Advanced Information and Knowledge Processing

Series Editors
Professor Lakhmi Jain
Lakhmi.jain@unisa.edu.au
Professor Xindong Wu
xwu@cems.uvm.edu

For other titles published in this series, go to
www.springer.com/series/4738

Sergei V. Chekanov

Scientific Data Analysis using Jython Scripting and Java

 Springer

Dr. Sergei V. Chekanov
Argonne National Laboratory (ANL)
9700 S. Cass Ave
Argonne
60439 IL, USA
chakanau@hep.anl.gov

AI&KP ISSN 1610-3947
ISBN 978-1-4471-2581-5   ISBN 978-1-84996-287-2 (eBook)
DOI 10.1007/978-1-84996-287-2
Springer London Dordrecht Heidelberg New York

British Library Cataloguing in Publication Data
A catalogue record for this book is available from the British Library

Printed on acid-free paper

Springer is part of Springer Science+Business Media (www.springer.com)

This book is dedicated to my family

Preface

Over the course of the past twenty years I have learned many things relevant to this book while working in high-energy physics. As everyone in this field in the yearly to mid-90s, I was analyzing experimental data collected by particle colliders using the FORTRAN programming language. Then, gradually, I moved to C++ coding following the general trend at that time. I was not too satisfied with this transition: C++ looked overly complicated and C++ source codes were difficult to understand. With C++, we were significantly constrained by particular aspects of computer hardware and operating system (Linux and Unix) on which the source codes were compiled and linked against existing libraries. Thus, to bring the analysis environment outside the high-energy community to the Windows platform, used by most people, was almost impossible.

I began serious development of ideas that eventually led to the jHepWork Java analysis environment in 2004, when I was struck by the simplicity and by the power of the Java Analysis Studio (JAS) program developed at the SLAC National Accelerator Laboratory (USA). One could run it even on the Windows platform, which was incredible for high-energy physics applications; We never really used Windows at that time, since high-energy physics community had wholly embraced Unix and Linux as the platform of choice, together with its build-in GNU C++ and FORTRAN compilers. More importantly, JAS running on Windows had exactly the same interface and functionality as for Unix and Linux! It was a few months after that I made the decision to focus on a simplified version of this Java framework which, I thought, should befit from Java scripting, will be simpler and more intuitive. Thus, it should be better suited for general public use. I have called it "jHepWork" ("j" means Java, "HEP" is the abbreviation for high-energy physics, and "work" means a sedentary lifestyle in front of a computer monitor).

Indeed, I was able to simplify the language and semantic of the JAS analysis environment by utilizing more appropriate short names for classes and methods, which are more suited for scripting languages. The entire project had grown tremendously after inclusion of many new GNU-licensed packages and extending the functionality of JAS in many areas, such as 3D graphics, serialized I/O and numerical packages. At present, jHepWork covers an impressive list of Java-written packages ranged from basic mathematical functions to neural networks and cellular automation. And, eventually, a little of JAS has left inside jHepWork! One important thing, however, has remained: As JAS, jHepWork was still an open-source software that can be downloaded freely from the Web.

For this project, Python was chosen as the main programming language because it is elegant and easy to learn. It is a great language for teaching scientific computation. For developers, this is an ideal language for fast prototyping and debugging. However, since the whole project was written in Java, it is Jython (Python implemented in Java) that was eventually chosen for the jHepWork project.

This book is intended for general audience, for those who use computing power to make sense of surrounding us data. This book is a good source of knowledge on data analysis for students and professionals of all disciplines. Especially, this book is for scientists and engineers, and everyone who devoted themselves to the quest of where we find ourselves in the Universe and what we find ourselves made of.

This book is also for those who study financial market; I hope it will be useful for them because the methods discussed in this book are undoubtedly common to any scientific research. However, I have to admit that this book may have little interest for a commercial use since financial-market analysts, unlike researches in basic scientific fields, could afford costly commercial products.

This book is about how to understand experimental data, how to reduce complexity of data, derive some meaningful conclusions and, finally, how to present results using Java graphical packages. It concentrates on computational aspects of these topics: as you will see, due to the simplicity of Python, one could catch ideas of many examples of this book just by looking at the code snippets without even explaining them in words. This book is also about how to simulate more or less realistic data samples which can mimic real situations. Such simulated data are used in this book in order to give simple and intuitive examples of data analysis techniques using Java scripting.

In this book I did not go deep inside of particular statistical or physics topics, since the aim was to give concrete numeral receipts and examples using Jython scripting language interfaced with Java numerical packages. My aim was also to give an introduction to many data-analysis subjects with sample code snippets based on Jython and jHepWork Java libraries. In cases when I could not cover the subject in detail, a sufficient number of relevant references was given, so the reader can easily find necessary information for each chapter using external sources.

Thus, this book presents practical approaches for data analysis, focusing on programming techniques. Each chapter describes the conceptual and methodological underpinning for statistical tools and their implementation in Java, covering essentially all aspects of data analysis, from simple multidimensional arrays and histograms to clustering analysis, curve fitting, metadata and neural networks. This book includes a comprehensive coverage of various numerical and graphical packages currently implemented in Java that are part of the jHepWork project.

The book was written by the primary developer of the software, and aimed to present a reliable and complete source of reference which lays the foundation for future data-analysis applications using Java scripting. The book includes more than

200 code snippets which are directly runnable and used to produce all graphical plots given in the text. A detailed description and several real-life data-analysis examples which develop a genuine feeling for data analysis techniques and their programming implementation are given in the last chapter of this book.

Finally, I am almost convinced myself that this book is self-contained and does not depend on knowledge of any computing package, Java, Python or Jython (although knowledge of Python and Java is desirable for professionals).

Chicago-Hamburg-Minsk   Sergei V. Chekanov

Acknowledgements

Several acknowledgements are in order. Much of this project grew out of fruitful collaboration with many of my colleagues who devoted themselves to high energy physics.

My scientific career in experimental high-energy physics was most influenced by Prof. Dr. V.I. Kuvshinov and Prof. Dr. E.W. Kittel, who were my supervisors almost fifteen years ago. I've learned experimental computation in the yearly 90x from Dr. L.F. Babichev and Dr. W.J. Metzger. I've learned experimental physics and its computational aspects from Dr. M. Derrick, Prof. Dr. E. Lohrmann, Dr. J. Repond, Dr. R. Yoshida, Dr. S. Magill, Dr. C. Glasman, Prof. Dr. J. Terron, Dr. J. Proudfoot, Dr. A. Vanyashin and many others.

The author is grateful to many authors writing free scientific software for their dedication to science and open-source analysis tools. I would like to thank many of my collegues for checking and debugging the jHepWork package, especially J. Dale, E. May, L. Lee, T. Johnson, P. Di Stefano and many others.

I would like to thank my parents and sister for their support, guidance, and love. Not least, I would like to express my eternal gratitude for my dear wife and children for their love and patience to a husband and father who wrote this book at home and thus was only half (mentally) present after coming from his work. Without their patience and understanding, this book would not have been possible.

Acknowledgments

Contents

Conventions and Acronyms

In this book, we will use the following typographical convention: A box with a code inside usually means interactive Jython commands typed in the Jython shell. All such commands start with the symbol >>>, which is the usual invitation in Python to type a command. This is shown in the example below:

```
>>> print 'Hello, jHepWork'
```

Working interactively with the Jython prompt has the drawback that it is impossible to save typed commands. In most cases, the code snippets are not so short, although they are still much shorter than in any other programming language. Therefore, it is desirable to save the typed code in a file for further modification and execution. In this case, we use Jython macro files, i.e. we write a code using the jHepWork (or any other) editor, save it in a file with the extension ".py", and run it using the keyboard shortcut [F8] or the button "run" from the jHepWork tool bar menu. In the book, such code examples are also shown inside the box, but code lines do not start with the Jython invitation symbol >>>. In such situations, the example codes will be shown as:

```
print 'Hello, jHepWork'
```

If a code snippet should be used as a Jython module for inclusion into other programs, then we should write our program inside a file. A Python code always imports an external module using its file name. Since the file names are important, we always indicate which exactly file name should be used on the top of the box with a code. For example, if a program code is considered as a module to be imported by another code example, we will show it as:

```
Module 'hello.py'
print 'Hello, jHepWork'
```

with the box title indicationing the file name. For instance, we call the module above as:

```
>>> import hello
Hello, jHepWork
```

when using the Jython prompt (recall the >>> symbol!). The code imports the file 'hello.py' and executes it, printing the string 'Hello, jHepWork'. In other cases, one can use any file names for the code snippets.

We use typewriter font for Jython and Java classes and methods. For file names and directories, we also use the same font style after adding additional parentheses.

We remind also that the directory name separators are backward slashes for Windows, and slashes for Linux and Mac computers. In this book, we use the latter convention. For example, the directory with examples will be shown as:

```
... /macro/examples/
```

while for Windows computers, the same directory should be shown as:

```
C:...\macro\examples\
```

The dots in this example are used to indicate the jHepWork installation directory.

This is all. We will try to avoid using abbreviations. When we use abbreviations, we will explain their meaning directly in the text.

Introduction

Introduction to Data Analysis and Why This Book Is Special

Data analysis is a systematic process of understanding surrounding us world by means of several phases in a scientific research:

- Data gathering, digitization and transformation to a necessary format. Usually, data comes from experimental apparatus;
- Reduction of data volume, structuring and cleaning erroneous entries where possible;
- Data description, which can usually be done via statistical analysis of data; At this stage, producing data summaries is an important computational task to proceed further;
- Data mining that focuses on knowledge discovery and predictions. This stage aims to identify and classify patterns in data. The data mining is usually machine-based exploration of data;
- Comparison of data with other data sets and finding interdependence or similarities;
- Confronting data with numerical or analytical models. Numerical data modeling and simulation of experimental apparatus can be used if an analytical description is impossible;
- Data visualization and extraction of relevant results.

As you can see, the topic of data analysis is very broad and cannot easily be covered in a single book. We do not plan to do this.

The approach of this book is different. There are plenty of books which go into the depth of certain data-analysis subjects. In this book, I give numerical recipes and complete code snippets which illustrate essentially all phases in data analysis discussed above. Not only this: we will not only illustrate data-analysis computational techniques, but also will show how to simulate data that can be used for our analysis examples. In addition, we perform data analysis computations using real-life data ranged from particle physics to astrophysics and biology.

Data analysis is a difficult research topic. It requires a good knowledge of not only your specific research field, but also computer programming. On top of this,

S.V. Chekanov, *Scientific Data Analysis using Jython Scripting and Java,*
Advanced Information and Knowledge Processing,
DOI 10.1007/978-1-84996-287-2_1, © Springer-Verlag London Limited 2010

the knowledge of mathematics and statistics is essential. To make the data analysis examples as simple as possible from the computational point of view, I fully embrace the scripting approach in the course of this book. This leads to short and clear analysis codes, so one could concentrate on the logic of analysis flow rather than on language-specific details.

Until now, if one needs to analyze large data volumes, most likely one has to use either C++ or FORTRAN, thus one should write some rather low-level code, compilation and execution of which require a certain computer platform. At this moment, this is the only book which teaches how to combine the power of a high-level scripting language with Java numerical libraries, and how to make use of truly platform-independent and multi-threaded computational environment for data analysis. I hope this book will help to unleash the tremendous potential of this new approach and will encourage to use it by a wide audience.

The main emphasis of this book has been put on numerical methods and codding techniques, thus we are going to equip you with the necessary knowledge for data analysis from the computational point of view. In this book, you will learn how to write analysis codes, and numerous code snippets give you some ideas that can easily be incorporated into your own research application.

Who Is This Book for

I have written this book for undergraduate and graduate students, academics, professors, professionals of any field and any age. The book could be used as a textbook for students.

I assume that the reader is not familiar with any particular programming language, but some basic understanding of statistics and mathematics would be very helpful to understand the material of this book. This is why I have spent so much time in this book showing how to write analysis codes, rather than explaining the basics of statistics and data mining. Note that if the reader will decide to write his/her own Java libraries to be deployed as jar files for a new project, some experience with the Java programming language will be required.

Chapter 1
Jython, Java and jHepWork

1.1 Introduction

The data analysis approach implemented in the jHepWork data-analysis framework [1] and to be discussed in this book was designed for everyone who does not have a big desire to go into the depth of low-level hardcore programming. Yes, this book is for you, scientists, engineering and students, and the rest of us, whose brain is already busy with professional research or hobby. jHepWork was designed to enable researches to spend their time thinking about problems and their solutions, rather than diving into a low-level codding using programming languages, such as C/C++ or FORTRAN, which are more oriented toward a computing machine, rather than to a person who has to understand and interpret the code. jHepWork analysis macros for data manipulations are based on Jython, an implementation of the high-level language Python. Thus, one can fully benefit from variety of programming possibilities offered by Python, including its syntax clarity and high-level libraries. But Jython is not prerequisite for this framework: Java can also be used to access the mathematical and graphical libraries of jHepWork.

With time, any computational framework based on a simple-to-learn programming language naturally gets large and difficult to handle; this is a quite inevitable feature of the modern life. Properly chosen computation language is essential to maintain simplicity of user communication with exponentially growing programs. And this is where Java comes to its power: Java virtual machine and its popular integrated development environments can help to develop programs, tell about errors or mistyped classes and, in general, provide a layer of intelligent activity between a human, who writes a code or interprets its algorithmic logic, and a machine designed for program execution. This is rather different from low-level languages like C/C++ or FORTRAN which are often used for numerical calculations. For such languages, a researcher is usually on his own with a text editor and a programing language itself which typically requires good programming skills and several manuals on a bookshelf.

jHepWork is by no means a simple framework, although it is based on Java and high-level Python language. It has more than ten thousands Java classes and methods

S.V. Chekanov, *Scientific Data Analysis using Jython Scripting and Java,*
Advanced Information and Knowledge Processing,
DOI 10.1007/978-1-84996-287-2_2, © Springer-Verlag London Limited 2010

designed for data manipulation, analysis and data visualization (excluding those provided by the native Java API and Python itself). The jHepWork library core for statistical and graphical analysis is based on the jHPlot library, which contains more than 1200 Java classes and methods. However, you will be surprised to find out about how easy to work with this program. Partially, this is because of the Python language implemented in Java (Jython) and, partially, because of Java itself. We will explain this point in this chapter.

1.1.1 Books You May Read Before

Generally, the text of this book is self-contained. But to understand the material deeper, you may need to look at other sources. First of all, there are plenty of good books [2–6] about Python and Jython, which are more complete for language-specific topics than the material given in this book.

Secondly, a great deal of supplementary information can be found in Java books [7–11]. These books are especially useful if you will decide to understand many issues on much deeper level than that given in this book. The truth is that Java forms the backbone of the jHepWork numerical and graphical libraries. This means that one can write your data-analysis software in pure Java language using exactly the same libraries as those used for the scripting examples in this book! Of course, in this case, you should use the proper Java syntax.

Thirdly, as you read, you may need to look at external sources to understand the material better, especially when we come to statistical interpretation of data. We will supply the reader with the necessary references, so he or she can choose the most appropriate (and affordable) books to discover the world of data analysis and data mining.

1.1.2 Yes, It Is Pure Java

You may want to skip this section if you are not too interested in any further discussion of Java and C++.

Nowadays, the advantages of Java over C++ in many areas seems to be overwhelming. First of all, Java is the most popular object-oriented programing language. According to SourceForge.net and Freshmeat.net statistics, the number of open-source applications written in Java exceeds those written in C++. According to the TIOBE Programming Community Index (http://www.tiobe.com), the popularity of Java in industry is at the level of 20%, while C++ has 10% popularity at the time when this book was written.

Java retains the C++ syntax, but significantly simplifies the language. This is an incomplete list of advantages of Java over C++: (1) Java is multi-platform with the philosophy of "write once, run anywhere"; (2) Better structured, clean, efficient, simpler (no pointers); (3) Stable, robust and well supported: Java programs written (or compiled) many years from now can be compiled (or executed) without modifications even today. This is true even for JAVA source code with graphic widgets. In contrast, C++ programs usually require continues time-consuming maintenance in order to follow up the development of C++ compilers and graphic desktop environment; (4) Java has the reflection technology, which is not present in C++. The reflection allows an application to discover information about created objects, thus a program can design itself at runtime. In particular, this is considered to be an essential feature for building integrated-development environments (IDEs); (5) Java has several intelligent IDEs, which are indispensable tools for large software projects. Some of them, like Eclipse or Netbeans, are free. We probably should note that one can also use free IDEs for C/C++, but they are not as intelligent as those for Java and usually miss many important features; (6) Automatic garbage collection. Having in hands this feature, a programmer does not need to perform a low-level memory management; (7) Extensive compile-time and run-time checking; (8) Java is truly multi-threaded and this significantly simplifies the development of applications which should run in parallel on multi-core machines; (9) Programs written in Java can be embedded to the Web. This is important for distributed analysis environment (Java WebStart, plugins, applets), especially when data-analysis tools are not localized in one single place but scattered over the Web (nowadays, this is the most common situation).

We will stop here assuming that you are convinced.

1.1.3 Some Warnings

We should immediately warn you: the jHepWork numerical and graphical libraries can be considered neither as most efficient nor error free. The code of jHepWork does not always follow the coding recommendations for Java developers including naming conventions and code layout. We even admit that some parts were not designed with highest-possible performance for code execution in mind. The reason is simple: it was not written by professional programmers. The numerical libraries were written by many people at different time, most of them were students and scientists who had to develop numerical and data-visualization algorithms for their own research programs, since commercial software companies either could not offer similar programs or their products were too expensive. Many contributed packages were discontinued many years ago and were brought into life only recently after their inclusion into jHepWork. In addition, some packages were written using Java 1.1, and this had also some impact on the coding style of certain libraries.

Thus, a professional programmer may immediately find unprofessionally written pieces of code. This is true even for some examples shown in this book. The reason for this was not because we were not aware of such codding issues. In some cases, we did not find appealing reasons to keep very strict coding standard at the expense of simplicity. For example, in most cases, we import all classes inside a package using the statement:

```
>>>   from PackageName import *
```

instead of importing only certain classes as

```
>>>   from PackageName import class1,class2
```

We did not enforce the latter case to keep the examples of this book short and concise, so we could fit the code snippets into the pages of this book. Also, it is possible
that you may not like to type long lists of imported classes during a code prototyping (personally, I do not like this style), since this can be done at later stage during
code deployment.

A professorial programmer might find some other odds, like why some object
containers are designed to store only double values (like the P1D class to be discussed below), while it is more practical to store integer values when necessary.
Again, the motivation was not because of omissions. The reason was that the reader
may not want to dive into extra complexity of dealing with different types, since
integers are only a subset of float values. There are plenty of other classes which are
well suited for storing integer values (we will discuss them in this book).

The main motivation for the jHepWork project was to develop an accessible and
friendly tool to be used in scientific search, with a syntax oriented towards scientists
rather than programmers, not towards a particular operating system. The design of
this project was mainly motivated by the simplicity: there are many programming
languages which are required to learn for many years before starting to write useful
scientific and engineering projects. The approach discussed in this book is very
different: generally, the reader does not need to know any programming language to
start writing analysis codes using jHepWork libraries. However, if it happens that the
reader knows either Java or Python (or both) already, he or she will find this book
also interesting, since jHepWork is not just a simplified entry to the world of the
Java and Python. The program and this book go much beyond a simple introduction
to programming, as they supply with a significant amount of information on how to
professionally analyze data.

The reader may also notice that a little attention has been paid to how to write
and use Java or Jython classes. Of course, classes are necessary for any object-
oriented language. The reason for this was following: for the majority of scientific
data-analysis projects, the logic of scripting programs is linear, i.e. an analysis code
typically consists of a well-defined sequence of statements to be evaluated one by
one, from the top to the bottom of the code. It is very unlikely that data-analysis logic
will contain highly parallel algorithmic branches as those for the usual graphical-
user-interface (GUI) development.[1] Certainly, the classes are necessary when one

---

[1]We should probably tell you that this may not be totally true in the future when multi-core machines will be rather common and one will be faced with the question of how to parallelize analysis
codes to gain high performance. We will discuss this topic later on in this book.

develops Java libraries to be used by a scripting language. But, in this book, we mainly concentrate on the scripting examples based on the existing Java libraries of jHepWork, rather than discussing how to write classes for numerical computation to be deployed as external libraries.

1.1.4 Errors

This book may contain typos, omissions or even errors. jHepWork can also contain bugs. If you notice any errors or if you have suggestions regarding the book and code examples, I would be happy to hear from you. You can send your comments to:

jhepwork@jworks.org

One can also post bug reports to the jHepWork forum accessible from the main Web page:

http://jwork.org/jhepwork/

1.2 Introduction to Scientific Computing

Let us say a few words about scientific analysis environment. Scripting in a scientific research is essential. There are plenty of programs heavily based on graphical user interfaces (GUI), where a researcher should go over many mouse clicks before reaching a designed result (which, usually, is a graph or some statistical summary). Typical examples are Microsoft Excel, Origin and many other commercial products. The scripting approach is somewhat different: it requires from a developer to type only short commands and store them in files so one can easier reproduce the results by executing such macro files later. During the program development, an analysis framework should help to find a proper method for a particular class instance and to supply with a comprehensive description of its methods which can fit to the program logic. It should control your code while typing and correct it when needed!

In this respect, jHepWork is similar to Wolfram Research Mathematica or Maple. However, unlike these commercial products, jHepWork is significantly more data oriented. Being Java-based, jHepWork is also more GUI oriented since all the power of Java graphical widgets to build user interfaces is in your hands. In addition, jHep-Work uses Python, which is very popular programming language for science and engineering. Finally, jHepWork is free.

1.2.1 Book Examples and the Power of Jython

You will be surprised to know that even the most realistic data-analysis examples given in this book have rather short source-code snippets. I will promise that all our example codes typically fit to 2/3 of the printed page of this book at most. This came to be possible using the Python syntax and its high-level built-in data structures. This was also possible due to known Python capabilities to glue different programming languages. In case of its Java implementation—Jython, one can seamlessly integrate Python, Java and jHepWork libraries.

As you walk through the examples, you may decide to type all the listed codes in by hand, since this is the best way to get familiar with the coding techniques. But, even although Jython examples of this book are short, you may still avoid typing them when following the book pages. The example source codes of this book (for each section separately) can be downloaded from:

http://jwork.org/jhepwork/

or from the mirrors:

http://projects.hepforge.org/jhepwork/
http://sourceforge.net/projects/jhepwork

Look at the section called "Documentation" which gives the location of the tar file with all examples.

1.2.2 The History of jHepWork

You can easily skip this section if you are not interested in the history of this project and jump directly to the next section.

jHepWork libraries have their roots in high-energy physics in the late 1990s when first effort was done in accessing visibility of using Java for high-energy physics [12]. Later, the AIDA project was formulated, with the goal to define abstract interfaces for common physics analysis objects, such as histograms, ntuples, fitters, input and output. The adoption of these interfaces allowed some simplification in using different tools without having to learn new interfaces or change their code. The AIDA interface was implemented in Java and then was included into the core of the Java Analysis Studio (JAS), which also contained a built-in editor and other software tools.

JAS has become a powerful modular application framework into which various analysis components can be plugged. The framework allowed to use various scripting languages, such as Jython and Peanut. JAS and JAIDA have become the core of the FreeHEP library [13], which was mainly aimed for future International Linear Collider project. While the initial focus of this project was high-energy physics, many self-contained libraries are generic and very common in science and engineering.

Being very powerful as a Java application, JAS was not ideally suited for scientific scripting due to long names of factories used to create objects. JAS had a rather basic editor without syntax highlighting, help assist and without a robust 3D support for data visualization. In 2005, a new project has started based on the same JAIDA libraries, with the main goal to improve graphics and to use short class names, so it would require less typing and, naturally, can lead to a higher productivity. The main goal was to utilize short names for Java classes, which could naturally match to the concise syntax of the Python language. Mixing Jython and Java objects for scientific scripting was expected to form a natural semantic flow and could lead to short codes for data-analysis programs. In addition to the FreeHEP high-energy physics libraries, the project was expected to include other freely available Java libraries aimed for statistical analysis, data mining and consequent visualization. The name of this project was jHepWork [1].

1.2.3 Why Jython?

Python [14] became increasingly popular programming language in science and engineering [6], since it is interactive, object-oriented, high-level, dynamic and portable. It has simple and easy to learn syntax which reduces the cost of program maintenance. Jython [15] is an implementation of Python in Java. In contrast to the standard Python (or CPython) written in C, Jython is fully integrated with the Java platform, thus Jython programs can make full use of extensive built-in and third-party Java libraries. Therefore, Jython programs have even more power than the standard Python implemented in C. Finally, the Jython interpreter is freely available for both commercial and non-commercial use.

jHepWork is a full-featured object-oriented data-analysis framework for scientists that takes advantage of the Jython language. It is truly multi-platform product, implemented 100 percent in Java.

Jython macros are used for input and output (I/O), mathematical manipulations with data, to create histograms, visualize data, perform statistical analysis, curve fitting and so on. The program includes many tools for interactive scientific plots in 2D and 3D. It can be used to develop a range of data-analysis applications focusing on analysis of complicated data sets. Data structures and data manipulation methods, integrated with Java and the JAIDA FreeHEP libraries, combine a remarkable power with a very clear syntax. It offers a full-featured, extensible multi-platform integrated development environment (IDE) implemented in Java.

You may ask this question: what is the point in using Jython and Java if CPython is also portable and can be installed on Linux or Windows platforms? This answer is this: CPython calls libraries implemented in C/C++ or FORTRAN, but these libraries, by definition, should be compiled separately for each platform (in fact, as CPython itself). Thus, CPython cannot provide a genuine multi-platform framework. In the case of Jython, libraries developed using Java are truly multi-platform and do not require separate deployment for each computer platform.

Programs written using jHepWork are usually rather short due the simple Python syntax and high-level constructs implemented in Python and in the core jHepWork libraries. As a front-end data-analysis environment, jHepWork helps to concentrate on interactive experimentation, debugging, rapid script development and finally on work-flow of scientific tasks, rather than on low-level programming.

The main web pages of jHepWork are:

http://jwork.org/jhepwork/
http://projects.hepforge.org/jhepwork/

These web pages contain numerous examples and documentation API.

jHepWork consists of two major libraries, jeHEP (jHepWork IDE) and jHPlot (jHepWork data-analysis library). Both are licensed under the GNU General Public License (GPL).

1.2.4 Differences with Other Data-analysis Packages

How does jHepWork compare with other commercial products? Throughout the years there have been many commercial products for data analysis, but it is important to realize that they are typically platform specific. On top of this, commercial products are either rather costly or do not provide a user with the source code, or both.

We would not be too wrong in saying that it is very hard to find a commercial product with the same functionality in certain analysis areas, and with such variety of methods existing in jHepWork. For example, only one single class used to build, manipulate and display an one-dimensional histogram has about eighty methods (plus dozens of methods inherited from other classes). Usually, commercial software is not competitive enough for such specialized tasks as histogramming and processing of large data samples. Together with a high price, this was one of the reasons why commercial products have never penetrated into the software environment of high-energy physics in which the data reduction and data mining were the most important tasks. Here, we should also probably add that Linux and Unix are the most common platforms in universities and laboratories and this also has a certain impact on the number of data-analysis packages to be used in scientific research.

Below we will compare jHepWork with another free object-oriented package currently used in high-energy physics, the so-called ROOT [16, 17]. The ROOT analysis framework is written in C++, and at the time when this book was written, ROOT was a *de facto* standard program in high-energy physics laboratories. Compared to ROOT, jHepWork:

- is Java-based and thus inherits the well-know robustness. For example, the source code of this project developed 5 years ago can easily be compiled without any changes even today. Even jar libraries compiled many years from now can run without problems on modern Java Virtual Machines. C++ programs, such as ROOT, require a constant support, especially if they include a graphical interface; A typical lifetime of unsupported C++ code based on graphic widgets is several years on Linux-based platforms;
- being Java-based, is truly multi-platform. jHepWork does not require compilation and installation. This is especially useful for plugins distributed via the Internet in form of bytecode jar libraries;

- can be integrated with the Web in form of applets or Java Web-start applications, thus it is better suited for a distributed analysis environment via the Internet. This is an essential feature for large scientific communities or collaborations working on a single project;
- Java has an automatic garbage collection. This has a significant advantage over C++/C as the user does not need to perform a memory management;
- being Java-based, jHepWork is designed from the ground up to support programming with multiple threads. It is truly multi-threaded language. This makes parallel programming easier and leads to a more efficient use of modern computers with multi-core processors. Unlike C++, the Java virtual machine takes care of low-level threads according to the host multi-core computers;
- being Java based, it comes with the reflection technology, i.e. the ability to examine or modify behavior of applications at runtime. This feature is missing in C++ and, therefore, in ROOT;
- has a full-featured IDE with syntax highlighting, syntax checker, code completion and analyzer;
- is designed for calculations based on Jython, thus it is seamlessly integrated with hundreds of Java classes for advanced 2D graphics and imaging;
- is used for calculations based on Jython scripts that can be compiled into Java bytecode files and packed into jar libraries without modifications of Jython scripts. In contrast, ROOT/CINT scripts have to be written using a proper C++ syntax, without CINT shortcuts, if they will be compiled into shared libraries. This makes the ROOT/CINT scripts to be almost identical to the standard C++ code with long programming statements;
- includes an advanced help system with a code completion based on the Java reflection technology. With increasingly large number of classes and methods in ROOT/C++, it is difficult to access information on classes and methods without advanced IDE. Using the jHepWork IDE, it is possible to access the full description of classes and methods while editing Jython scripts;
- essentially all jHepWork objects, including histograms, can be saved into files and restored using Java serialization mechanism. One can store collections of objects using Jython maps or lists. One can even serialize dialog widgets and other GUI components.

Finally, as mentioned before, Java is considered to be the most popular programming language. One can find more detailed information about differences between Java and C++ on this Web site [18].

1.2.5 How Fast It Is?

Sometime one can hear that Java is slower than C++. The subject itself is controversial, since the answer totally depends on the nature and the goal of an application. Nowadays, most people agree that after introduction of Just-in-Time compiler (JIT), Java is as fast as C++. Probably, in some areas, Java is still slower than C++, but the nature of such controversy is already a sign that the performance gap is now quite small and there is no alarming difference in speed between Java and C++ programs. And, anyway, the proper comparisons with C++ is usually unfair: Java does tremendous amount of run-time checks, such as array bound checking, thread synchronization, run-time checking, garbage collection etc. to make sure that a code runs without problems and without putting extra stress on a programmer.

The JIT compilation converts Java bytecode into a native machine code at run-time. The conversion step could be slow, but for most numerical calculations involving large loops over objects, this does not matter so much. With the recent development of Java engines, the speed penalty is not that significant, especially for projects based on massively parallel processing where Java's multi-thread support is the strongest argument.

Secondly, while programming in C++, one can often pass objects to functions by value, which leads to an overhead. Java always passes references to objects instead of objects themselves, therefore, independent of how you program in Java, your code will always be rather efficient.

Thirdly, Java is a process virtual machine, not a system virtual machine is usually used to run C/C++ applications which are difficult to run using the host computers. Thus, Java avoids a significant overhead due to running non-native operating systems used for C/C++ or FORTRAN applications.

jHepWork is also based on Jython, not only on Java. Jython scripts are about four-five times slower than equivalent Java programs for operations on primitive data types (remember, all Jython data types are objects). This means that CPU intensive tasks should be moved to Java jar libraries.

One should bear in mind that jHepWork was designed for data manipulations and visualizations in which the program speed is not essential, as it is assumed that jHepWork scripts are used for manipulations with high-level data objects (like histograms). For such front-end data analysis, the bottleneck is human interaction with graphical objects using the mouse, keyboard or network latency in case of remote data or programs.

In practice, results obtained with Jython programs can be obtained much faster than those designed in C++/Java, because the development is so much easier that a user often winds up with a much better algorithm based on the Jython syntax and jHepWork high-level objects than he/she would in case of C++ or Java. For CPU intensive tasks, like large loops over primitive data types, reading files etc., one should use high-level structures of Jython and jHepWork or user-specific libraries which can be developed using the jHepWork IDE. This is the basic idea. The rest of this book will spell it out more carefully.

1.2.6 Jython and CPython Versions

The jHepWork described in this book is based on the Jython version 2.5.1, which supports all language level features of CPython 2.5. This Jython release is believed to provide a much cleaner and consistent code base than the previous releases.

1.3 Installation

The good news is that you do not need to install anything to make all the examples discussed in this book to work. But to run jHepWork, you need to have the Java JDK (Java development kit) installed. You can also use Java Runtime Environment (JRE), which is very likely already installed on your computer but, in this case, there will be some limitations (for example, you will not be able to compile Java source codes).

Java software is available at:

http://java.sun.com/javase/downloads/

Installing the JDK or JRE is rather simple for any platform (Windows, Solaris, Mac and Linux). Once installed, check Java by typing `java -version` on the prompt. If Java is installed, you should see

```
java version "1.6.X"
Java(TM) 2 Runtime Environment, Standard Edition
Java HotSpot(TM) Client VM (build 1.6.X, mixed mode)
```

or a similar message ("X" indicates a subversion number). You will need at least Java 1.6 or above.

jHepWork does not require installation. Download the package from the following location:

http://jwork.org/jhepwork/

or

http://projects.hepforge.org/jhepwork/
http://sourceforge.net/projects/jhepwork/

The package for download has the name "jhepwork-VERSION.zip", where "VERSION" is a version number. Unzip this file to a folder. You will see several files and the directories `'lib'`, `'macros'` and `'doc'`. For Windows, just click on the file `'jhepwork.bat'` which brings up the jHepWork IDE windows. For Linux, Unix and Mac, run the script `'jhepwork.sh'`.

For Mac, Linux and UNIX, one can put the file `'jhepwork.sh'` to the `'$HOME/bin'` directory, so one can start jHepWork from any directory. In this case, one should set the variable JEHEP_HOME (defined inside the script `'jhepwork.sh'`) to the directory path where the file `'jehep.jar'` is located.

First time you execute the `'jhepwork.sh'` or `'jhepwork.bat'`, you will see many messages such as:

```
*sys-package-mgr*: processing new jar, 'jhplot.jar'
*sys-package-mgr*: processing new jar, 'jminhep.jar'
. . . . . . . . . . . . . . . . . . . . . . . . . . . . . . . . . . . . . . . . . . . . . . . .
```

This is normal: one should wait until the end of Java libraries scan. Jython cashes the jar libraries, i.e. it creates a new directory `'cachedir'` inside the directory `'lib/jython'` with the description of all classes located in jar files defined in the CLASSPATH variable. Next time when you execute the start-up script, jHepWork IDE will start very fast as the package cache is ready (of course, if you did not modify the Java CLASSPATH before starting the jHepWork IDE).

1.4 Introduction to the jHepWork IDE

Feel free to skip this section and jump to Chap. 2, since for the readers with some programming experience, this section could be too obvious. For those who are just entering the computational world, I'll try to explain here several tricks which could be useful for source code editing and execution. jHepWork comes with a light-weight integrated development environment (IDE) which includes a source code editor with a code completion and a code analyzer, a Jython shell ("Jython-Shell"), a Bean shell ("BeanShell") and a panel with the file manager. The script `'jhepwork.sh'` (for Linux/UNIX/Mac) or `'jhepwork.bat'` (Windows) starts the jHepWork IDE. After initialization, you will see the jHepWork IDE with a source code editor as shown in Fig. 1.1.

It should be noted that the source-code development using the jHepWork libraries can be done using any text editor, while the execution of Jython scripts or compiled

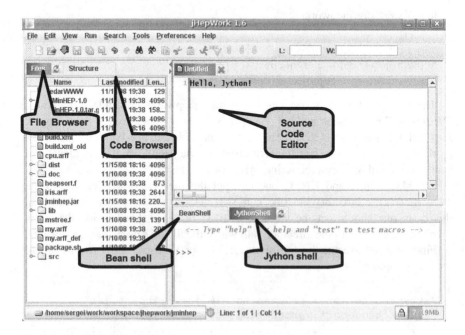

Fig. 1.1 jHepWork IDE workbench

Java codes can be done using a shell prompt after specifying the CLASSPATH environmental variable. This part should be considered for advanced users and will be discussed later.

The jHepWork workbench has three main windows:

- The Source Code Editor (central window);
- The Tool Bar menu (above the text area);
- File and code browser window (left window);
- Bean-shell and Jython-shell window (bottom windows).

When jHepWork is started for the first time, it creates files with preferences located in the directory:

    $HOME/.jehep

for Linux/Mac, or

    $HOME\jehep.ini

for Windows. There are several preference files inside this directory: the main file, 'jehep.pref', with all source-code editor preferences, a user dictionary file, a JabRef preference file and other files. If you need to reset all settings to their default values, just remove the directory with the preference files (or just the file 'jehep.pref').

1.4.1 Source Code Editor

The source code editor can be used to edit files, and it has all the features necessary for effective programming: syntax highlighting, syntax checker and a basic code completion. For bookmarks, one should click on the right margin of the source code editor. A blue mark should appear that tells that the bookmark is set at a given line. One can click on it with the mouse in order to jump to the bookmarked text location.

The file browser is used to display files and directories. By clicking on the selected file name one can open it in a new tab of the text editor. For most types of files (LaTeX, C++, Java, Python), the code browser shows the structure of the currently opened document.

1.4.2 jHepWork Java Libraries and Python Packages

Although this topic is probably for advances users, we feel that it is necessary to describe the jHepWork library structure here. Generally, the program contains Java jar libraries and Python-based libraries. After the installation, the main jHepWork directory contains the following subdirectories:

- 'lib'—contains Java libraries. When you start jHepWork, this directory is scanned by jHepWork and all libraries inside this directory are put to the Java CLASSPATH environmental variable. The directory lib contains several sub-directories with jar files: freehep—contains only FreeHEP Java libraries; the directory 'jython' contains Jython libraries; system contains libraries necessary to run jHepWork, including third-party libraries. User-specific libraries can be put under the user directory (which will also be scanned by the jHepWork startup script).

- 'python'—contains Python libraries. By default, they are not imported by Python modules, so it is up to you to import them into your programs. There are several ways to do this: (1) put the directory name in the file 'registry' located in the directory 'lib/jython'. You should define the variable python.path as:

```
python.path=[dir]/python/packages
```

where [dir] is the installation directory (it should contain the file 'jehep.jar'). Or, alternatively, one can set the location of Python libraries at the beginning of your Jython code as:

```
import sys
sys.path.append('[dir]/python/packages')
```

If you run a Jython module inside the jHepWork IDE, one can specify the current installation directory using the variable SystemDir which always points to the installation directory. Thus the line above will look as:

```
import sys
sys.path.append(SystemDir+'/python/packages')
```

I will come back to this point later in the text where several Python-based scientific libraries will be discussed.

- 'macros'—contains Python modules necessary to run jHepWork. It also contains examples in the sub-directory 'examples'. When you start jHep-Work, the directory 'macros/system' is put inside the Jython class path automatically by the script 'sysjehep.py' located inside the subdirectory 'system'.

A user can put new macros in the 'macros/user' directory. There are already several macros in this directory: For example, one can replace a string with another string in the current text just by calling the method replace(str1,str2), where str1 and str2 are input strings; In fact, all user macros are rather similar to those used by the jEdit[2] editor, as long as the `textArea` class is used.

The Jython class path is defined via the module `'sysjehep.py'` located in the `'macros/system'` directory. This variable specifies the location of Jython modules which can be visible for the "import" statement. The module `'sysjehep.py'` is loaded automatically every time you start jHepWork (or reload the Jython console). One can specify the location of the user modules in this file. By default, every file which is put into the installation directory or to `'macros/system'` or `'macros/user'`, should be visible for jHepWork macros and normally you do not need to import them.

1.4.3 Jython and Bean Shell Consoles

The BeanShell and JythonShell can help quickly prototype small pieces of codes without using text editors. By default, the BeanShell window is active. To activate the PythonShell, window, click on the PythonShell tab.

BeanShell and JythonShell can be used to run external programs. Just type "!" in front of the program name you want to execute. For example:

```
>>> !latex <file>   #   latex for the file <file>
>>> !make           #   compile C++/Fortran
>>> !ls             #   list files (linux/unix/max)
>>> !dir            #   list files (windows)
```

One can use the command history (keyboard key: [Up] and [Down]) and Java and Python code completion using the key combination [Ctrl] + [Space]. Use the help options to learn more about additional features for interactions with the native OS. Both shells contain their own help system.

There are several predefined variables available for the both consoles. They have been exported by the scripts located in the `'macros/system'` directory: `'sysjehep.py'` (for Jython) `'sysjehep.bsh'` (for BeanShell).

There are several predefined objects useful for text manipulations: Assume a user is editing a file in the text editor. Using the JythonShell, one can access properties of this file using the `Editor` class and its methods. This class holds currently opened document and allows a manipulation with its content. For example, if one types `print Editor.DocDir()` in the JythonShell, one can display the full path to the currently opened document. One can access all public methods of the `Editor` static class using the jHepWork code assist for the `Editor` class:

```
>>> Editor.  [Ctrl]+[Shift] # show all methods
```

---

[2]jEdit is one of the most popular Java-based text editors.

or

```
>>> dir(Editor) # show all methods
'DocDir', 'DocMasterName', 'DocMasterNameShort',
'DocName', 'DocStyle' ..
```

(below we print only a few first methods). As you can see, having in hand this class, one can access a broad variety of methods for text manipulation. More information about this class can be found using the code assist to be discussed in Sect. 1.4.10.

Below we give several examples of how to access the information about an opened file using several predefined variables. Use the Jython shell to type the following commands:

```
>>> print 'Java classpath',ClassPath
>>> print 'File name=',DocName -
>>> print 'Directory=',DocDir
>>> print 'File separator=',fSep
>>> print 'Name with complete path',DocMasterName
>>> print 'Name without extension',DocMasterNameShort
```

All these variables are automatically imported by the IDE.

But how one can find the predefined variables available in jHepWork while working with Jython macros files? This should be easy: Click on JythonShell and type the following command:

```
>>> dir()
[ClassPath', 'DicDir', 'DocDir', 'DocMasterName',
 'DocMasterNameShort', 'DocName', 'DocStyle',
 'Editor', 'ObjectBrowser', 'ProjDir', 'SetEnv',
 'SystemDir', 'SystemMacrosDir', 'UserMacrosDir
 ...
```

(again I show only a few first variables). The best approach to learn about them is just to print them out. Analogously, one can print all such variables using the BeanShell commands (but using the BeanShell syntax).

It is useful in many cases to clean up messages from the interactive consoles. To reload either the File Browser or Jython/Bean shell, one should use the Reload buttons located directly on the small blue tabs of the console windows.

jHepWork uses the following aliases for the BeanShell macros:

[CLASSPATH]	java class path;
[FILE_SHORT]	returns the filename without the extension;
[FILE_SHORT_NODIR]	returns the file name without the extension and the path;
[FILE]	returns the full name of the file including the path;
[FILE_NODIR]	returns the full name of the file excluding the path;

```
[DIR_FILE]              returns file directory;
[DIR_SYS]               returns system directory.
```

For example, typing `print [FILE]` prints the name of the currently opened file. Such substitutions can be used in macros. For example, if a macro contains `[FILE]`, it will be automatically replaced by the current file name.

1.4.4 Accessing Methods of Instances

A user can view the available methods by typing `obrowser` in the BeanShell. This will open an object browser window with all objects. If one needs to add some object, one should type `obrowser.add(obj)`, where "obj" is an instance of a Java class.

```
>>> obrowser
>>> obj=new JLabel('OK')
>>> obrowser.add(obj)
```

For a similar task in the JythonShell prompt, a user should use the code assist, see Sect. 1.4.10, or the Python `dir('obj')` method. The jHepWork code assist cannot be used in BeanShell. In this case, one should use the `obrowser` class.

One can manually execute the `obrowser` class by running the script `'obrowser.bsh'` or `'obrowser.py'` from the directory `macros/system`.

1.4.5 Editing Jython Scripts

First, a new file with the extension `'.py'` should be created. There are several ways to do this: (1) Select `[File]-[New]-[jHPlot script]` in the main menu. A new template Jython script should appear; (2) Use the menu `[File]-[New]-[Text document]`. Then, save it as a Jython file with the extension `'.py'`; (3) Click on any Jython file with the extension `'.py'` in the File Browser.

If a script with the user analysis program is ready, one should save it. But before this, you may check for syntax errors without actual execution of the macro file. Look at the menu `[Run]` and `[Check Jython syntax]`. In the case of errors, the IDE will point to a line with an error. For the syntax checker, it is impossible to identify run-time errors, i.e. errors which may happen during the execution of the script. Once you know that there are no errors, save the file again using the menu `[File]`.

1.4.6 Running Jython Scripts

To run a Jython script, use the `[run]` button from the main tool-bar of jHepWork. One can also use the keyboard key `[F8]` for fast script execution. In case of an error, the jHepWork main editor will move the cursor to the appropriate line with the detected error. Press any key to remove the line highlighting (red color). One can also execute a Jython file line-by-line using the [run] menu of the main tool-bar.

During the execution, all program outputs will be redirected to the JythonShell window. It is appropriate mention here one feature: JythonShell is not designed for very heavy output, it is mainly a debugging tool. If you want to print a lot of messages, then you should be careful: you may wipe out significant resources by doing this and your code execution will be rather slow. For example, consider a simple Jython program: printing integer values:

```
for i in range(1000000):
    print 'Test=',i
```

Save these lines in the file 'test.py'. Then open it in the jHepWork editor and run it by clicking on the button [run]. You will immediately see that the memory monitor in the right corner of the IDE will show an increase in the memory usage. So, try to avoid such situations. If you need to debug loops like this, use a small number of iterations. If you still want to run the loop over many iterations, you should not print many debugging messages for every single iteration inside a loop. For example, one can print a status line every 100 iteration as:

```
for i in range(1000000):
    if i%100 == 0: print 'Test=',i
```

As you may guess, % means a remainder of $i/100$, which should be 0 for printing the string 'Test=',i.

1.4.7 Running a BeanShell Scripts

One can run BeanShell scripts and Java source codes in the same way. However, jHepWork has much less advanced error handling in this case. As in the Jython case, the [F8] key can be used fast execution of a BeanShell script.

During the execution, the output from the scripts will be redirected to the Bean-Shell window.

1.4.8 Compiling and Running Java Code

Java files can be edited in the same way. Java source code can be compiled into bytecodes using the menu from the [Run] tool-bar menu as [Run]-[Javac current file]. Similarly, one can also compile all Java source codes located in the same directory or build a jar library.

1.4.9 Working with Command-line Scripts

For advanced users, we should be more specific about how to run Jython scripts from jHepWork without the main text editor. For this, one can set the 'CLASSPATH' variable to the libraries located in the 'lib' directory. One can find an example bash script 'a_run.sh' in the directory macros/examples. To run a Jython file, say file.py, run the command:

```
bash#: a_run.sh file.py
```

from the Linux/UNIX or Mac prompt.

It is however more convenient to use the ANT tool [19], since it is a multi-platform Java program. It reads a configuration file 'build.xml' to run and compile programs using the command line. jHepWork has a special 'build.xml' file which allows to run and compile Jython macros from the command line without the main text editor. Go to the directory 'macros/cmd' and type ant. If the ANT tool is installed, one should see command-line options to run or compile the scripts. For example, to run the 'file.py' script without the main text editor using Linux, Unix or Mac, type:

```
bash#: ant run -DM file
```

If you want to compile this script into a jar library, type:

```
bash#: ant jarpy  -DM file
```

If one needs to compile standard Java source files with the purpose of creating a jar library, one should put the source files to the 'src' directory and execute the command 'ant jarjava'.

In all such cases, it is assumed that a user is working in the 'cmd' directory. If one needs to work in some other directory, modify the variable jhepwork in the 'build.xml' file.

1.4.10 jHepWork Code Assist

If a user is working with the Jython shell, all methods associated with a particular object are shown in a drop down menu after typing a dot after the name of an object and pressing [Space] after holding down the [Ctrl] key. For example, if a Jython or Java object obj was instantiated, type:

```
>>> obj.  # press [Ctrl]+[Space] for the help
```

to bring up a table with all methods associated with a class instance obj.

However, the help system works differently for the code editor. In this case, one can get the information on available public methods using the jHepWork code assist. When a user types a name of some object followed by a dot, [F4] key can be used to check the methods associated with this object. For example:

```
>>> from jhplot import P0D
>>> obj=P0D()      # create a Java object
>>> obj.           # press [F4] for  P0D methods
>>> obj="string"   # create a Jython object
>>> obj.           # press [F4] for "str" methods
```

Pressing the [F4] key after each object, followed by a dot, brings up a table with all methods that belong to a particular Jython or Java class. In the above example, the table shows the methods of the class P0D.

One can also search for a particular method using the pop-up table of the Code Assist. For this, the standard Java regular expressions [20, 21] can be used. To sort rows with methods, one should click on the column headers. Then one can push a selected method to the text editor using a double-click or mouse menu. The selected method will be inserted into the code editor at the line after the dot.

The Code Assist also allows you to look at the full API documentation of the jHPlot classes or methods. Click on a selected line in the Code Assist table and use the right mouse button to get the associated JavaDoc information. One can also push a selected method into the editor window by clicking on a pop-up mouse menu.

Many (not all!) classes of jHepWork have their own help, usually based on the original work of the developers. To access the API help documentation, use the method doc(). For example, after creation of HPlot() canvas as c1= HPlot(), one can look at the API documentation of its methods by executing the line c1.doc().

The code assist of the BeanShell window is somewhat different. Please read the corresponding help using this shell.

1.4.11 Other Features

The jHepWork GUI has several other features which should be useful while working with a text or data:

- The menu [Search] can be used for searching particular strings or substrings in an opened document;
- The menu [Run] is useful for compilation of Java source codes, building Java jar libraries, execution of Jython scripts and checking Jython syntax;
- The menu [Tool] is designed for working with LaTeX files. It provides a bibliographic manager (based on the JabRef program), LaTeX tools for common LaTeX symbols. It can be used also to start a graphical canvas for plotting and a basic image editor based on the popular ImageJ program.

It is worth exploring the menu toolbar of the IDE. Especially, if you would like to see jHepWork examples, look at the menu [Tool] and select [jHPlot examples]. This brings up a dialog with the list of available Jython examples. One can run the example macro files to get a clue on how to use the jHPlot package with numerical and graphical libraries.

Before jumping to the data analysis sections, I would like to give one advice. The jHepWork IDE is only an experimental tool and, probably, it was not tested as careful as any other professional IDE. It was designed as a light-weight source-code editor with IDE-like features to help for new users to get started. Therefore, if you are not too satisfied with the jHepWork IDE, you are likely to be a professional programmer. In this case, use ether Eclipse or NetBeans (well, if you are a professional programmer, you must know about them!). Both Eclipse and Netbean can be used for editing jHepWork scripts if you will set the $CLASSPATH to the jHepWork jar libraries inside the directory 'lib' (again, I do not need to explain this to advanced users).

Finally, you may even use popular code editors, like jEdit, Emacs or VI (Linux and Windows) or Notepad (Windows). How to run the jHepWork scripts in a console without using the IDE has been described before.

1.5   Third-party Packages and the License

1.5.1   Contributions and Third-party Packages

Note that some third-party libraries are not licensed by the GPL license but free only for *non-commercial* purposes. Therefore, generally, you can safely use the entire software for *non-commercial* purposes only. If you will use it for *commercial* purposes, you should contact the authors.

jHepWork is based on several reused classes that have been rewritten and adopted for the use together in jHepWork. This book has numerous references on other third-party packages if the discussed package contains or relies on a particular library. The author apologize in advance if some reference is missing; such omissions were wholly unintentional.

All packages listened below are subject to their licenses (as well as the packages cited in the following sections). The vast majority of the included packages are GNU-licensed or have very permissive open-source licenses.

Many projects from the list given below are not supported any longer by the original authors, but they were very useful at the time when jHepWork was under heavy development. jHepWork (version 2.2) contains:

- FreeHEP java libraries
  http://java.freehep.org;
- jEdit TextArea components (from the jEdit 2003 version, written by Slava Pestov);

- some classes from the jMySpell project
  http://jmyspell.javahispano.net/en/index.html by DreamTagnerine;
- classes from early versions of JabRef project
  http://jabref.sourceforge.net/;
- classes for color syntax highlighting taken from the Jext project
  http://www.jext.org/;
- LatexTools Beanshell macro developed for jEdit (this is still a test version);
- Debuxter package
  http://dexter.sourceforge.net/;
- classes from the JyConsole project by Artenum
  http://www.artenum.com;
- some classes of the jpEdit project
  http://www.jpedit.co.uk/;
- drawing classes from jPlot
  http://www.cig.ensmp.fr/~vanderlee/jplot/ by Jan van der Lee;
- the Surface Plotter package
  http://www.fedu.uec.ac.jp/~yanto/java/surface/ by Yanto Suryono;
- a rewrite of the Browser3d package (I could not identify its developers);
- a rewrite of the popular JaxoDraw package
  http://jaxodraw.sourceforge.net/;
- the Jakart Common Math library
  http://jakarta.apache.org/commons/math/;
- self-contained mathematics and statistics components addressing the most common problems. In addition, T. Flanagan's Java Library
  http://www.ee.ucl.ac.uk/~mflanaga/java/updates.html;
- a self-contained framework for clustering analysis by S.Chekanov, the JMinHEP package
  http://hepforge.cedar.ac.uk/jminhep/;
- the core engine of the JOONE package http://www.jooneworld.com/ and the Encog
  http://code.google.com/p/encog-java/;
- the JGraph package and the JGrapht package
  http://www.jgraph.com/
  http://jgrapht.sourceforge.net/;
- the jFreeChart package
  http://www.jfree.org/;
- the classes from the ObjectBrowser project by J. Hrivnac;
- jHepWork includes the GlobalDocs program
  http://globaldocs.zeevbelkin.com/;
- the Colt package
  http://dsd.lbl.gov/~hoschek/colt/;
- refactored classes from the PTPlot 5.6
  http://ptolemy.berkeley.edu/java/ptplot/;
- several classes from the high performance jMathTools package
  http://jmathtools.sourceforge.net/;

- look and feel based upon LiquidInf and JGoodies and therefore subject to their license;
- the Protocol Buffers library
  http://code.google.com/p/protobuf/;
- VLJTable table class from VLsolutions
  http://www.vlsolutions.com;
- classes from the ImageJ java program
  http://rsb.info.nih.gov/ij/;
- the package Sympy, a Python library for symbolic mathematics
  http://code.google.com/p/sympy/;
- several icons are from Gnome and Eclipse and therefore subject to their licenses;
- 3D-XplorMath project by the 3D-XplorMath Consortium
  http://3d-xplormath.org/.

1.5.2 Disclaimer of Warranty

jHepWork is *not commercial* product, although it is professionally written and many libraries have been tested by a large scientific community. I cannot guarantee that it is fault free in all possible foreseeable situations. Therefore, you use this package at your own risk.

The author and publisher make no warranties, express or implied, that the programs contained in this book are free of errors, or are consistent with any particular standard of merchantability. They should not be relied on for solving a problem whose incorrect solution could result in injury to a person or loss of property. If you do use the program in such a manner, it is at your own risk. The authors and publisher disclaim all liability for direct or consequential damages resulting from your use of the programs.

1.5.3 jHepWork License

jHepWork is licensed by the GNU General Public License (GPL). However, it contains some third-party libraries integrated to jHepWork which are free for *non-commercial* purposes.

Here is the GNU License:

Copyright (C) 2006 S. Chekanov. The jHepWork project.

This program is free software; you can redistribute it and/or modify it under the terms of the GNU General Public License as published by the Free Software Foundation; either version 2 of the License, or any later version.

This program is distributed in the hope that it will be useful, but without any warranty; without even the implied warranty of merchantability or fitness for a particular purpose. See the GNU General Public License for more details.

You should have received a copy of the GNU General Public License along with this program; if not, write to the Free Software Foundation, Inc., 59 Temple Place— Suite 330, Boston, MA 02111-1307, USA.

References

1. Chekanov, S.: HEP data analysis using JHEPWORK and JAVA. In: Proceedings of the HERA-LHC workshop (CERN-DESY) 2007–2008 (2008), p. 763. URL http://arxiv.org/abs/0903.3861
2. Pilgrim, M.: Dive into Python. Apress, New York (2004)
3. Guzdial, M.: Introduction to Computing and Programming in Python. A Multimedia Approach. Prentice Hall, New York (2005)
4. Martelli, A.: Python in a Nutshell (In a Nutshell (O'Reilly)). O'Reilly Media, Sebastopol (2006)
5. Lutz, M.: Learning Python, 3rd edn. O'Reilly Media, Sebastopol (2007)
6. Langtangen, H.: Python Scripting for Computational Science. Springer, Berlin/Heidelberg (2008)
7. Richardson, C., Avondolio, D., Vitale, J., Schrager, S., Mitchell, M., Scanlon, J.: Professional Java, JDK 5th edn. Wrox, Birmingham (2005)
8. Arnold, K., Gosling, J., Holmes, D.: Java(TM) Programming Language, 4th edn. Java Series. Addison-Wesley, Reading (2005)
9. Flanagan, D.: Java in a Nutshell, 5th edn. O'Reilly Media, Sebastopol (2007)
10. Eckel, B.: Thinking in Java, 4th edn. Prentice Hall PTR, Englewood Cliffs (2006)
11. Bloch, J.: Effective Java, 2nd edn. The Java Series. Prentice Hall PTR, Englewood Cliffs (2008)
12. Johnson, A.: A Java based analysis environment JAS (1996)
13. FreeHEP Java Libraries. URL http://java.freehep.org/
14. Python Programming Language. URL http://www.python.org/
15. The Jython Project. URL http://www.jython.org/
16. Brun, R., Rademakers, F., Canal, P., Goto, M.: Root status and future developments. ECONF C0303241 (2003) MOJT001
17. Brun, R., Rademakers, F.: ROOT: An object oriented data analysis framework. Nucl. Instrum. Methods A **389**, 81 (1997). URL http://root.cern.ch/
18. Wikipedia, Comparison of Java and C++. URL http://en.wikipedia.org/wiki/Comparison_of_Java_and_C
19. The Apache Ant. URL http://ant.apache.org/
20. Java regular expressions, ReGex package. URL http://java.sun.com/j2se/1.6.0/docs/api/
21. Stubblebine, T.: Regular Expression Pocket Reference: Regular Expressions for Perl, Ruby, PHP, Python, C, Java and .NET (Pocket Reference (O'Reilly)). O'Reilly Media, Sebastopol (2007)

Chapter 2
Introduction to Jython

In this chapter, we will give a short introduction to the Jython programming language. We have already pointed out that Jython is an implementation of the Python programming language, unlike CPython which is implemented in C/C++.

While these implementations provide almost identical a Python-language programming environment, there are several differences. Since Jython is fully implemented in Java, it is completely integrated into the Java platform, so one can call any Java class and method using the Python-language syntax. This has some consequences for the way you would program in Jython. During the execution of Jython programs, the Jython source code is translated to Java bytecode that can run on any computer that supports the Java virtual machine.

We cannot give a comprehensive overview of Jython or Python in this chapter: This chapter aims to describe a bare minimum which is necessary to understand the Jython language, and to provide the reader with sufficient information for the following chapters describing data-analysis techniques using the jHepWork libraries.

2.1 Code Structure and Commentary

As for CPython, Jython programs can be put into usual text files with the extension '.py'. A Jython code is a sequence of statements that can be executed normally, line-by-line, from the top to the bottom. Jython statements can also be executed interactively using the Jython shell (the tab 'JythonShell' of the jHepWork IDE).

Comments inside Jython programs can be included using two methods: (1) To make a single-line comment, put the sharp "#" at the beginning of the line; (2) To comment out a multi-line block of a code, use a triple-quoted string.

It is good idea to document each piece of the code you are writing. Documentation comments are strings positioned immediately after the start of a module, class

S.V. Chekanov, *Scientific Data Analysis using Jython Scripting and Java,*
Advanced Information and Knowledge Processing,
DOI 10.1007/978-1-84996-287-2_3, © Springer-Verlag London Limited 2010

or function. Such comment can be accessed via the special attribute __doc__. This attribute will be considered later on when we will discuss functions and classes.

Jython statements can span multiple lines. In such cases, you should add a back slash at the end of the previous line to indicate that you are continuing on the next line.

2.2 Quick Introduction to Jython Objects

As for any dynamically-typed high-lavel computational language, one can use Jython as a simple calculator. Let us use the jHepWork IDE to illustrate this. Start the jHepWork and click on the "JythonShell" tab panel. You will see the Jython invitation ">>>" to type a command. Let us type the following expression:

```
>>>   100*3/15
```

Press [Enter]. The prompt returns "20" as you would expect for the expression (100*3)/15. There was no any assignment in this expression, so Jython assumes that you just want to see the output. Now, try to assign this expression to some variable, say W:

```
>>> W=100*3/15
```

This time, no any output will be printed, since the output of the expression from the right side is assigned directly to the variable W. Jython supports multiple assignments, which can be rather handy to keep the code short. Below we define three variables W1, W2 and W3, assigning them to 0 value:

```
>>> W1=W2=W3=0
```

One can also use the parallel assignments using a sequence of values. In the example below we make the assignment W1=1, W2=2 and W3=3 as:

```
>>> W1,W2,W3=1,2,3
```

At any step of your code, you can check which names in your program are defined. Use the built-in function dir() which returns a sorted list of strings

```
>>> dir()
['W','W1','W2','W3','__doc__', '__name__' ...
```

(we have truncated the output in this example since the actual output is rather long). So, Jython knows about all our defined variables including that defined using the character W. You will see more variables in this printed list which are predefined by jHepWork.

One can print out variables with the print() method as:

```
>>>  print W1
1
```

One can also append a comment in front:

```
>>>  print 'The output =',W1
The output = 1
```

You may notice that there is a comma in front the variable W1. This is because 'The output' is a string, while 'W1' is an integer value, so their types are distinct and must be separated in the print statement.

How do we know the variable types? For this we can use the type() method which determines the object type:

```
>>>  type(W1)
<type 'int'>
```

The output tells that this variable holds an integer value (the type "int").

Let us continue with this example by introducing another variable, 'S' and by assigning a text message to it. The type of this variable is "str" (string).

```
>>>  S='The output ='
>>>  type (S)
<type 'str'>
```

So, the types of the variables 'W1' and S are different. This illustrates the fact that Jython, as any high-level language, determines types based on assigned values during execution, i.e. a variable may hold a binding to any kind of object. This feature is very powerful and the most useful for scripting: now we do not need to worry about defining variable types before making assignments to a variable. Clearly, this significantly simplifies program development.

Yet, the mechanics behind such useful feature is not so simple: the price to pay for such dynamical features is that all variables, even such simple as 'W1' and 'S' defined before, are objects. Thus, they are more complicated than simple types in other programming languages, like C/C++ or FORTRAN. The price to pay is slower execution and larger memory consumption. On the other hand, this also means that you can do a lot using such feature! It should also be noted that some people use the word "value" when they talk about simple types, such as numbers and strings. This is because these objects cannot be changed after creation, i.e. they are immutable.

First, let us find out what can we do with the object 'W1'. We know that it holds the value 10 (and can hold any value). To find out what can be done with any objects in the JythonShell window is rather simple: Type 'W1' followed by a dot and press the [Space] key by holding down [Ctrl]:

```
>>> W1. [Ctrl]-[Space]
```

You will see a list of methods attributed to this object. They usually start as __method__, where 'method' is some attribute. For example, you will see the method like __str__, which transforms an object of type integer to a string. So, try this

```
>>> SW=str(W1); type(SW)
<type 'str'>
```

Here we put two Jython statements on one line separated by a semi-column. This, probably, is not very popular way for programming in Jython, but we use it to illustrate that one can this syntax is also possible. In some cases, however, a program readability can significantly benefit from this style if a code contains many similar and short statements, such as a1=1; a2=2. In this case, the statements have certain similarity and it is better to keep them in one single logical unit. In addition, we will use this style in several examples to make our code snippets short.

The last expression in the above line does not have "=", so Jython assumes that what you really want is to redirect the output to the interactive prompt. The method type() tells that "SW" is a string. As before, you may again look at the methods of this object as:

```
>>> SW. [Ctrl]-[Space]
```

This displays a list of the methods attributed to this object. One can select a necessary method and insert it right after the dot.

In addition to the JythonShell help system, one can discover the attributes of each Jython object using the native Jython method dir():

```
>>> dir(SW)
```

In the following sections we will discuss other methods useful to discover attributes of Java objects.

2.2.1 Numbers as Objects

Numbers in Jython are immutable objects called values, rather than simple types as in other programming languages (C/C++, Fortran or Java). There are two main types: integers (no fractional part) and floats (with fractional part). Integers can be represented by long values if they are followed by the symbol 'L'. Try this:

```
>>> Long=20L
<type 'long'>
```

The only limit in the representation of the long numbers is the memory of Java virtual machine.

Let us take a look again at the methods of a real number, say "20.2" (without any assignment to a variable).

```
>> 20.2. [Ctrl]-[Space]
```

or, better, you can print them using the method "dir()"

```
>>>  dir(20.2)
```

Again, since there is no any assignment, Jython just prints the output of the dir() method directly to the same prompt. Why they are needed and what you can do with them? Some of them are rather obvious and shown in Table 2.1.

Table 2.1 A short overview of the Jython operators for values

Jython operators for values		
$abs(x)$	__abs__	absolute value
$pow(x, y)$ or y**x	__pow__	raise x to the power y
$-x, +x$	__neg__, __pos__	negative or positive
+, -	__radd__, __rsub__	add or subtract
*, /	__rmul__, __rdiv__	add or subtract
$x < y, x > y$	__com__	less or larger. Returns 0 (false) or 1 (true)
$cmp(x, y)$	__com__	compare numbers. Returns 0 (false) or 1 (true)
$x <= y, x >= y$	–	comparison: less (greater) or equal
$x == y, x != y$	–	comparison: equal or not equal
$str(x)$	__str__	convert to a string
$float(x)$	__float__	convert to float
$int(x)$	__int__	convert to integer
$long(x)$	__long__	convert to long

There are more methods designed for this object, but we will not go into further discussion. Just to recall: any number in Jython is an object and you can manipulate with it as with any object to be discussed below. For example, Jython integers are objects holding integer values. This is unlike C++ and Java where integers are primitive types.

Is it good if simple entities, such as numbers are, have properties of objects? For interactive manipulation with a code and fast prototyping, probably we do not care so much, or even can take advantage of this property. But, for numerical libraries, this feature is unnecessary and, certainly, is too heavy for high-performance calculations. We will address this issue later in the text.

2.2.2 Formatted Output

In the above examples we have used the `print` command without setting control over the way in which printed values are displayed. For example, in the case of the expression "1.0/3.0", Jython prints the answer with 17 digits after the decimal place!

Obviously, as for any programming language, one can control the way the values are displayed: For this, one can use the % command to produce a nicely formatted output. This is especially important of one needs to control the number of decimal places the number is printed to. For example, one can print 1.0/3.0 to three decimal places using the operator %.3f inside the string:

```
>>> print 'The answer is %.3f'%(1.0/3)
The answer is 0.333
```

As you can see, Jython replaces the character "f" with the variable value that follows the string. One can print more than one variable as shown in the example below:

```
>>> print 'The answer is %.3f and %.1f'% (1.0/3, 2.0/3)
The answer is 0.333 and 0.7
```

One can also use the operator % to control the width of the displayed number, so one can make neatly aligned tables. For example, the string "10.1f" forces the number to be printed such that it takes up to ten characters. The example below shows how to do this using the new-line character to print the second number on a new line. As one can see, we align this second number with that printed on the first line:

```
>>> print 'The answer: %.3f \n  %13.1f'% (1.0/3, 2.0/3)
The answer: 0.333
            0.7
```

2.2.3 Mathematical Functions

To perform mathematical calculations with values, one should use the Jython `math` module which comes from the standard specification of the Python programming language. Let us take a look at what is inside of this module. First, we have to import this module using the "`import math`" statement:

```
>>> import math
```

Use the usual approach to find the methods of this module:

```
>>> dir(math)
['acos', 'asin', 'atan', 'atan2', 'ceil',
 'classDictInit', 'cos', 'cosh', 'e', 'exp',
 'fabs', 'floor', 'fmod', 'frexp', 'hypot',
 'ldexp', 'log', 'log10', 'modf', 'pi', 'pow',
 'sin', 'sinh', 'sqrt', 'tan', 'tanh']
```

Most of us are familiar with all these mathematical functions that have the same names in any programming language. To use these functions, type the module name `math` followed by the function name. A dot must be inserted to separate the module and the function name:

```
>>> math.sqrt(20)
4.47213595499958
```

As before for JythonShell, one can pick up a necessary function as:

```
>>> math.   [Ctrl]-[Space]
```

It should be noted that, besides functions, the `math` module includes a few well-known constants: π and e:

```
>>> print "PI=", math.pi
PI= 3.141592653589793
>>> print "e=", math.e
e= 2.718281828459045
```

If you have many mathematical operations and want to make a code shorter by skipping the "`math`" attribute in front of each function declaration, one can explicitly import all mathematical functions using the symbol "`*`":

```
>>> from math import *
>>> sqrt(20)
4.47213595499958
```

2.2.4 Complex Numbers

Python has a natural support for complex numbers. Just attach "J" or "j" for the imaginary part of a complex number:

```
>>> C=2+3j
>>> type(C)
<type 'complex'>
```

Once a complex number is defined, one can perform mathematical manipulations as with the usual numbers. For example:

```
>>> 1j*1j
(-1+0j)
```

Mathematical operations with complex numbers can be performed using the 'cmath' module, which is analogous to the 'math' module discussed above. The example below demonstrates how to calculate hyperbolic cosine of a complex value:

```
>>> import cmath
>>> print cmath.cosh( 2+3j )
(-3.724+0.511j)
```

The output values for the real and imaginary part in the above example were truncated to fit the page width.

2.3 Strings as Objects

Strings are also treated as values since they are immutable. To define a string, one should enclose it in double (") or single (') quote. The escape character is a backslash, so one can put a quote character after it. The newline is given by *n* directly after the backslash character. Two strings can be added together using the usual "+" operator.

As mentioned above, an arbitrary value, val, can be converted into a string using the method str(val). To convert a string into int or float value, use the methods `int(str)` or `float(str)`. Below we illustrate several such conversions:

```
>>> i=int('20')
>>> type(i)
<type 'int'>
>>> f=float('20.5')
>>> type(f)
<type 'float'>
```

As before, all the methods associated with a string can be found using [Ctrl]-[Space] or the `dir()` method:

```
>>> dir('s')
...
'capitalize', 'center', 'count','decode',
'encode', 'endswith', 'expandtabs', 'find',
'index', 'isalnum', 'isalpha', 'isdecimal',
'isdigit', 'islower', 'isnumeric', 'isspace'...
```

(we display only a few first methods). Some methods are rather obvious and do not require explanation. All methods that start from the string "`is`" check for a particular string feature.

Below we list more methods:

`len(str)`	gives the number of characters in the string `str`
`string.count(str)`	counts the number of times a given word appears in a string
`string.found(str)`	numeric position of the first occurrence of word in the string
`str.lower()`	returns a string with all lower case letters
`str.upper()`	returns a string with all upper case letters

Strings can be compared using the standard operators: `==`, `!=`, `<`, `>`, `<=`, and `>=`.

2.4 Import Statements

There are several ways that can be used to import a Java or Python package. One can use the '`import`' statement followed by the package name. In case of Python, this corresponds to a file name without the extension '`.py`'. The import statement executes the imported file, unlike lower-level languages, like C/C++, where the `import` statement is a preprocessor statement. The consequence of this is that the `import` statement can be located in any place of your code, as for the usual executable statement. We have seen already how to import the Python package "`math`".

Here is an example illustrating how to import the Java Swing package (usually used to build a GUI):

```
>>> from javax.swing import *
```

In the above example we use the wildcard character "*" to import all packages from Java Swing. In this book, you will see that we use "*" wildcard almost for every example, since we what to keep our examples short. This is often considered as a bad style since it "pollutes" the global namespace. However, if you know that the code is not going to be very long and complicated, we should not worry too much about this style.

Let us give another example showing how to import of a Java class. We remind that the code below works only for Python implemented in Java (Jython):

```
>>> from javax.swing import JFrame
```

This time we have imported only a single class (JFrame) from Java, unlike the previous example with the "polluted" namespace.

Another way to import classes is to use the 'import' statement without the string 'from'. For example:

```
>>> import javax.swing
```

In this case, we should use the qualified names, i.e.:

```
>>> f=javax.swing.JFrame('Hello')
```

Although it takes more typing, we have avoided polluting the global namespace of our code.

2.4.1 Executing Native Applications

In Sect. 1.4.3 we have shown that native applications can be run using JythonShell by appending "!" in front of an external command. In addition, one can also use Jython 'os.system' package to run an external program.

The code below shows how to run an external command. In this example, we bring up the Acroread PDF file viewer (it should be found on the current PATH if this program installed on your system):

```
>>> import os
>>> rc=os.system('acroread')
```

```
>>> if rc == 0:
>>> ... print 'acroread started successfully'
```

The statement 'if' checks whether the execution has succeeded or not. We will discuss the comparison tests in the next section. The same operation will look like !acroread when using the Jython shell.

2.5 Comparison Tests and Loops

2.5.1 The 'if-else' Statement

Obviously, as in any programming language, one can use the 'if-else' statement for decision capability of your code. The general structure of comparison tests is

```
if [condition1]:
      [statements to execute if condition1 is true]
elif [condition2]:
      [statements to execute if condition2 is true]
....

else:
      [rest of the program]
```

The text enclosed in square brackets represents some Jython code. After the line with the statement 'if', the code is placed farther to the right using white spaces in order to define the program block. Either space characters or tab characters (or even both!) are accepted as forms of indentation. In this book, we prefer two spaces for indentation. It should also be noted that the exact amount of indentation does not matter, only the relative indentation of nested blocks (relative to each other) is important.

The indentation is good Python feature: The language syntax forces to use the indentation that you would have used anyway to make your program readable. Thus even a lousy programmer is forced to write understandable code!

Now let us come back to the comparison tests. The [condition] statement has several possibilities for values 'a' and 'b' values as shown in Table 2.2:

Let us illustrate this in the example below:

```
>>> a=1; b=2;
>>> if a*b>1:
>>> .. print "a*b>1"
>>> else:
>>> .. print "a*b<=1"
```

Comparison tests	
$a == b$	a is equal to b
$a! = b$	a is not equal to b
$a > b$	a is greater than b
$a >= b$	a is greater than or equal to b
$a < b$	a is less than b
$a <= b$	a is less than or equal to b
$a == b$	a is equal to b
$a! = b$	a is not equal to b

In case if you will need more complex comparisons, use the boolean operators
such as 'and' and 'or':

```
>>> a=1; b=0
>>> if a>0 and b=0:
>>> ..print 'it works!'
```

One can also use the string comparisons =   (equal) or ! = (not equal). The comparison statements are case sensitive, i.e. 'a' == 'A' is false.

2.5.2 Loops. The "for" Statement

The need to repeat a statement or a code block is essential feature of any numerical calculation. There is, however, one feature you should be aware of: Python should be viewed as an "interface" type of language, rather than that used for heavy repeated operations like long loops over values. According to the author's experience, if the number of iterations involving looping over values is larger than several thousands, such part of the code should be moved to an external library to achieve a higher performance and a lower memory usage compared to the Python code operating with loops over objects. In case of Java, such libraries should be written in Java.

In this section we will be rather short. One can find more detailed discussion about this topic in any Python textbook.

The simplest loop which prints, say, 10 numbers is shown below:

```
>>> for i in range(10):
>>> ... print i
```

This 'for' loop iterates from 0 to 9. Generally, you can increment the counter by any number. For example, to print numbers from 4 to 10 with the step 2, use this example:

```
>>> for i in range(4,10,2):
>>> ... print i
```

2.5.3 The 'continue' and 'break' Statements

The loops can always be terminated using the 'break' statement, or some iterations can be skipped using the 'continue' statement. All such control statements
are rather convenient, since help to avoid various 'if' statements which makes the
Python code difficult to understand. This is illustrated in the example bellow:

```
>>> for i in range(10):
>>> ... if (i == 4): continue
>>> ... if (i == 8): break
>>> ... print i
```

In this loop, we skip the number 6 and break the loop after the number 8:

2.5.4 Loops. The 'while' Statement

One can also construct a loop using the 'while' statement, which is more flexible
since its iteration condition could be more general. A generic form of such loop is
shown below:

```
while CONDITION:
... <Code Block as long as CONDITION is true>
```

Let us give a short example which illustrates the while loop:

```
>>> a=0
>>> while a<10:
>>> ... a=a+1
```

The while loop terminates when a=10, i.e. when the statement after the
'while' is false. As before, one can use the control statements discussed above
to avoid overloading the execution block with various "if" statements.

One can also create an infinite loop and then terminate it using the "break"
statement:

```
>>> a=0
>>> while 1:
>>> ... print "infinite loop!"
>>> ... a=a+1;
>>> ... if a>10:
>>>            break
>>> ... print i
```

In this example, the 'break' statement together with the 'if' condition controls the number of iterations.

2.6 Collections

Data-analysis computations are usually based on object collections, since they have being designed for various repetitive operations on sequential data structures—exactly what we mostly do when analyzing multiple measurements. In addition, a typical measurement consists of a set of observations which have to be stored in a data container as a single unit.

Unlike to other languages, we consider Python collections to be useful mainly for storing and manipulation with other high-level objects, such as collections with a better optimized performance for numerical calculations. In this book, we will use the Jython collections to store sets of jHepWork histograms, mathematical functions, Java-based data containers and so on.

Of course, one can use Jython collections to keep numerical values, but this approach is not going to be very efficient: An analysis of such values requires Python loops which are known to be slow. Secondly, there are no too many pre-build Jython libraries for object manipulation.

Nevertheless, in many parts of this books we will use collections which contain numerical values: this is mainly for pedagogical reasons. Besides, we do not care too much about the speed of our example programs when analyzing tens of thousands events.

2.6.1 Lists

As you may guess, a list is an object which holds other objects, including values. The list belongs to a *sequence*, i.e. an ordered collection of items.

2.6.2 List Creation

An empty list can be created using squared brackets. Let us create a list and check its methods:

```
>>> list=[]
>>> dir(list) #  or list. + [Ctrl]+[Space]
```

One can also create a list which contains integer or float values during the initialization:

```
>>> list=[1,2,3,4]
>>> print list
[1, 2, 3, 4]
```

The size of this list is accessed using the len(list) method. The minimum and maximum values are given by the min(list) and max(list) methods, respectively. Finally, for a list which keeps numerical values, one can sum-up all list elements as sum(list).

One can create a mixed list with numbers, strings or even other lists:

```
>>> list=[1.0,'test',int(3),long(2),[20,21,23]]
>>> print list
[1.0, 'test', 3, 2L, [20, 21, 23]]
```

One can obtain each element of the list as list[i], where 'i' is the element index, 0<i<len(list). One can select a slice as list[i1:i2], or even select the entire list as list[:]. A slice which selects index 0 through 'i' can be written as list[:i]. Several lists can be concatenated using the plus operator '+', or one can repeat the sequence inside a list using the multiplication '*'.

As before, one can find the major methods of the list using [Ctrl]+[Space] keys. Some methods are rather obvious:

To add a new value, use the method append():

```
>>> list.append('new string')
```

A typical approach to fill a list in a loop would be:

```
>>> list=[]
>>> for i in range(4,10,2):
>>> ...    list.append(i)
```

(here, we use a step 2 from 4 to 10). The same code in a more elegant form looks like:

```
>>> list=range(4, 10, 2)
>>> print list
[4, 6, 8]
```

If one needs a simple sequence, say from 0 to 9 with the step 1, this code can be simplified:

```
>>> list=range(10)
>>> print 'List from 0 to 9:',list
List from 0 to 9: [0,1,2,3,4,5,6,7,8,9]
```

One can create a list by adding some condition to the range statement. For example, one create lists with odd and even numbers:

```
>>> odd =range(1,10)[0::2]
>>> even=range(1,10)[1::2]
```

Another effective "one-line" approach to fill a list with values is demonstrated below:

```
>>> import math
>>> list = [math.sqrt(i) for i in range(10)]
```

Here we created a sequence of sqrt(i) numbers with $i = 0..9$.

Finally, one can use the 'while' statement for adding values in a loop. Below we make a list which contains ten zero values:

```
>>> list=[]
>>> while len(list)<10:
>>> ... list.append(0)
```

2.6.3 Iteration over Elements

Looping over a list can be done with the 'for' statement as:

```
>>> for i in list:
>>> ...print i
```

or calling its elements by their index 'i':

```
>>> for i in range(len(list)):
>>> ...print list[i]
```

2.6.3.1 Sorting, Searches, Removing Duplicates

The list can be sorted with the `sort()` method:

```
>>> list.sort()
>>> print list
[1.0, 2L, 3, [20, 21, 23], 'new string', 'test']
```

To reverse the list, use the method `reverse()`.

To insert a value, use the `insert(val)` method, while to remove an element, use the `remove(val)` method. Finally, one can delete either one element of a list or a slice of elements. For example, to remove one element with the index i1 of a list use this line of the code: `'del list[i1]'`. To remove a slice of elements in the index range i1–i2, use `'del list[i1:i]'`. To empty a list, use `'del list[:]'`. Finally, `'del list'` removes the list object from the computer memory.

It should be noted that the list size in the computer memory depends on the number of objects in the list, not on the size of objects, since the list contains pointers to the objects, not objects themselves.

Advanced statistical analysis will be considered in Sect. 7.4, where we will show how to access the mean values, median, standard deviations, moments and etc. of distributions represented by Jython lists.

Jython lists are directly mapped to the Java ordered collection `List`. For example, if a Java function returns `ArrayList<Double>`, this will be seen by Jython as a list with double values.

To search for a particular value `'val'`, use

```
>>> if val in list:
>>> ...print 'list contains', val
```

For searching values, use the method `index(val)`, which returns the index of the first matching value. To count the number of matched elements, the method `count(val)` can be used (it also returns an integer value).

2.6.4 Removal of Duplicates

Often, you may need to remove a duplicate element from a list. To perform this task, use the-called dictionary collection (will be discussed below). The example to be given below assumes that a list object has been created before, and now we create a new list (with the same name) but without duplicates:

```
>>> tmp={}
>>> for x in list:
>>> ...tmp[x] = x
>>> list=tmp.values()
```

This is usually considered to be the fastest algorithm (and the shortest). However, this method works for the so-called hashable objects, i.e. class instances with a "hash" value which does not change during their lifetime. All Jython immutable built-in objects are hashable, while all mutable containers (such as lists or dictionaries to be discussed below) are not. Objects which are instances of user-defined Jython or Java classes are hashable.

For unhashable objects, one can first sort objects and then scan and compare them. In this case, a single pass is enough for duplicate removal:

```
>>> list.sort()
>>> last = list[-1]
>>> for i in range(len(list)-2, -1, -1):
>>> ...if last==list[i]:
>>>        del list[i]
>>> ...else:
>>>        last=list[i]
```

The code above is considered to be the second fastest method after that based on the dictionaries. The method above works for any type of elements inside lists.

2.6.4.1 Examples

Lists are very handy for many data-analysis applications. For example, one can keep names of input data files which can be processed by your program in a sequential order. Or, one can create a matrix of numbers for linear algebra. Below we will give two small examples relevant for data analysis:

A matrix. Let us create a simple matrix with integer or float numbers:

```
>>> mx=[
...    [1, 2],
...    [3, 4],
...    [5, 6],
...    ]
```

One can access a row of this matrix as mx[i], where 'i' is a row index. One can swap rows with columns and then access a particular column as:

```
>>> col=[[x[0] for x in mx], [x[1] for x in mx]]
>>> print col
[[1, 3, 5], [2, 4, 6]]
```

In case of an arbitrary number of rows in a matrix, use the map container for the same task:

```
>>> col=map(None,*mx)
>>> print col
[[1, 3, 5], [2, 4, 6]]
```

Advanced linear-algebra matrix operations using a pure Jython approach will be considered in Sect. 7.5.4.

Records with measurements. Now we will show that the lists are very flexible for storing records of data. In the example below we create three records that keep information about measurements characterized by some identification string, a time stamp indicating when the measurement is done and a list with actual numerical data:

```
>>> meas=[]
>>> meas.append(['test1','06-08-2009',[1,2,3,4]])
>>> meas.append(['test2','06-09-2009',[8,1,4,4,2]])
>>> meas.append(['test3','06-10-2009',[9,3]])
```

This time we append lists with records to the list holding all event records. We may note that the actual numbers are stored in a separate list which can have an arbitrary length (and could also contain other lists). To access a particular record inside the list meas use its indexes:

```
>>> print meas[0]
>>> ['test1', '06-08-2009', [1, 2, 3, 4]]
>>> print meas[0][2]
[1, 2, 3, 4]
```

2.6.5 Tuples

Unlike lists, tuples cannot be changed after their creation, thus they cannot grow or shrink as the lists. Therefore, they are *immutable*, similar to the values. As the Jython lists, they can contain objects of any type. Tuples are very similar to the lists and can be initiated in a similar way:

```
>>> tup=()                      # empty tuple
>>> tup=(1,2,"test",20.0) # with 4 elements
```

Of course, now operations that can change the object (such as append()), cannot be applied, since we cannot change the size of this container.

In case if you need to convert a list to a tuple, use this method:

```
>>> tup=tuple([1,2,3,4,4])
```

Below we will discuss more advanced methods which add more features to manipulations with the lists and tuples.

2.6.6 Functional Programming. Operations with Lists

Functional programming in Jython allows to perform various operations on data structures, like lists or tuples. For example, to create a new list by applying the formula:

$$\frac{b[i] - a[i]}{b[i] + a[i]} \tag{2.1}$$

for each element of two lists, a and b, you would write a code such as:

```
>>> a=[1,2,3]
>>> b=[3,4,5]
>>> c=[]
>>> for i in range(len(a)):
>>> ... c.append( b[i]-a[i] / (a[i]+b[i]) )
```

To circumvent such unnecessary complexity, one can reduce this code to a single line using functional programming:

```
>>> a=[1.,2.,3.]
>>> b=[3.,4.,5.]
>>> c= map(lambda x,y: (y-x)/(y+x),a,b)
>>> print c
[0.5, 0.33, 0.25]
```

The function map creates a new list by applying (2.1) for each element of the input lists. The statement lambda creates a small anonymous function at runtime which tells what should be done with the input lists (we discuss this briefly in Sect. 2.10).

As you can see, the example contains much lesser code and, obviously, programming is done at a much higher level of abstraction than in the case with the usual loops over list elements.

To build a new list, one can also use the 'math' module. Let us show a rather practical example based on this module: assume we have made a set of measurements, and, in each measurement, we simply counting events with our observations. The statistical error for each measurement is the square root of the number of events, in case of counting experiments like this. Let us generate a list with statistical errors from the list with the numbers of events:

```
>>> data=[4,9,25,100]
>>> import math
>>> errors= map(lambda x: math.sqrt(x),data)
>>> print errors
[2.0, 3.0, 5.0, 10.0]
```

The above calculation requires one line of the code, excluding the standard 'import' statement and the 'print' command.

Yet, you may not be totally satisfied with the 'lambda' function: sometime one needs to create a rather complicated function operating on lists. Then one can use the standard Jython functions:

```
>>> a=[1.,2.,3.]
>>> b=[3.,4.,5.]
>>> def calc(x,y):
>>> ... return (y-x)/(x+y)
>>> c= map(calc,a,b)
>>> print c
[0.5, 0.33, 0.25]
```

The functionality of this code is totally identical to that of the previous example. But, this time, the function calc() is the so-called "named" Jython function. This function can contain rather complicated logic which may not fit to a single-line 'lambda' statement.

One can also create a new list by selecting certain elements. In this case, use the statement filter() which accepts an one-argument function. Such function must return the logical true if the element should be selected. In the example below we create a new list by taking only positive values:

```
>>> a=[-1,-2,0,1,2]
>>> print "filtered:",filter(lambda x: x>0, a)
filtered: [1, 2]
```

As before, the statement 'lambda' may not be enough for more complicated logic for element selection. In this case, one can define an external (or named) function as in the example below:

```
>>> a=[-1,-2,0,1,2]
>>> def posi(x):
>>> ... return x > 0
>>> print "filtered:",filter(posi, a)
filtered: [1, 2]
```

Again the advantage of this approach is clear: we define a function posi(), which can arbitrary be complicated, but the price to pay is more codding.

Finally, one can use the function reduce() that applies a certain function to each pair of items. The results are accumulated as shown below:

```
>>> print "accumulate:",reduce(lambda x, y: x+y,[1,2,3])
>>> accumulate: 6
```

The same functional programming methods can be applied to the tuples.

2.6.7 Dictionaries

Another very useful container for analysis of data is the so-called dictionary. If one needs to store some objects (which, in turn, could contain other objects, such as more efficiently organized collections of numbers), it would be rather good idea to annotate such elements. Or, at least, to have some human-readable description for each stored element, rather than using an index for accessing elements inside the container as for lists or tuples. Such a description, or the so-called "key", can be used for fast element retrieval from a container.

Dictionaries in Jython (as in Python) are designed for one-to-one relationships between keys and values. The keys and the corresponding values can be any objects. In particular, the dictionary value can be a string, numerical value or even other collection, such as a list, a tuple, or other dictionary.

Let us give an example with two keys in form of strings, 'one' and 'two', which map to the integer values '1' and '2', respectively:

```
>>> dic={'one':1, 'two':2}
>>> print dic['one']
1
```

In this example, we have used the key 'one' to access the integer value '1'. One can easily modify the value using the key:

```
>>> dic['one']=10
```

It should be noted that the keys cannot have duplicate values. Assigning a value to the existing key erases the old value. This feature was used when we removed duplicates from the list in Sect 2.6.3.1. In addition, dictionaries have no concept of order among elements.

One can print the available keys as:

```
>>> print dic.keys()
```

The easiest way to iterate over values would be to loop over the keys:

```
>>> for key in dic:
>>> ... print key, 'corresponds to', dic[key]
```

Before going further, let us rewrite the measurement example given in the previous section when we discussed the lists. This time we will use record identifications as keys for fast retrieval:

```
>>> meas={}
>>> meas['test1']=['06-08-2009',[1,2,3,4]]
>>> meas['test2']=['06-09-2009',[8,1,4,4,2]]
>>> meas['test3']=['06-10-2009',[9,3]]
>>> print meas['test2']
['06-09-2009', [8, 1, 4, 4, 2]]
```

In this case, one can quickly access the actual data records using the keys. In our example, a single data record is represented by a list with the date and additional list with numerical values.

Let us come back to the description of the dictionaries. Here are a few important methods we should know about:

dic.clear()   clean a dictionary;
dic.copy()   make a copy;
has_key(key)   test, is a key present?;
keys()   returns a list of keys;
values()   returns a list of values in the dictionary.

One can delete entries from a dictionary in the same way as for the list:

```
>>> del dic['one']
```

One can sort the dictionary keys using the following approach: convert them into a list and use the `sort()` method for sorting:

```
>>> people = {'Eve':10, 'Tom': 20, 'Arnold': 50}
>>> list = people.keys()
>>> list.sort()
>>> for p in list:
>>> ...  print p,'is ',people[p]
Arnold is   50
Eve is   10
Tom is   20
```

2.7 Java Collections in Jython

It was already said that the concept of collections is very important for any data analysis, since "packing" multiple records with information into a single unit is a very common task.

There are many situations when it is imperative to go beyond the standard Python-type collections implemented in Jython. The strength of Jython is in its complete integration with Java, thus one can call Java collections to store data. Yes, the power of Java is in your hands!

To access Java collections, first you need to import the classes from the package `java.util`. Java collections usually have the class names started with capital letters, since this is the standard convention for class names in the Java programming language. With this observation in mind, there is a little chance for mixing Python collections with Java classes during the code development. In this section, we will consider several collections from the Java platform.

2.7.1 List. An Ordered Collection

To build an ordered collection which contain duplicates, use the class `List` from the Java package `java.util`. Since we are talking about Java, one can check what is inside of this Java package as:

```
>>> from java.util import *
>>> dir()
[ .. 'ArrayList','Currency','Date',List,Set,Map]
```

Here we printed only a few Java classes to fit the long list of classes to the page width. One can easily identify the class `ArrayList`, a class which is usually used to keep elements in a list. One can check the type of this class and its methods using either `dir()` or the JythonShell code assist:

```
>>> from java.util import *
>>> jlist=ArrayList()
>>> type(jlist)
<type 'java.util.ArrayList'>
>>> dir(jlist):
[... methods ...]
>>> jlist. #  [Ctrl]+[Space]
```

As you can see, the `type()` method indicates that this is a Java instance, so we have to use the Java methods of this instance for further manipulation. Let us add elements to this list and print them:

```
>>> e=jlist.add('test')
>>> e=jlist.add(1)
>>> jlist.add(0,'new test')
>>> e=jlist.add(2)
>>> print jlist
[new test, test, 1, 2]
>>> print jlist.get(0)
new test
>>> print jlist.toArray()
array(java.lang.Object,['new test', 'test', 1, 2])
```

You may notice that when we append an element to the end of this list, we assign the result to the variable `'e'`. In Jython, it returns '1' for success (or `true` for Java). We also can add an object `obj` at the position characterized with the index i using the method `add(i,obj)`. Analogously, one can access elements by their integer positions. For example, one can retrieve an object back using the method `get(i)`. The list of elements can be retrieved in a loop exactly as we usually do for the Jython lists. Let us show a more complete example below:

```
––––––––––––––––––––– Java list example –––––––––––––––––––––

from java.util import *

jlist=ArrayList()
# append integers
for i in range(100):
    jlist.add(i)
print jlist.size()

# replace at 0 position
jlist.set(0,100)
s=jlist
print type(s)
```

```
# range between 0-50
newlist=jlist.subList(0,50)
for j in newList:
    print j
```

Run the above code and make sense of its output.

Probably, there are not too strong reasons to use Java `List` while working with Jython, since the native Jython list discussed in the previous section should be sufficient for almost any task. However, it is possible that you will need to use Java lists in order to integrate your application natively into the Java platform after moving your code into a pure Java codding.

2.7.1.1 Sorting Java Lists

One can do several manipulations with the `List` using the Java `Collection` class. Below we show how to sort a list using the natural ordering of its elements, and how to reverse the order:

```
>>> from java.util import *
>>> jlist=ArrayList()
>>> jlist.add('zero'); jlist.add('one'); jlist.add('two')
>>> Collections.sort(jlist)
>>> print jlist
>>> [one, two, zero]
>>> Collections.reverse(jlist)
>>> print jlist
>>> [zero, two, one]
```

The next question is how to sort a list with more complicated objects, using some object attribute for sorting. Consider a list containing a sequence of other lists as in the case shown below:

```
>>> from java.util import *
>>> jlist=ArrayList()
>>> jlist.add([2,2]); jlist.add([3,4]); jlist.add([1,1])
>>> print jlist
[[2, 2], [3, 4], [1, 1]]
```

Here there is a small problem: how can we tell to the method `sort()` that we want to perform a sorting using a first (or second) item in each element-list? Or, more generally, if each element is an instance of some class, how can we change ordering objects instantiated by the same class?

One can do this by creating a small class which implements the `Comparator` interface. We will consider Jython classes in Sect. 2.11, so at this moment just accept this construction as a simple prescription that performs a comparison of two objects. The method `compare(obj1,obj2)` of this class compares objects and returns a negative value, zero, or a positive integer value depending on whether the object is less than, equal to, or greater than the specified object. Of course, it is up to you to define how to perform such object comparison. For the example above, each object is a list with two integers, so one can easily prototype a function for object comparison. Let us write a script which orders the list in increasing order using the first element of each list:

—————————————— Sorting Java lists ——————————————

```
from java.util import *

jlist=ArrayList()
jlist.add([2,2]); jlist.add([3,4]); jlist.add([1,1])

class cmt(Comparator):
    def compare(self, i1,i2):
        if i1[0]>i2[0]: return 1
        return 0

Collections.sort(jlist,cmt())
print jlist
```

After running this script, all elements will be ordered and the print method displays `[[1, 1],[2, 2],[3, 4]]`.

We will leave the reader here. One can always find further information about the Java lists from any Java textbook.

2.7.2 Set. A Collection Without Duplicate Elements

The `Set` container from the package `java.util` is a Java collection that cannot contain duplicate elements. Such set can be created using general-purpose implementations based on the `HashSet` class:

```
>>> from java.util import *
>>> s=HashSet()
>>> e=s.add('test')
>>> e=s.add('test')
>>> e=s.add(1)
>>> e=s.add(2)
>>> print s
[1, 2, test]
```

As you can see from this example, the string `'test'` is automatically removed from the collection. Operations with the Java sets are exactly the same as those with the `ArrayList`. One can loop over all elements of the set collection using the same method as that used for the `ArrayList` class, or one can use a method by calling each element by its index:

```
>>> for i in range(s.size()):
>>> ...print s[i]
```

As in the case with the Java lists, you may face a problem when go beyond simple items in the collection. If you want to store complicated objects with certain attributes, what method should be used to remove duplicates? You can do this as well but make sure that instances of the class used as elements inside the Java set use hash tables (most of them do). In case of the example shown in Sect. 2.7.1.1, you cannot use `HashSet` since lists are unhashinable. But with tuples, it is different: Tuples have hash tables, so the code snippet below should be healthy:

```
>>> from java.util import *
>>> s=HashSet()
>>> e=s.add( (1,2) )
>>> e=s.add( (2,4) )
>>> e=s.add( (1,2) )
>>> print s
[(2, 4), (1, 2)]
```

As you can see, the duplicate entry (1,2) is gone from the container. In case if you need to do the same with Python lists, convert them first into tuples as shown in Sect. 2.6.5.

2.7.3 SortedSet. Sorted Unique Elements

Next, why not to keep all our elements in the Java set container in a sorted order, without calling an additional sorting method each time we add a new element? The example below shows the use of the `SortedSet` Java class:

```
>>> from java.util import *
>>> s=TreeSet()
>>> e=s.add(1)
>>> e=s.add(4)
>>> e=s.add(4)
>>> e=s.add(2)
>>> print s
[1, 2, 4]
```

the second value "4" is automatically removed and the collection appears in the sorted oder.

2.7.4 Map. Mapping Keys to Values

As it is clear from the title, now we will consider the Java Map collection which maps keys to specific objects. This collection is analogous to a Jython dictionary, Sect. 2.6.7. Thus, a map cannot contain duplicate keys as we have learned from the Jython dictionaries.

Let us build a map collection based on the HashMap Java class:

```
>>> from java.util import *
>>> m=HashMap()
>>> m.put('a', 1)
>>> m.put('b', 2)
>>> m.put('c', 3)
>>> print m
{b=2, c=3, a=1}
```

Now you can see that Java maps have the same functionality as the Jython dictionaries. As for any Java collection, the size of the Map is given by the method size(). One can access the map values using the key:

```
>>> print m['a']
1
```

Similar to the lists, one can print all keys in a loop:

```
>>> for key in m:
>>> ... print key, 'corresponds to', m[key]
b corresponds to 2
c corresponds to 3
a corresponds to 1
```

Here we print all keys and also values corresponding to the keys.

2.7.5 Java Map with Sorted Elements

This time we are interested in a map with sorted keys. For this one should use the class TreeMap and the same methods as for the HashMap class discussed before:

```
>>> from java.util import *
>>> m=TreeMap()
>>> m.put('c', 1)
>>> m.put('a', 2)
```

```
>>> m.put('b', 3)
>>> print m
{a=2, b=3, c=1}
```

Compare this result with that given in the previous subsection. Now the map is sorted using the keys.

2.7.6 Real Life Example: Sorting and Removing Duplicates

Based on the Java methods discussed above, we can do something more complicated. In many cases, we need to deal with a sequence of data records. Each record, or event, can consist of strings, integer and real numbers. So we are dealing with lists of lists. For example, assume we record one event and make measurements of this event by recording a string describing some feature and several numbers characterizing this feature. Such example was already considered in Sect. 2.6.4.1.

Assume we make many such observations. What we want to do at the end of our experiment is to remove duplicates based on the string with a description, and then sort all the records (or observations) based on this description. This looks like a real project, but not for Jython! The code below does everything using a several lines of the code:

—————— Sorting and removing duplicates ——————

```
from java.util import *

data=ArrayList()
data.add( ["star",1.1,30] )
data.add( ["galaxy",2.2,80] )
data.add( ["galaxy",3.3,10] )
data.add( ["moon",4.4,50] )

map=TreeMap()
for row in data:
    map.put(row[0],row[1:])

data.clear()
for i in map:
    row=map[i]
    row.insert(0,i)
    data.add(row)

print data
```

Let us give some explanations. First, we make a data record based on the list 'data' holding all our measurements. Then we build a TreeMap class and use the first element to keep the description of our measurement in form of a "key".

The rest of our record is used to fill the map values (see `row[1:]`). As you already know, when we fill the `TreeMap` object, we remove duplicate elements and sort the keys automatically. Once the map is ready, we remove all entries from the list and refill it using a loop over all the keys (which are now ordered). Then we combine the key value to form a complete event record. The output of the script is given below:

```
[['galaxy',3.3,10],['moon',4.4,50],['star',1.1,30]]
```

We do not have extra record with the description 'galaxy' and, expectedly, all our records are appropriately sorted.

2.8 Random Numbers

A generation of random numbers is an essential phase in scientific programming. Random numbers are used for estimating integrals, generating data encryption keys, data interpretation, simulation and modeling complex phenomena. In many examples of this book, we will simulate random data sets for illustrating data-analysis techniques.

Let us give a simple example which shows how to generate a random floating point number in the range [0, 1] using the Python language:

```
>>> from random import *
>>> r=Random()
>>> r.randint(1,10) # a random number in range [0.10]
```

Since we do not specify any argument for the `Random()` statement, a random seed from the current system time is used. In this case, every time you execute this script, a new random number will be generated.

In order to generate a random number predictably for debugging purpose, one should pass an integer (or long) value to an instance of the `Random()` class. For the above code, this may look as: `r=Random(100L)`. Now the behavior of the script above will be different: every time when you execute this script, the method `randint(1,10)` will return the same random value, since the seed value is fixed.

Random numbers in Python can be generated using various distributions depending on the applied method:

```
>>> r.random()              # in range [0.0, 1.0)
>>> r.randint(min,max)      # int in range [min,max]
>>> r.uniform(min,max)      # real number in [min,max]
>>> r.betavariate(a,b)      # Beta distribution (a>0,b>0)
>>> r.expovariate(lambda)   # Exponential distribution
>>> r.gauss(m,s)            # Gaussian distribution
>>> r.lognormvariate(m,s)   # Log normal distribution
```

```
>>> r.normalvariate(m,s)    # Normal distribution
>>> r. gammavariate(a, b)   # Gamma distribution.
>>> r.seed(i)               # set seed (i integer or long)
>>> state=r.getstate()      # returns internal state
>>> setstate(state)         # restores internal state
```

In the examples above, 'm' denotes a mean value and 's' represents a standard
deviation for the output distributions.

Random numbers are also used for manipulations with Jython lists. One can
randomly rearrange elements in a list as:

```
>>> list=[1,2,3,4,5,6,7,8,9]
>>> r.shuffle(list)
>>> print list
[3, 4, 2, 7, 6, 5, 9, 8, 1] # random list
```

One can pick up a random value from a list as:

```
>>> list=[1,2,3,4,5,6,7,8,9]
>>> r.choice(list)   #  get a random element
```

Similarly, one can get a random sample of elements as:

```
>>> list=[1,2,3,4,5,6,7,8,9]
>>> print r.sample(list,4) # random list
>>> [4, 2, 3, 6]
```

Of course, the printed numbers will be different in your case.

2.9 Time Module

The time module is rather popular due to several reasons. First, it is always a good
idea to find our the current time. Secondly, it is an essential module for more serious
tasks, such as optimization and benchmarking analysis programs or their parts. Let
us check the methods of the module time:

```
>>> import time
>>> dir(time)    # check what is inside
['__doc__', 'accept2dyear', 'altzone', 'asctime',
 'classDictInit', 'clock', 'ctime', 'daylight',
 'gmtime', 'locale_asctime', 'localtime', 'mktime',
 'sleep', 'strftime', 'struct_time', 'time',
 'timezone', 'tzname']
```

You may notice that there is a method called __doc__. This looks like a method to keep the documentation for this module. Indeed, by printing the documentation of this module as

```
>>> print time.__doc__
```

you will see a rather comprehensive description. Let us give several examples:

```
>>> time.time()  # time in seconds since the Epoch
>>> time.sleep() # delay for a number of seconds
>>> t=time.time()
>>> print t.strftime('4-digit year: %Y, 2-digit year: \
                      %y, month: %m, day: %d')
```

The last line prints the current year, the month and the day with explanatory annotations.

To find the current day, the easiest is to use the module datetime:

```
>>> import datetime
>>> print "The date is", datetime.date.today()
>>> The date is 2008-11-14
>>> t=datetime.date.today()
>>> print t.strftime("4-digit year: \
    %Y, 2-digit year: %y, month: %m, day: %d")
```

To force a program to sleep a certain number of seconds, use the sleep() method:

```
>>> seconds = 10
>>> time.sleep(seconds)
```

2.9.1 Benchmarking

For tests involving benchmarking, i.e. when one needs to determine the time spent by a program or its part on some computational task, one should use a Jython module returning high-resolution time. The best is to use the module clock() which returns the current processor time as a floating point number expressed in seconds. The resolution is rather dependent on the platform used to run this program but, for our benchmarking tests, this is not too important.

To benchmark a piece of code, enclose it between two time.clock() statements as in this code example:

```
>>> start = time.clock();   \
    [SOME CODE FOR BENCHMARKING];  \
    end = time.clock()
>>> print 'The execution of took (sec) =', end-start
```

Let us give a concrete example: We will benchmark the creation of a list with integer numbers. For benchmarking in an interactive mode, we will use the exec() statement. This code benchmarks the creation of a list with the integer numbers from 0 to 99999.

```
>>> code='range(0,100000)'
>>> start=time.clock();List=exec(code);end=time.clock()
>>> print 'Execution of the code took (sec)=',end-start
Execution of the code took (sec) = 0.003
```

Alternatively, one can write this as:

```
>>> List=[]
>>> code='for x in range(0,100000): List.append(x)'
>>> start=time.clock();exec(code);end=time.clock()
>>> print 'Execution of the code took (sec)=',end-start
```

2.10 Python Functions and Modules

Jython supports code reuse via functions and classes. The language has many built-in functions which can be used without calling the import statement. For example, the function dir() is a typical built-in function. But how one can find out which functions have already been defined? The dir() itself cannot display them. However, one can always use the statement dir(module) to get more information about a particular module. Try to use the lines:

```
>>> import __builtin__
>>> dir(__builtin__)
...'compile', 'dict', 'dir', 'eval' ..
```

This prints a rather long list of the built-in functions available for immediate use (we show here only a few functions).

Other ("library") functions should be explicitly imported using the import statement. For example, the function sqrt() is located inside the package 'math', thus it should be imported as 'import math'. One can always list all functions of a particular package by using the `dir()` function as shown in Sect. 2.2.3.

It is always a good idea to split your code down into a series of functions, each of which would perform a single logical action. The functions in Jython are declared using the statement `def`. Here is a typical example of a function which returns `(a-b)/(a+b)`:

```
>>>def func(a,b):
>>>  ... "function"
>>>  ... d=(a-b)/(a+b)
>>>  ... return d
>>>print func(3.0,1.0)
0.5
>>> print func.__doc__
function
```

To call the function `func()`, a comma-separated list of argument values is used. The `'return'` statement inside the function definition returns the calculated value back and exits the function block. If no return statement is specified, then `'None'` will be returned. The above function definition contains a string comment `'function'`. A function comment should always be on the first line after the `def` attributed. One can print the documentation comment with the method `__doc__` from a program from which the function is called.

One can also return multiple values from a function. In this case, put a list of values separated by commas; then a function returns a tuple with values as in this example:

```
>>>def func(a,b,c=10):
>>>  ... d1=(a-b)/(a+b)
>>>  ... d2=(a*b*c)
>>>  ... return d1,d2
>>>print func(2,1)
(0, 20)
>>> >print func(2.,1.0)
(0.5, 30.0)
```

The example shows another features of Jython functions: the answer from the function totally depends on the type of passed argument values. The statement `func(2,1)` interprets the arguments as integer values, thus the answer for `(a-b)/(a+b)` is zero (not the expected 0.5 as in case of double values). Thus, Jython functions are *generic* and any type can be passed in.

One can note another feature of the above example: it is possible to omit a parameter and use default values specified in the function definition. For the above example, we could skip the third argument in the calling statement, assuming `c=10` by default.

All variable names assigned to a function are local to that function and exist only inside the function block. However, you may use the declaration 'global' to force a variable to be common to all functions.

```
>>>def func1(a,b):
>>> ... global c
>>> ... return a+b+c
>>>def func2(a,b):
>>> ... global c
>>> ... c=a+b
>>>
>>>print func2(2,1)
None
>>>print func1(2,1)
6
```

Thus, once the global variable 'c' is assigned a value, this value is propagated to other functions in which the 'global' statement was included. The second function does not have the 'return' statement, thus it returns 'None'.

We should note that a function in Jython can call other functions, including itself.

In Jython, one can also create an anonymous function at runtime, using a construct called 'lambda' discussed in Sect. 2.6.6. The example below shows two function declarations with the same functionality. In one case, we define the function using the standard ("named") approach, and the second case uses the "lambda" anonymous declaration:

```
>>> def f1 (x): return x*x
>>> print f1(2)
4
>>> f1=lambda x: x*x
>>> print f1(2)
4
```

Both function definitions, f1 and f2 do exactly the same operation. However, the "lambda" definition is shorter.

It is very convenient to put functions in files and use them later in your programs. A file containing functions (or any Jython statement!) should have the extension '.py'. Usually, such file is called a "module". For example, one can create a file 'Func.py' and put these lines:

```
──────────────── File 'Func.py' ────────────────
def func1(a,b):
    "My function 1"
    global c
    return a+b+c

def func2(a,b):
```

```
    "My function 2"
    global c
    c=a+b
```

This module can be imported into other modules. Let us call this module from the JythonShell prompt with the following commands:

```
>>> import Func
>>>print Func.func2(2,1)
None
>>>print Func.func1(2,1)
6
```

We can access functions exactly as if they are defined in the same program, since the import statement executes the file 'Func.py' and makes the functions available at runtime of your program.

Probably, we should remind again that one can import all functions with the statement 'from Func import *', as we usually do in many examples of this book. In this case, one can call the functions directly without typing the module name.

Another question is where such modules should be located? How can we tell Jython to look at particular locations with module files? This can be done by using a predefined list sys.path from the 'sys' module. The list sys.path contains strings that specify the location of Jython modules. One can add an additional module location using the append() method: In this example, we added the location '/home/lib' and printed out all directories containing Jython modules:

```
>>> import sys
>>> sys.path.append('/home/lib')
>>> print sys.path
```

Here we have assumed that we put new functions in the directory '/home/lib'.

Now we are equipped to go further. We would recommend to read any book about Python or Jython to find more detail about Jython modules and functions.

2.11 Python Classes

As for any object-oriented language, one can define a Jython class either inside a module file or inside the body of a program. Moreover, one can define many classes inside a single module.

Classes are templates for creation of objects. Class attributes can be hidden, so one can access the class itself only through the methods of the class. Any Python book or tutorial should be fine in helping to go into the depth of this subject.

A Jython class is defined as:

```
>>> class ClassName[args]:
>>> ... [code block]
```

where [code block] indicates the class body and bounds its variables and methods.

The example below shows how to create a simple class and how to instantiate it:

```
>>> class Func:
>>> ... 'My first class'
>>> ... a='hello'; b=10
>>>
>>> c=Func()
>>> print c.a, c.b
hello 10
```

The class defined above has two public variables, a and b. We create an instance of the class 'Func' and print its public attributes, which are just variables of the type string and integer. As you can see, the class instance has its own namespace which is accessible with the dot. As for functions and modules, classes can (and should) have documentary strings.

The created instance has more attributes which can be shown as a list using the built-in function dic(): Try this line:

```
>>> dir(Func)
['__doc__', '__module__', 'a', 'b']
```

The command displays the class attributes and the attributes of its class base. Note that one can also call the method dir(obj), where 'obj' is an instance of the class (c in our example), rather than explicitly using the class name.

But what about the attributes which start from the two leading underscores? In the example above, both variables, a and b, are public, so they can be seen by a program that instantiates this class. In many cases, one should have private variables seen by only the class itself. For this, Jython has a naming convention: one can declare names in the form __Name (with the two leading underscores). Such convention offers only the so-called "name-mangling" which helps to discourage internal variables or methods from being called from outside a class.

In the example above, the methods with two leading underscores are private attributes generated automatically by Jython during class creation. The variable __doc__ keeps the comment line which was put right after the class definition, and the second variable __module__ keeps a reference to the module where the class is defined.

```
>>> print c.__doc__
My first class
>>> print c.__module__
None
```

The last call returns 'None' since we did not put the class in an external module file.

2.11.1 Initializing a Class

The initialization of a Jython class can be done with the __init__ method, which takes any number of arguments. The function for initialization is called immediately after creation of the instance:

```
>>> class Func():
>>>     'My class with initialization'
>>>     def __init__(self, filename=None):
>>>       self.filename=filename
>>>     def __del__(self):
>>>       # some close statement goes here
>>>     def close(self):
>>>       # statement to release some resources
```

Let us take a closer look at this example. You may notice that the first argument of the __init__ call is named as self. You should remember this convention: every class method, including __init__, is always a reference to the current instance of the class.

In case if an instance was initialized and the associated resources are allocated, make sure they are released at the end of a program. This is usually done with the __del__ method which is called before Jython garbage collector deallocates the object. This method takes exactly one parameter, self. It is also a good practice to have a direct cleanup method, like close() shown in this example. This method can be used, for example, to close a file or a database. It should be called directly from a program which creates the object. In some cases, you may wish to call close() from the __del__ function, to make sure that a file or database was closed correctly before the object is deallocated.

2.11.2 Classes Inherited from Other Classes

In many cases, classes can be inherited from other classes. For instance, if you have already created a class 'exam1' located in the file 'exam1.py', you can use this class to build a new ("derived") class as:

```
>>> from exam1 import exam1
>>> class exam2(exam1):
>>> ... [class body]
```

As you can see, first we import the ancestor class 'exam1', and then the ancestor of the class is listed in parentheses immediately after the class name. The new class 'exam2' inherits all attributes from the 'exam1' class. One can change the behavior of the class 'exam1' by simply adding new components to 'examp2' rather than rewriting the existing ancestor class. In particular, one can overwrite methods of the class 'exam1' or even add new methods.

2.11.3 Java Classes in Jython

The power of Jython, a Java implementation of the Python language, becomes clear when we start to call Java classes using Python syntax. Jython was designed as a language which can create instances of Java classes and has an access to any method of such Java class.

This is exactly what we are going to do while working with the jHepWork libraries. The example below shows how to create the Java Date object from java.util and use its methods:

```
>>> from java.util import Date
>>> date=Date()
>>> date.toGMTString()
'09 Jun 2009 03:48:17 GMT'
```

One can use the code assist to learn more about the methods of this Java class (Type) the object name followed by a dot and use [Ctrl]+[Space] in JythonShell for help. Similarly, one can call dir(obj), where obj is an object which belongs to the Java platform. For jHepWork IDE code editor, use a dot and the key [F4].

In this book, we will use Java-based numerical libraries from jHepWork, thus most of the time we will call Java classes of this package. Also, in many cases, we call classes from the native Java platform. For example, the AWT classes 'Font' and 'Color' are used by many jHepWork objects to set fonts and colors. For example, Sect. 3.3.1 shows how to build a Java instance of graphical canvas based on the Java class HPlot.

2.11.4 Topics Not Covered

In this book, we will try to avoid going into the depths of Python classes. We cannot cover here many important topics, such as inheritance (the ability of a class to inherit propertied from another class) and abstract classes. We would recommend any Python or Jython textbook to learn more about classes.

As we have mentioned before, we would recommend to develop Java libraries to be linked with Jython, rather than building numerical libraries using pure-Jython classes; for the latter approach, you will be somewhat locked inside the Python language specification, plus this may result in slow overall performance of your application. Of course, you have to be familiar with the Java language in order to develop Java classes.

2.12 Used Memory

To know how much memory used by the Java virtual machine for an application is important for code debugging and optimization. The amount of memory currently allocated to a process can be found using the standard Java library as in the example below:

```
>>> from java.lang import Runtime
>>> r=Runtime.getRuntime()
>>> Used_memory = r.totalMemory() - r.freeMemory()
>>> 'Used memory in MB = ', Used_memory/(1024*1024)
```

We will emphasize that this only can be done in Jython, but not in CPython which does not have any knowledge about the Java virtual machine.

We remind that if you use the jHepWork IDE, one can look at the memory monitor located below the code editor.

2.13 Parallel Computing and Threads

A Jython program can perform several tasks at once using the so-called threads. A thread allows to make programs parallelizable, thus one can significantly boost their performance using parallel computing on multi-core processors.

Jython provides a very effective threading compared to CPython, since JAVA platform is designed from the ground up to support multi-thread programming. A multi-threading program has significant advantage in processing large data sets, since one can break up a single task into pieces that can be executed in parallel. At the end, one can combine the outputs. We will consider one such example in Sect. 16.4.

To start a thread, one should import the Jython module 'threading'. Typically, one should write a small class to create a thread or threads. The class should contain the code to be executed when the thread is called. One can also put an initialization method for the class to pass necessary arguments. In the example below, we create ten independent threads using Jython. Each thread prints integer numbers. We create instances of the class shown above and start the thread using the method start() which executes the method run() of this class.

```
————————————— A thread example —————————————

from threading import Thread

class test(Thread):
  def __init__ (self,fin):
    Thread.__init__(self)
    self.fin = fin
  def run(self):
    print 'This is thread No='+str (self.fin)

for x in xrange ( 10 ):
    current=test(x)
    current.start()
    print 'done!'
```

Here we prefer to avoid going into detailed discussion of this topic. Instead, we will illustrate the effectiveness of multi-threading programs in the following chapters when we will discuss concrete data-analysis examples.

2.14 Arrays in Jython

This is an important section: Here we will give the basics of objects which can be used for effective numerical calculations and storing consecutive values of the same type.

Unfortunately, the Java containers discussed in Sect. 2.7 cannot be used in all cases. Although they do provide a handy interface for passing arrays to Java and jHepWork objects to be discussed later, they do not have sufficient number of built-in methods for manipulations.

Jython lists can also be used for data storage and manipulation. However, they are best suited for general-purpose tasks, such as storing complex objects, especially if they belong to different types. They are rather heavy and slow for numerical manipulations with numbers.

Here we will discuss Python/Jython arrays that can be used for storing a sequence of values of a certain type, such as integers, long values, floating point numbers etc. Unlike lists, arrays cannot contain objects with different types.

The Jython arrays directly mapped to Java arrays. If you have a Java function which returns an array of double values, and declared as double[] in a Java code, this array will be seen by Jython as an array.

Table 2.3 Characters used to specify types of the Jython arrays

Jython array types	
Typecode	Java type
z	boolean
c	char
b	byte
h	short
i	int
l	long
f	float
d	double

To start working with the arrays, one should import the module `jarray`. Then, for example, an array with integers can be created as:

```
>>> from jarray import *
>>> a=array([1,2,3,4], 'i')
```

This array, initialized from the input list `[1,2,3,4]`, keeps integer values, see the input character `'i'` (integer). To create an array with double values, the character `'i'` should be replaced by `'d'`. Table 2.3 shows different choices for array types. The length of arrays is given by the method `len(a)`.

Arrays can be initialized without invoking the lists. To create an array containing, say, ten zeros, one can use this statement:

```
>>> a=zeros(10, 'i')
```

here, the first argument represents the length of the array, while the second specifies its type.

A new value `'val'` can be appended to the end of an array using the `append(val)` method if the value has exactly the same type as that used during array creation. A value can be inserted at a particular location given by the index `'i'` by calling the method `insert(i,val)`. One can also append a list to the array by calling the method `fromlist(list)`.

The number of occurrences of a particular value `'val'` in an array can be given by the method `count(val)`. To remove the first occurrence of `'val'` from an array, use the `remove(val)` method.

2.14.1 Array Conversion and Transformations

Many Java methods return Java arrays. Such arrays are converted to Jython arrays when Java classes are called from a Jython script.

Very often, it is useful to convert arrays to Jython list for easy manipulation. Use the method `tolist()` as below:

```
>>> from jarray import *
>>> a=array([1,2,3,4], 'i')
>>> print a.tolist()
[1, 2, 3, 4]
```

One can reverse all elements in arrays using the `reverse()` method. Finally, one can also transform an array into a string applying the `tostring()` method.

There are no too many transformations for Jython arrays: in the following chapters, we will consider another high-level objects which are rather similar to the Jython arrays but have a large number of methods for numerical calculations.

2.14.2 Performance Issues

We have already noted that in order to achieve the best possible performance for numerical calculations, one should use the built-in methods, rather than Python-language constructs.

Below we show a simple benchmarking test in which we fill arrays with one million elements. We will consider two scenarios: In one case, we use a built-in function. In the second case, we use a Python-type loop. The benchmarking test was done using the time module discussed in Sect. 2.9. The only new component in this program is the one in which we format the output number: here we print only four digits after the decimal point.

```
———————— Benchmarking Jython arrays ————————
import time
from  jarray import *

start=time.clock()
a=zeros(1000000, 'i')
t=time.clock()-start
print 'Build-in method (sec)= %.4f' % t

start=time.clock()
a=array([], 'i')
for i in range(0,1000000,1):
    a.append(0)
```

```
t=time.clock()-start
print 'Python loop  (sec) %.4f' % t
```

Run this small script by loading it in the editor and using the "[run]" button. The performance of the second part, in which integers are sequentially appended to the array, is several orders of magnitudes slower than for the case with the built-in array constructor zeros().

Generally, the performance of Jython loops is not so dramatically slow. For most examples to be discussed later, loops are several times slower than equivalent loops implemented in built-in functions.

2.15 Exceptions in Python

Exception is an unexpected error during program execution. An exception is raised whenever an error occurs.

Jython handles the exceptions using the "try"-"except"-"else" block. Let us give a short example:

```
>>> b=0
>>> try:
>>> ... a=100/b
>>> except:
>>> ... print "b is zero!"
```

Normally, if you will not enclose the expression a=100/b in the "try"-"except" block, you will see the message such as:

```
>>> a=100/b
Traceback (innermost last):
  File "<input>", line 1, in ?
  ZeroDivisionError: integer division or modulo by zero
```

As you can see, the exception in this case is ZeroDivisionError.

Another example of the exceptions is "a file not found" which happens while attempting to open a non-existing file (see the next chapter describing Jython I/O). Such exception can be caught in a similar way:

```
>>> try:
>>> ... f=open('filename')
>>> except IOError, e:
>>> ... print e
```

This time the exception is IOError, which was explicitly specified. The variable e contains the description of the error.

Exceptions can be rather different. For example, NameError means unknown name of the class or a function, TypeError means operation for incompatible types and so on. One can find more details about the exceptions in any Python manual.

2.16 Input and Output

2.16.1 User Interaction

A useful feature you should consider for your Jython program is interactivity, i.e. when a program asks questions at run-time and a user can enter desired values or strings. To pass a value, use the Jython method input():

```
>>> a=input('Please type a number: ')
>>> print 'Entered number=',a
```

In this example, the input() method prints the string 'Please type a number:' and waits for the user response.

But what if the entered value is not a number? In this case, we should handle an exception as discussed in Sect. 2.15.

If you want to pass a string, use the method raw_input() instead of input().

The above code example works only for the stand-alone Jython interpreter, outside the jHepWork IDE. For the jHepWork IDE, this functionality is not supported. In fact, you do not need this feature at all: When working with the IDE, you are already working in an interactive mode. However, when you run Jython using the system prompt, the operations input() or raw_input() are certainly very useful.

2.16.2 Reading and Writing Files

File handling in Jython is relatively simple. One can open a file for read or write using the open() statement:

```
>>> f=open(FileName, option)
```

where 'FileName' represents a file name including the correct path, 'option' is a string which could be either 'w' (open for writing, old file will be removed), 'r' (open for reading) or 'a' (file is opened for appending, i.e. data written to it is added on at the end). The file can be closed with the close() statement.

Let us read a file 'data.txt' with several numbers, each number is positioned on a new line:

```
>>> f=open('data.txt','r')
>>> s=f.readline()
>>> x1=float(s)
>>> s=f.readline()
>>> x2=float(s)
>>> f.close()
```

At each step, readline() reads a new line and returns a string with the number, which is converted into either a float or integer.

The situation is different if several numbers are located on one line. In the simplest case, they can be separated by a space. For such file format, we should split the line into pieces after reading it. For example, if we have two numbers separated by white spaces in one line, like '100 200', we can read this line and then split it as:

```
>>> f=open('data.txt', 'r')
>>> s=f.readline()
>>> x=s.split()
>>> print s
['100','200']
```

As you can see, the variable 'x' is a list which contains the numbers in form of strings. Next, you will need to convert the elements of this list into either float or integer numbers:

```
>>> x1=float( x[0] )
>>> x2=float( x[1] )
```

In fact, the numbers can also be separated by any string, not necessarily by white spaces. Generally, use the method split(str), where 'str' is a string used to split the original string.

There is another powerful method: readlines(), which reads all lines of a file and returns them as a list:

```
>>> f=open('data.txt')
>>> for l in f.readlines():
>>> ... print l
```

To write numbers or strings, use the method write(). Numbers should be coerced into strings using the str() method. Look at the example below:

```
>>> f=open('data.txt', 'w')
>>> f.write( str(100)+'\n' )
```

```
>>> f.write( str(200) )
>>> f.close()
```

here we added a new line symbol, so the next number will be printed on a new line.

One can also use the statement 'print' to redirect the output into a file. This can be done with the help of the >> operator. (Note: by default, this operator prints to a console.). Let us give one example that shows how to print ten numbers from zero to nine:

```
>>> f=open('data.txt', 'w')
>>> for i in range(10):
>>> ... print >> f, i
>>> f.close()
```

One can check the existence of the file using the Jython module 'os':

```
>>> import os
>>> b=os.path.exists(fileName)
```

where 'b=0' (false in Java) if the file does not exist, and 'b=1' (true in Java) in the opposite case.

2.16.3 Input and Output for Arrays

Jython arrays considered in the previous section can be written into an external (binary) file. Once written, one can read its content back to a new array (or append the values to the existing array).

```
>>> from  jarray import *
>>> a=array([1,2,3,4],'i')
>>> f=open('data.txt','w')
>>> a.tofile(f)        # write values to a file
>>> f.close()
>>> # read values
>>> f=open('data.txt','r')
>>> b=array([],'i')
>>> b.fromfile(f,3) # read 3 values from the file
>>> print b
array('i',[1, 2, 3])
```

It should be noted that the method fromfile() takes two arguments: the file object and the number of items (as machine values).

2.16.4 Working with CSV Python Module

The CSV ("Comma Separated Value") file format is often used to store data structured in a table. It is used for import and export in spreadsheets and databases and to exchange data between different applications. Data in such files are either separated by commas, tabs, or some custom delimiters.

Let as write a table consisting of several rows. We will import Jython csv file and write several lists with values using the code below:

```
──────────────── Writing a CSV file ────────────────
import csv

w=csv.writer(open('test.csv', 'w'),delimiter=',')
w.writerow(['London', 'Moscow', 'Hamburg'])
w.writerow([1,2,3])
w.writerow([10,20,30])
```

Execute this script and look at the current directory. You will see the file 'test.csv' with the lines:

```
London,Moscow,Hamburg
1,2,3
10,20,30
```

This is expected output: each file entry is separated by a comma as given in the delimiter attribute specified in our script. One can put any symbol as a delimiter to separate values. The most popular delimiter is a space, tab, semi-column and the symbol ' | '. The module also works for quoted values and line endings, so you can write files that contain arbitrary strings (including strings that contain commas). For example, one can specify the attribute quotechar=' | ' to separate fields containing quotes.

In the example below we read a CSV file and, in case of problems, we print an error message using the Jython exception mechanism discussed in Sect. 2.15:

```
──────────────── Reading a CSV file ────────────────
import csv

r = csv.reader(open('test.csv', 'rb'), delimiter=',')
try:
  for row in r:
    print row
except csv.Error, e:
    print 'line %d: %s' % (reader.line_num,e)
```

Let us convert our example into a different format. This time we will use a double quote (useful when a string contains comma inside!) for each value and tab for value

separations. The conversion script is based on the same Jython `csv` module and will
look as:

```
─────────────────── Converting CSV file ───────────────────
import csv


reader=csv.reader(open('test.csv',"rb"),delimiter=',')
writer=csv.writer(open('newtest.csv',"wb"),\
                  delimiter='\t',\
                  quotechar='"', quoting=csv.QUOTE_ALL)
for row in reader:
    writer.writerow(row)
```

The output file will look as:

```
"London"    "Moscow"    "Hamburg"
"1"         "2"         "3"
"10"        "20"        "30"
```

But what if we do not know which format was used for the file you want to
read in? First of all, one can always open this file in an editor to see how it looks
like, since the CSV files are human readable. One can use the jHepWork editor by
printing this line in the JythonShell prompt:

```
>>> view.open('newtest.csv', 0  )
```

which opens the file `'newtest.csv'` in the IDE. Alternatively, one can determine the file format automatically using the `Sniffer` method for safe opening of
any CSV file:

```
─────────────── Reading a CSV file using sniffer ───────────────
import csv

f=open('newtest.csv')
dialect = csv.Sniffer().sniff(f.read(1024))
f.seek(0)
reader = csv.reader(f, dialect)
for row in csv.reader(f, dialect):
        print row
```

This time we do not use the exception mechanism, since it is very likely that your
file will be correctly processed.

We will come back to the CSV file format in the following chapters when we
will discuss Java libraries designed to read the CSV files.

2.16.5 Saving Objects in a Serialized File

If you are dealing with an object from the Python-language specification, you may want to store this object in a file persistently (i.e. permanently), so another application can read it later. In Jython, one can serialize (or pickle) an object as:

```
>>> import pickle
>>> f=open('data.pic','w')
>>> a=[1,2,3]
>>> pickle.dump(a,f)
>>> f.close()
```

One can restore the object back as:

```
>>> import pickle
>>> f=open('data.pic','r')
>>> a=pickle.load(f)
>>> f.close()
```

In this example, we save a list and then restore it back from the file 'data.pic'. One cannot save Java objects using the same approach. Also, any object which has a reference to a Java class cannot be saved. We will consider how to deal with such special situations in the following chapters.

2.16.6 Storing Multiple Objects

To store one object per file is not too useful feature. In many cases, we are dealing with multiple objects. Multiple objects can be stored in one serialized file using the shelve module. This Jython module can be used to store anything that the pickle module can handle.

Let us give one example in which we store two Jython objects, a string and a list:

```
>>> import shelve
>>> sh=shelve.open('data.shelf')
>>> sh['describe']='My data'
>>> sh['data']=[1,2,3,4]
>>> sh.close()
```

The example above creates two files, 'data.shelf.dir' and 'data.shelf.dat'. The first file contains a "directory" with the persistent data. This file is in a human-readable form, so if you want to learn what is stored inside of the data file, one can open it and read its keys. For the above example, the file contains the following lines:

```
'describe', (0, 15)
'data', (512, 22)
```

The second file, 'data.shelf.dat', contains the actual data in a binary form.

One can add new objects to the "shelf" file. In the example below, we add a Jython map to the existing file:

```
>>> import shelve
>>> sh=shelve.open('data.shelf')
>>> sh['map']={'x1':100,'x2':200}
>>> sh.close()
```

Let us retrieve the information from the shelve storage and print out all saved objects:

```
>>> import shelve
>>> sh=shelve.open('data.shelf')
>>> for i in sh.keys():
>>>   ...print i, ' = ',sh[i]
>>> sh.close()
```

The output of this code is:

```
describe  =  My data
data   =   [1, 2, 3, 4]
map   =   {'x2': 200, 'x1': 100}
```

Finally, one can remove elements using the usual del method.

As you can see, the "shelve" module is very useful since now one can create a small persistent database to hold different Jython objects.

2.16.7 Using Java for I/O

In this section, we show how to write and read data by calling Java classes. Let us give an example of how to write a list of values into a binary file using the DataOutputStream Java class. In the example below we also use the Java class BufferedOutputStream to make the output operations to be more efficient. In this approach, data are accumulated in the computer memory buffer first, and are only written when the memory buffer is full.

```
──────────── Writing data using Java ────────────
from java.io import *
```

```
fo=FileOutputStream('test.d')
out=DataOutputStream(BufferedOutputStream( fo ))

list=[1.,2.,3.,4]
for a in list:
    out.writeFloat(a)

out.close()
fo.close()
```

The output of this example is binary data. The DataOutputStream class allows to write any of the basic types of data using appropriate methods, such
as boolean (writeBoolean(val)), double (writeDouble(val)), integers
(writeInt(val)), long (writeLong(val)) and so on.

Now let us read the stored float numbers sequentially. We will do this in an infinite loop using the 'while' statement until we reach the end of the file (i.e.
until the "end-line" exception is thrown). Then, the break statement exits the infinite loop. Since we know that our data are a sequence of float numbers, we use
the method readFloat(). One can play with other similar methods, such as
readInt() (read integer values), readDouble() (read double values).

```
———————————————— Reading data using Java ——————————————

from java.io import *

fo=FileInputStream('test.d')
inf=DataInputStream(BufferedInputStream(fo))

while 1:
  try:
      f=inf.readFloat()
      print f
  except:
      print 'end of file'
      break

inf.close()
fo.close()
```

We will continue the discussion of high-level Java classes for I/O which allow us
to store objects or sequences of objects in Chap. 11.

2.16.8 Reading Data from the Network

Files with data may not be available from a local file storage, but exist in network-
accessible locations. In this case, one should use the module 'urllib2' that can

read data from URLs using HTTP, HTTPS, FTP file protocols. Here is an example of how to read the HTML Jython web page with Jython news:

```
>>> from urllib2 import *
>>> f = urlopen('http://www.jython.org/Project/news.html')
>>> s=f.read()
>>> f.close()
>>> print s
```

This code snippet is very similar to the I/O examples shown above, with the only one exception: now we open a file using the `urlopen` statement. The web access is an unauthenticated. One can always check the response headers as `f.info()`, while the actual URL can be printed using the string `f.geturl()`. As usual, one can also use the method `readlines()` to put all HTML-page lines into a Jython list.

One can also use a jHepWork module for downloading files from the Web. It has one advantage: it shows a progress bar during file retrievals. This will be discussed in Sect. 12.2.

If authentication is required during file access, a client should retry the request with the appropriate name and password. The module `'urllib2'` also provides such functionality, but we will refrain from further discussion of this advanced topic.

2.17 Real-life Example. Collecting Data Files

Here we will consider a rather common data-analysis task: we collect all files located in a file system, assuming that all such files have the extension `'.dat'`. The files will be located in the root directory `'/home'`, which is the usual user-home directory on the Linux/UNIX platform. Our files contain numbers, each of which is positioned on a new line. We will persuade the following task: we will try to sum up all numbers in the files and calculate the sum of all numbers inside these files.

A snippet of a module `'walker.py'` which returns a list of files is given below. The module accepts two arguments: the root directory for scanning and the extension of the files we are interested in. The function builds a list of files with the appended full path. We will call the function `walk()` recursively until all directories are identified:

———————————— File `'walker.py'` ————————————

```
import os

def walker (dir,extension):
  files=[]
  def walk( dir, process):
   for f in os.listdir( dir ):
    fpath = os.path.join( dir, f)
    if os.path.isdir(fpath) and not os.path.islink(fpath):
```

```
              walk( fpath, process )
        if os.path.isfile( fpath ) :
            if fpath.endswith(extension) :
                files.append(fpath)

    walk(dir,files)
    return files
```

Let us test this module. For this, we will write a small program which: (1) imports the module 'walker.py'; (2) lists all descendant files and subdirectories under the specified directory and fills the file list with all files which have the extension '.dat'; (3) then it loops over all files in the list and reads the numbers positioned on every new line; (4) Finally, all numbers are summed up. The code which does all of this is given below:

———————————————— File collector ————————————————
```
import os
from walker import *

files= walker('/home/','.dat')

sum=0
lines=[]
for file in files:
    ifile = open(file,'r')
    lines=lines+ifile.readlines()
    ifile.close()
    for i in range(len(lines)):
        sum=sum+float(lines[i])
print "Sum of all numbers=", sum
```

The described approach is not the only one. The module which lists all files recursively can look much sorter using the os.walk function:

———————————————— Building a file list ————————————————
```
def getFileList(rootdir):
    fileList = []
    for root, subFolders, files in os.walk(rootdir):
        for f in files:
            fileList.append(os.path.join(root,f))
    return fileList

print getFileList('/home/')
```

This code builds a list of files in the directory "/home/".

In Sect. 12.9 we will show another efficient code based on the jHepWork Java class which can also be used in pure-Java applications. As in the example above, it builds a list of files recursing into all subdirectories.

The above code can significantly be simplified if we know that all input files are located inside a single directory, thus there is no need for transversing all subdirectories.

```
>>> list=[]
>>> for f in os.listdir('/home/'):
>>>    if not file.endswith('.dat'):   continue
>>>    list.append(f)
```

Finally, there is a simpler approach: import the module 'glob' and scan all files:

```
>>> import glob
>>> list=glob.glob('/home/*.dat')
```

The asterisk (*) in this code indicates that we are searching for a pattern match, so every file or directory with the extension '.dat' will be put into a list, without recursing further into subdirectories. One can specify other wildcard characters, such as '/home/data?.dat', that matches any single character in that position in the name starting from 'data'. Another example: '/home/*[0-9].dat' string considers all files that have a digit in their names before the extension '.dat'.

Often, in order to process data stored in many files, it is useful to divide a list with file names into several lists with equal number of files in each list. In this way, one can process files in parallel using multiple computers or multiple processors. This task can easily be achieved with the code given below:

```
─────────────────── File list splitter ───────────────────
def splitlist(seq, size):
    newlist = []
    splitsize = 1.0/size*len(seq)
    for i in range(size):
      k1=round(i*splitsize)
      k2=round((i+1)*splitsize)
      newlist.append(seq[int(k1):int(k2)])
      newlist.append(seq[k])
    return newlist
```

The code accepts a list of files and an integer size which specifies how many lists need to be generated. The function returns a new list in which each entry represents a list of files. The number of entries in each sublist is roughly equal.

2.18 Using Java for GUI Programming

Undoubtedly, the major strength of Jython is in its natural integration with Java, a
language used to build Jython. This opens infinite opportunities for a programmer.
Assuming that you had already a chance to look at one of these Java books [1–5],
you can start immediately use Java libraries to write a Jython code.

Below we show a small example of how to write a graphical user interface which
consists of a frame, a button and a text area. While the code still uses the Python
syntax, it calls classes from the Java platform.

```
————————————— Swing GUI using Jython —————————————
from java.awt import *
from javax.swing import *

fr = JFrame('Hello!')
pa1 = JPanel()
pa2 = JTextArea('text',6,20)

def act(event):
   pa2.setText('Hello, jHepWork')

bu=JButton('Hello', actionPerformed=act)
pa1.add(bu)

fr.add(pa1,BorderLayout.SOUTH)
fr.add(pa2,BorderLayout.NORTH)
fr.setDefaultCloseOperation(JFrame.DISPOSE_ON_CLOSE)
fr.pack()
fr.setVisible(1)
```

In this example, we call Java swing components directly, like they are usual Python
classes. The main difference with Python is in the class names: Java classes always
have names starting with capital letters.

When comparing this code with Java, one should note several important differences: there is no need to use the 'new' statement when creating Java objects. Also,
there is no need to put a semicolon at the end of each Java method or class declaration. We should also recall that Java boolean values are transformed into either "1"
(true) or "0" (false) in Jython programs.

So, let us continue with our example. Create a file, say, 'gui.py', copy the
lines from the example below and run this file in the jHepWork editor. You will
see a frame as shown in Fig. 2.1. By clicking on the button, the message "Hello,
jHepWork" should be shown.

In the following chapters, we try to follow our general concept: a Jython macro
is already a sufficiently high-level program style, so we will avoid detailed discussion of GUI-type of program development. In this book, we aim to show how to
develop data analysis programs for which GUI-type of features are less frequent,

Fig. 2.1 A Java Swing frame
with a button "Hello"

compare to "macro"-type of programming. Since Jython macros allow manipulations with objects without dealing with low-level features of programing languages, in some sense, they are already some sort of "user-interfaces". In addition, Jython macros have much greater flexibility than any GUI-driven application, since they can quickly be altered and rerun.

Yet, GUI is an important aspect of our life and we will discuss how to add GUI features to data-analysis applications in appropriate chapters.

2.19 Concluding Remarks

This concludes our introduction to the world of Python, Jython and Java. If there is one message I have tried to convey here is that the combination of all these three languages (actually, only two!) gives you an extremely powerful and flexible tool for your research. There are dozens of books written for each language and I would recommend to have some of them on your table if you want to study the topic in depth. To learn about Jython, you can always pick up a Python book (version 2.5 at least). In several cases, you may look at Jython and Java programming books, especially if you will need to do something very specific and non-standard using Java libraries. But, I almost guarantee, such situations will be infrequent if you will learn how to use the jHepWork libraries to be discussed in the following chapters.

References

1. Richardson, C., Avondolio, D., Vitale, J., Schrager, S., Mitchell, M., Scanlon, J.: Professional Java, JDK 5th edn. Wrox, Birmingham (2005)
2. Arnold, K., Gosling, J., Holmes, D.: Java(TM) Programming Language, 4th edn. Java Series. Addison-Wesley, Reading (2005)
3. Flanagan, D.: Java in a Nutshell, 5th edn. O'Reilly Media, Sebastopol (2005)
4. Eckel, B.: Thinking in Java, 4th edn. Prentice Hall PTR, Englewood Cliffs (2006)
5. Bloch, J.: Effective Java, 2nd edn. The Java Series. Prentice Hall PTR, Englewood Cliffs (2008)

Chapter 3
Mathematical Functions

Functions allow programming code reuse, and thus are essential in any programming language. We have already discussed how to define general-purpose functions in the Python language in Sect. 2.10. Now we will turn to the question of how to build *mathematical* functions. First, we will remind how to construct mathematical functions in Jython and then we will discuss Java implementations of mathematical functions in jHepWork. At the end of this chapter, we will discuss symbolic manipulations with functions.

For many examples of this chapter, we will "glue" Python-type statements with Java libraries. Therefore, we will stop using the word Python[1] when refer to the code snippets, and will switch to the word "Jython" as the main programming language used throughout this book.

3.1 Jython Functions

As we already know, Jython functions can be declared with the def() statement. In case of mathematical functions, it is very likely you will need to import the module 'math' before or during creation of Jython functions, so a general form of a function definition would be:

```
>>> def FunctionName( arg1, arg2, .. ) :
>>> ... import math
>>> ... -- math statements --
>>> ...    return value
```

In this example, arg1, arg2 etc. is a comma-separated list of arguments. The arguments could be values, lists, tuples, strings or any Jython object.

---

[1]This usually implies the CPython implementation of the Python programming language.

S.V. Chekanov, *Scientific Data Analysis using Jython Scripting and Java,* 85
Advanced Information and Knowledge Processing,
DOI 10.1007/978-1-84996-287-2_4, © Springer-Verlag London Limited 2010

We have already discussed how to build functions in Jython in Sect. 2.10. Many mathematical functions are already defined in the module 'math', and one can list all such functions using the dir(math) statement (after importing this module first). Below we will consider several useful examples illustrating how to construct an arbitrary function:

Absolute value:

```
>>> def abs(x)  :
>>> ... 'absolute value'
>>> ... if x<0: return -x
>>> ... return x
>>>
>>> print abs(-200)
200
```

Factorial:

```
>>> def factor(x)  :
>>> ... 'calculate factorial'
>>> ... if x<0:
>>> ...     raise ValueError, 'Negative number!'
>>> ... if x<2: return 1
>>> ... return long(x)*factor(x-1)
>>>
>>> print factor(10)
3628800
```

One could build rather complicated mathematical functions, as shown in this example with two arguments, x and y:

```
>>> def myfun(x,y)  :
>>> ... 'calculate complicated function'
>>> ... from math import *
>>> ... return cos(x)*sin(y)+(2**x)+y**4
>>>
>>> print myfun(0.3,0.8)
2.3260608626777355
```

Jython mathematical functions become rather inefficient when they have to be displayed, since any drawing involves loops with multiple calls to the same function but with different arguments. Below we will discuss implementations of mathematical functions in Java numerical libraries included into jHepWork. Being more efficient and flexible, they are also tightly integrated into the jHepWork graphical canvas used for object visualization.

3.2 1D Functions in jHepWork

For one-dimensional (1D) functions, we have to deal with a dependence of one value, say y, on another, usually called x. jHepWork includes the F1D class to describe, evaluate and display such functions. A 1D function can be instantiated as:

```
>>> from jhplot import *
>>> f1=F1D('definition')
```

where the string `'definition'` should be replaced by a mathematical formula. This string can contain any combinations of +, -, *, / operations, parenthesis () and predefined mathematical functions. For numerical values, the scientific notations "e", "E", "d", "D" can be used. The only independent variable should be specified as x. The function definition can contain predefined mathematical functions listened in Tables 3.1 and 3.2, as well as the constants given in Table 3.3.

To evaluate a function at a fixed point, use the `eval(x)` method. For evaluation of a list of numerical values, pass a Jython list instead. In this case, the `eval(x)` method returns an array of values y calculated at specific values x from the input list.

Table 3.1 Mathematical functions used to build the F1D objects

Function	Defined as
x^y	Power
**	as before
exp(x)	Exponential
sqrt(x)	$\sqrt{x}$
log(x)	Natural Logarithm
log10(x)	Logarithm base 10
cos(x)	Cosine
sin(x)	Sine
tan(x)	Tangent
cosh(x)	Hyperbolic cosine
sinh(x)	Hyperbolic sine
tanh(x)	Hyperbolic tangent
acos(x)	Arc Cosine
asin(x)	Arc Sine
atan(x)	Arc Tangent
acosh(x)	Hyperbolic Arc Cosine
asinh(x)	Hyperbolic Arc Sine
atanh(x)	Hyperbolic Arc Tangent
rem(x)	Reminder
atan2(x,y)	Arc Tangent (2 parameters)

Table 3.2 Special functions that can be included into the F1D definition

Function	Defined as
j0	Bessel function of order 0 of the argument
j1	Bessel function of order 1 of the argument
jn	Bessel function of order n of the argument
y0	Bessel function of the 2nd kind of order 0
y1	Bessel function of the 2nd kind of order 1
yn	Bessel function of the 2nd kind of order n
fac(x)	factorial $x!$
gamma(x)	the Gamma function
erf(x)	the Error function
erfc(x)	the complementary Error function
normal(x)	the normal function
poisson(k,x)	sum of the first k terms of the Poisson distribution
poissonc(k,x)	the sum of the terms $k + 1$ to infinity of the Poisson distribution
igam(a,x)	incomplete Gamma function
igamc(a,x)	complemented incomplete i Gamma function
chisq(d,x)	area under the left hand tail (from 0 to x) of the χ^2 probability density function with "d" degrees of freedom
chisqc(d,x)	area under the right hand tail (from x to infinity) of the Chi square probability density function with "d" degrees of freedom

Let us give one example which makes this feature clear:

─────────────────────────── 1D function ───────────

```
from jhplot import *
f1=F1D('x^2+pi*sqrt(x)')
print f1.eval(20)
a=f1.eval([10,20,30])
print a.tolist()
```

The execution of this script gives:

```
414.049629462
[109.934, 414.049, 917.207]
```

As you may note, since eval(d) returns an array, we converted this array into a list using the tolist() method for shorter printout (still we truncated the output numbers to fit them to the page width).

In many situations, it is useful to create a "parsed" function object to pass it to some method or class inside a long loop. This can significantly boost performance since the string parsing is done outside loops. This example illustrates how to print ten numbers evaluated by the function $x * \exp(x)$:

Table 3.3 Constants used to define the F1D functions

Constants	Defined as
pi	PI
e	E
k	Boltzman Constant: 1.3807e-16
ec	Elementary Charge: 4.8032e-10
me	Electron Mass. Units g: 9.1095e-28
mp	Proton Mass. Units g: 1.6726e-24
gc	Gravitational Constant. Units dyne-cm^2/g^2: 6.6720e-08
h	Planck constant. Units erg-sec: 6.6262e-27
c	Speed of Light in a Vacuum. Units cm/sec: 2.9979e10
sigma	Stefan-Boltzman Constant. Units erg/cm^2-sec-deg^4: 5.6703e-5
na	Avogadro Number. Units 1/mol: 6.0220e23
r	Gas Constant. Units erg/deg-mol: 8.3144e07
g	Gravitational Acceleration at the Earths surface. Units cm/sec^2: 980.67

```
>>> from jhplot import *
>>> f=F1D('x*exp(x)')
>>> p=f.getParse()
>>> for i in range(10):
>>> ... print p.getResult(i)
```

Of course, one can also print the numbers using the eval() method (since the string parsing is done during creation of the F1D object). However, there are many situations in which a "parsed" function (i.e. the object "p" in this example) can be used as an argument for methods inside loops.

One can find all the methods associated with this class as explained in Sect. 1.4.10.

3.2.1 Details of Java Implementation

The object created in the above examples is an instance of the F1D Java class, thus it does not belong to the Python programming language, although we call it using the Python syntax. When we instantiate the object of the class F1D, we call the Java library package jhplot stored in the file jhplot.jar located in the 'lib/system' directory. How can we check this? As usual, call the method type():

```
>>> from jhplot import *
>>> f=F1D('x*exp(x)')
```

```
>>> type(f)
<type 'jhplot.F1D'>
```

Even more: one can also use Java methods to access the class instance and its name:

```
>>> c=f.getClass()
>>> print c
<type 'jhplot.F1D'>
>>> print c.getCanonicalName()
jhplot.F1D
>>> print c.getMethods()
... list of all method
```

As you can see, one can use several "get" methods to access the name of this object. This cannot be done if an object belongs to Jython. We can also print all methods associated with this object using the method getMethods(). We do not print a very long output from this method to save space.

Since we are dealing with the Java object, one can work with the F1D function using either Java code or BeanShell. The only difference is in the syntax of your program. We will discuss this topic later.

In the following sections we will learn now to build mathematical functions using several independent variables and how to create functions using Jython classes which will allow to use rather complicated logic in the function definition. But before we will give some elements of calculus based on the F1D class, focusing mainly on numerical integration and differentiation.

3.2.2 Integration and Differentiation

A F1D function can numerically be integrated in a region between min and max using the integral(N,min,max) method, where N is the number of points for integration. This method assumes the trapezoid rule for integration.

There are more options in which a particular integration method can be specified. Assuming that a function f1 is created, one can integrate it between a minimum value min and a maximum value max using the N number of integration points as:

```
>>> f1d.integral(type,N,min,max)
```

where type is a string which defines the type of integration. This string can take the following values:

"gauss4"   Gaussian integration formula (4 points)
"gauss8"   Gaussian integration formula (8 points)
"richardson" Richardson extrapolation

"simpson"   using Simpson's rule
"trapezium"   trapezium rule

The code below tests different integration algorithms using the methods shown above. We also benchmark the code (see Sect. 2.9.1) by printing the time spent by each algorithm (in milliseconds):

```
————————————— Integration of a function ——————
from jhplot import *
import time

f1=F1D('sin(1.0/x)*x^2')
methods=['gauss4', 'gauss8',  'richardson',\
         'simpson','trapezium']
for m in methods:
    start = time.clock()
    d=f1.integral(m,10000,1,10)
    t = time.clock()-start
    print m+' =',d,' time (ms)=',t*1000
```

The result of code execution is given below:

```
gauss4      = 49.1203116758  time (ms)= 42.245
gauss8      = 49.1203116758  time (ms)= 32.382
richardson  = 49.1203116758  time (ms)= 12.162
simpson     = 49.1203116758  time (ms)= 14.173
trapezium   = 49.1203116663  time (ms)=  4.234
```

While the time needed for the integration could be different for your tests (clearly, this depends on many factors), it appears that the fastest algorithm is that based on the trapezium method.

A function can numerically be differentiated using the method:

```
>>> a=f1d.differentiate(N,min,max)
```

The method returns an array with the result of differentiation. The size of this array is set using an integer value N.

In Sect. 3.9, we will again discuss the topic of integration and differentiation using symbolic mathematical calculations.

3.3 Plotting 1D Functions

The remaining issue is how to visualize the F1D functions. Below we will learn how to build a canvas suited for plotting F1D objects and discuss the main options for their visualization.

3.3.1 Building a Graphical Canvas

To plot the 1D functions, first we need to create a graphical canvas. This can be done by instantiation of a canvas object of the HPlot class:

```
>>> from jhplot import *
>>> c1=HPlot('Canvas')
```

This creates a default canvas with the title 'Canvas' and with the default frame size of 600 × 400 pixels. To display the canvas in a pop-up frame, execute the statement:

```
>>> c1.visible(1)
```

It should be reminded that "1" means boolean "true", while "0" means "false". One can also use the shortcut visible() instead of visible(1). If you do not want to pop-up the canvas frame and want to keep it in the computer memory, set the argument of the method visible() to zero.

There are more constructors for this class. For example, one can customize the frame size with the method:

```
>>> c1=HPlot('Canvas',800,600)
```

which creates a canvas with the frame size of 800 by 600 pixels. One can also resize the canvas using the mouse.

The created canvas can be divided into several plot regions (or "pads"). In this case, use the following initialization:

```
>>> c1=HPlot('Canvas',600,400,iX,iY)
```

which again creates a canvas of the size 600 by 400 pixels, but, in addition, the last two numbers are used to make two plot regions inside the canvas frame. The first integer number after 400 tells that we need iX plot regions in X, and the second number is used to set the number of pads in Y. For example, after setting iX=2 and iY=1, the above constructor creates two pads positioned horizontally.

A function can be plotted inside the canvas using the draw(obj) method, where obj is an instance of the F1D class. One can navigate to the current region using the method cd(i1,i2), where i1 and i2 specify the pad in X and Y. For example, if one needs to plot a function inside the first pad, use:

```
>>> c1.cd(1,1)
>>> c1.draw(f1d)
```

where f1d represents an object of the F1D function. If the function should be shown on the second region, use

```
>>> c1.cd(1,2)
>>> c1.draw(f1d)
```

By default, the HPlot canvas has the range between 0 and 1 for the X or Y axis. One should specify the necessary range using the method:

```
>>> c1.setRange(x1,x2,y1,y2)
```

where x1 is a minimum value in X, x2 is a maximum value in X, and y1 and y2 are the same but for the Y-axis. Alternatively, one can set "auto-range" using the method c1.setAutoRange().

Before plotting a function, one can show a global title of the entire canvas. This can be done by using the method setGTitle(str), where str represents a string. In the simplest case, it accepts one argument: a string with the title text. One can customize the text color and/or the font size as will be shown later. As usual, use the code assist (see Sect. 1.4.10) to learn about possible choices.

In addition to the global title, one can set titles for each drawing pad using the method setName(s), with "s" being a string with some text. We can also annotate X and Y axes. The methods for this are setNameX(s) and setNameY(s), where "s" is an annotation string.

Let us give a more concrete example. First, import the Color class from the Java AWT library and then set appropriate annotations:

```
>>> from java.awt import Color
>>> from jhplot import *
>>>
>>> c1.setGTitle('GlobalTitle',Color.red)
>>> c1.setNameX('X axis')
>>> c1.setNameY('Y axis')
>>> c1.setName('Pad title')
>>> c1.visible()        # make it visible
>>> c.setAutoRange()    # set autorange for X and Y
```

All the entries above are self-explanatory. One may add a background color for the canvas using:

```
>>> c1.setBackgroundColor(c)
```

where 'c' stands for the Java AWT color. Colors can be set using any of the static methods shown in Table 3.4.

Table 3.4 Colors from the Java AWT library	AWT Color	Color
	Color.black	black color
	Color.blue	blue color
	Color.cyan	cyan color
	Color.darkGray	dark gray
	Color.gray	gray color
	Color.green	green color
	Color.lightGray	light gray color
	Color.magenta	magenta color
	Color.orange	orange color
	Color.pink	pink color
	Color.red	red color
	Color.white	white color
	Color.yellow	yellow color

A color can also be defined in a more flexible way using the constructor Color(r,g,b), with specified red, green and blue values. Each value must be in the range [0–1]. Alternatively, one can use the same constructor, but specifying red, green, blue (integer) values in the range [0–255]. One can also define a transparency level (or "alpha" value) using the forth argument. There are other constructors for this class, so please refer to any Java textbook.

To check what colors are available and to define your own colors, use the jHep-Work dialog window:

```
>>> import utils; utils.ShowColors()
```

The above command brings up an Java color chooser frame which can be used to select a custom color using the mouse.

Custom fonts for the legends can be specified using the Font class from the same AWT package:

```
>>> from java.awt import Font
>>> font=Font('Lucida Sans',Font.BOLD, 12)
>>> c1.setLegendFont(font)
```

Here we created a font instance from the font collection existing in your computer environment. A custom font can be created with the statement Font('name', style,size) from the specified font name, style and font size. The style can be either Font.PLAIN (simple), Font.ITALIC (italic) or Font.BOLD (bold).

To find the names of fonts is more tricky, but possible. Use the methods getAvailableFontFamilyNames or getAllFonts. The lengthy statement below illustrates how to print out the available fonts installed on your system:

```
>>> from java.awt import *
>>> e = GraphicsEnvironment.getLocalGraphicsEnvironment()
>>> print e.getAllFonts()
>>> print e.getAvailableFontFamilyNames()
```

When using jHepWork, call the predefined command which lists the available fonts:

```
>>> import utils; utils.ShowFonts()
```

Use the above approach to find out necessary fonts to be used in Jython scrips to setup graphical options for drawing of titles, and other labels. In addition, one can use the GUI-dialogs of the HPlot canvas. For example, double-click on the global title to bring up a setup dialog used to change the global text, fonts and colors. In a similar way one can edit the margins of the pads.

3.3.2 Drawing 1D Functions

Once a canvas is ready, one can plot a 1D function discussed in Sect. 3.2 as:

```
>>> from jhplot import *
>>>
>>> c1=HPlot("Canvas")
>>> c1.visible()
>>> c1.setAutoRange()
>>> f1=F1D('2*exp(-x*x/50)+sin(pi*x)/x', -2.0,5.0)
>>> c1.draw(f1)
```

Obviously, -2.0 and 5.0 specify the range for the abscissa. There is an important difference with respect to the function definition given in Sect. 3.2: now we explicitly tell what abscissa range should be used for evaluation and drawing. One can set the range for the function later using the methods:

```
>>> f1.setMin(min)
>>> f1.setMax(max)
```

where min and max are a minimum and a maximum value for the abscissa. Naturally, the corresponding "getter" methods getMin() and getMax() return these values back. It should be noted that the specified abscissa range does not affect the method eval(x) which determines the value of a function at a certain x value.

Finally, you may need to determine the plotting resolution. By default, a function is evaluated at 500 points in the specified abscissa range. One can find this using the `getPoints()` method. One can change this value using the method `setPoints(i)`, where `'i'` is an integer number (should be sufficiently large). For a very large number, the price to pay is a larger memory consumption and slower plotting.

In order to plot several functions on the same canvas, simply repeat the `draw(obj)` statement for each function.

```
>>> f1=F1D('2*exp(-x*x/50)+sin(pi*x)/x', -2.0,5.0)
>>> f2=F1D('exp(-x*x/10)+cos(pi*x)/x', -2.0,5.0)
>>> c1.draw(f1)
>>> c1.draw(f2)
```

It should be noted that the method `draw(obj)` can also be used to visualize a Jython list of `F1D` functions as in the example below:

```
>>> from jhplot import *
>>>
>>> c1=HPlot('Canvas')
>>> c1.visible()
>>> c1.setAutoRange()
>>> f1=F1D('2*exp(-x*x/50)+sin(pi*x)/x', -2.0,5.0)
>>> f2=F1D('2*sqrt(x)')
>>> c1.draw([f1,f2])
```

For all these examples, the color used for drawing is the same, thus it is difficult to separate the plotted functions visually. One can draw the functions using various colors after importing the `Color` class from the Java AWT package as discussed before. Moreover, one can define the line width as:

```
>>> from java.awt import Color
>>> from jhplot import *
>>>
>>> f1=F1D('2*exp(-x*x/50)+sin(pi*x)/x', -2.0,5.0)
>>> f2=F1D('exp(-x*x/10)+cos(pi*x)/x', -2.0,5.0)
>>> f1.setPenWidth(2)
>>> f1.setColor(Color.green)
>>> f2.setColor(Color.red)
>>> c1.draw(f1)
>>> c1.draw(f2)
```

where the method `setPenWidth(i)` accepts an integer number for the line width in terms of the number of pixels. To draw dashed lines, use the `setPenDash()` method. One can change the dashed line length by specifying an integer value between 0 and 40. One can also use the `update()` method to redraw the plot.

Table 3.5 The most important methods for graphical representations of the F1D class. For the methods shown in this table, "b" indicates a boolean value (0 for true and 1 for false), while "i" is an integer parameter. The notation "d" indicates a float value. The attributes "c" and "f" correspond to the Color and Font classes of Java AWT. "text" represents a string

Methods	Definitions
setColor(c)	set line color
setPenWidh(i)	width of the line
setPenDash(i)	dashed style with "i" being the length
setLegend(b)	set (b=1) or not (b=0) the legend
setTitle(text)	set title text

The example above shows that a F1D function contains several methods for drawing. Some most important graphical options are listed in Table 3.5.

As mentioned before, instead of using all these graphical methods in Jython scripts, for function drawing, one can also edit function attributes using a GUI-driven dialog: Navigate the mouse to the pad with the graph and select [Edit settings] with the mouse button. Then select [Y item] and you will see a pop-up window with various attributes. In a similar way one can edit the global title and margins of the pads.

3.3.3 Plotting 1D Functions on Different Pads

To plot two or more functions on different plot regions, one should construct an appropriate canvas. Before calling the method draw(obj), the current plotting pad has to be changed using the cd(i1,i2) method. In the example below we make a canvas with two pads and then navigate to the necessary pad when we need to plot a function:

```
>>> from java.awt import Color
>>> from jhplot  import *
>>>
>>> c1=HPlot('Canvas',600,400,1,2)
>>> c1.visible()
>>> c1.setAutoRange()
>>> f1=F1D('2*exp(-x*x/50)+sin(pi*x)/x', -2.0, 5.0)
>>> f2=F1D('exp(-x*x/10)+cos(pi*x)/x', -2.0, 5.0)
>>> f1.setColor(Color.green)
>>> c1.draw(f1)
>>> c1.cd(1,2)      # go to the second pad (1,2)
>>> c1.draw(f2)
```

3.3.4 Short Summary of HPlot Methods

Table 3.6 shows the major methods of the HPlot class. Note that there are more than 300 methods associated with this class, which are divided into the "getter" (they start from the "get" string) and "setter" (starting from "set") groups of methods.

One can learn about the methods of the HPlot canvas using the jHepWork code assist, i.e. typing c1. and pressing [F4] key, where c1 is an instance of the HPlot class. One can also use the method dir(c1) to print the methods.

We will discuss the HPlot canvas in Sect. 10.2 in more detail. In addition, we can show how to use other graphical tools to display functions in Chap. 10.

3.3.5 Examples

Let us show a complete script example for visualization of several 1D functions using two plot regions. Save these lines in a file, say 'test.py', load this file into the jHepWork editor and click on the "[run]" button (or press [F8]) to execute the script.

—————————— Using HPlot canvas ——————————

```
from java.awt import Color
from jhplot  import *

c1 = HPlot('Canvas',600,400,2,1)
c1.visible()
c1.setGTitle('F1D Functions',Color.red)
c1.setNameX('X axis')
c1.setNameY('Y axis')

c1.cd(1,1)
c1.setAutoRange()
c1.setName('Local title')
f1 = F1D('2*exp(-x*x/50)+sin(pi*x)/x',-2.0,5.0)
f1.setPenDash(4)
c1.draw(f1)
f1 = F1D('exp(-x*x/50)+pi*x',-2.0,5.0)
f1.setColor(Color(10,200,50))
f1.setPenWidth(3)
c1.draw(f1)

c1.cd(2,1)
c1.setAutoRange()
f1 = F1D('20*x*cos(x)',-0,5.0)
f1.setColor(Color.red)
f1.setPenWidth(3)
c1.draw(f1)
```

Table 3.6 Most important methods of the `HPlot` class. For the methods shown in this table, "b" indicates a boolean value (0 or 1), while "a" is an integer parameter indicating axis (a=0 means X-axis, a=1 means Y-axis). The notation "d" indicates a float value. The attributes "c" and "f" correspond to the `Color` and `Font` classes of Java AWT. "text" represents a string

Methods	Definitions
destroy()	clean and destroy the canvas frame
draw(o)	draw some object, like F1D etc.
clear()	clean the current region
cd(X,Y)	go to a current region in X and Y
clear(X,Y)	clean one region given by X and Y
clearAll()	clean all regions
clearData()	clean data from the current region all graph settings are kept
clearAllData()	clean data from all regions all graph settings are kept
visible(b)	make canvas visible (b=1) or invisible (b=0)
visible()	make canvas visible
setAutoRange()	set autorange for all axes
setAutoRange(b)	set autorange (b=1) or not (b=0) for all axes
setAutoRange(a,b)	set (b=1) or not set (b=1) autorange for axis
setAxesColor(c)	set color for axes
setBox(b)	set or not a bounding box around the graph
setBoxOffset(d)	offset of the bounding box
setBoxFillColor(c)	fill color of the bounding box
setBoxColor(c)	color of the bounding box
setBackgroundColor(c)	background color of the graph
setGrid(a,b)	show grid (b=1) or not (b=1) for axis
setGridColor(c)	grid color
setGridToFront(b)	grid in front of drawing (b=1) or behind (b=0)
update()	update current plot defined by cd()
updateAll()	update plots in all regions
setGTitle(text,f,c)	set attributes for global title
setLegend(b)	set legend (b=1) or not (b=0)
setLegendFont(f)	set legend font
setLogScale(a,b)	set (b=1) or not set (b=0) log scale for axis
setLegendPosition(a,pos)	set legend position given by pos value
setTicsMirror(a,b)	set (b=1) or not set (b=0) mirror ticks for axis
setGrid(a,b)	show grid (b=1) or not (b=1) for axis
setRange(a,min,max)	min and max for axis
setRange(minX,maxX,minY,maxY)	min and max range for X and Y
setAntiAlias(b)	set (b=1) or not set antialiase for graphics
removeAxes()	remove all axes
export(FileName)	export to an image (png, eps, ps) Image format is given by the file extension
update()	update the current plot

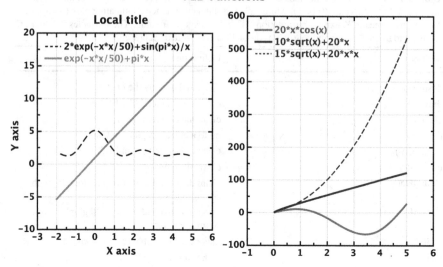

Fig. 3.1 Several F1D functions plotted on two pads of the same HPlot canvas

```
f1 = F1D('10*sqrt(x)+20*x',0,5.)
f1.setColor(Color.blue)
f1.setPenWidth(3)
c1.draw(f1)

f1 = F1D('15*sqrt(x)+20*x*x',0,5.)
f1.setColor(Color.blue)
f1.setPenDash(3)
c1.draw(f1)
```

The execution of this script creates a canvas with several plotted functions as shown in Fig. 3.1.

3.4 2D Functions

3.4.1 Functions in Two Dimensions

By considering functions with more than one independent variable, we are now beginning to venture into high-dimensional spaces. To build a two-dimensional (2D) function, jHepWork has the F2D class. It is defined in the same way as F1D. The only difference - it takes two independent variables, x and y.

```
>>> from jhplot import *
>>> f1=F2D( 'definition')
```

where 'definition' is a string that must be replaced by the actual mathematical formula with two independent variables, x and y. As for the F1D class, the function definition can contain any combination of predefined operators, functions and constants listened in Tables 3.1, 3.2 and 3.3.

To evaluate a 2D function at a fixed point (x, y), use the eval(x,y) method. One can evaluate a function for lists with x and y values using the method eval(x[],y[]) as in the example below:

```
>>> from jhplot import *
>>> f2=F2D('sqrt(x)*sin(y)/x + y^3')
>>> f2.eval(2,0.1)
0.07
>>> f2.eval([1,2],[2,4])
array([D, [array('d', [8.9,63.2]),array('d',[8.6,63.4])]])
```

The output numbers below are truncated to fit them to the width of this page.

As for the F1D function, use a "parsed" object if you need to pass it to some method inside long loops. Get it using the method getParse() which returns an object for evaluation.

The F2D functions can numerically be integrated for a region between minX-maxX (for x) and between minY-maxY (for y) using the method:

```
>>> f2.integral(N,minX,maxX,minY,maxY)
```

where N is the number of points for integration for each abscissa. The method assumes the trapezoid rule for integration.

3.4.2 Displaying 2D Functions on a Lego Plot

The F2D functions can be shown using a three-dimensional HPlot3D canvas. This canvas has very similar methods as those for the HPlot canvas, but allows to draw objects in 3D. What we should remember is that we have to prepare the function for drawing beforehand, which means one should specify the ranges for x and y axes during the initialization (or after, but before drawing).

During the initialization step, one can set the ranges as:

```
>>> from jhplot import *
>>> f2=F2D('sqrt(x)*sin(y)/x + y^3',minX,maxX,minY,maxY)
```

so the function will be plotted for x in the interval [minX-maxX], and in the interval [minY-maxY] for the y independent variable.

After the initialization, one can set the range using the setMinX(min), setMinX(max) methods for x, and analogously for y. The number of points for the evaluation can be set using setPoints(n) methods (the default number of points is 500).

Here is a typical example which shows how to plot either a single function or two functions on the same lego plot:

```
————————————————— Plotting 2D functions —————————————————
from java.awt import Color
from jhplot import *

c1  = HPlot3D('Canvas',600,700, 2,2)
c1.visible()
c1.setGTitle('F2D examples')

f1=F2D('cos(x*y)*(x*x-y*y)', -2.0, 2.0, -2.0, 2.0)
f2=F2D('cos(x+y)*x', -2.0, 5.0, -2.0, 5.0)
f3=F2D('sin(4*x)+x^2', -2.0, 2.0, -2.0, 2.0)
f4=F2D('x^2+y^2', -2.0, 2.0, -2.0, 2.0)

c1.cd(1,1)
c1.setScaling(8)
c1.setRotationAngle(30)
c1.draw(f1)

c1.cd(2,1)
c1.setAxesFontColor(Color.blue)
c1.setColorMode(3)
c1.setScaling(8)
c1.setElevationAngle(30)
c1.setRotationAngle(35)
c1.draw(f2)

c1.cd(1,2)
c1.setColorMode(4)
c1.setLabelFontColor(Color.red)
c1.setScaling(8)
c1.setRotationAngle(40)
c1.draw(f3)

c1.cd(2,2)
c1.setColorMode(1)
c1.setScaling(8)
c1.setElevationAngle(30)
c1.setRotationAngle(35)
c1.draw(f4,f2)
```

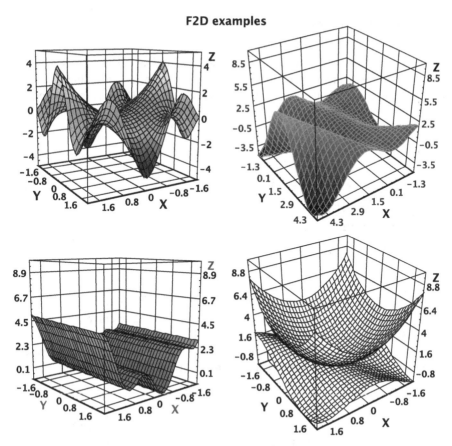

Fig. 3.2 F2D functions shown using the HPlot3D canvas in several regions

The resulting figure is shown in Fig. 3.2.

This example features one interesting property of the HPlot3D canvas: Instead of plotting one function as c1.draw(f1), we plot both functions on the same pad after passing two functions as arguments (see the last line). This is somewhat different from the HPlot behavior, where we could pass any number of functions in a form of the list. For the HPlot3D canvas, one can plot two functions at most. Also, the class does not accept lists of functions. Later we will show that one can plot 2D histograms or even mixing histograms with functions using the method draw(obj1,obj2).

For the three-dimensional canvas, the presentation style can be changed using setColorMode(i) method, where i=0 for wireframe, i=1 for hidden, i=2 for color spectrum, i=3 for gray scale, i=4—for dualshades. The methods setScaling(), setElevationAngle() and setRotationAngle() are self-explanatory. The plots can be rotated with the mouse and the title can be modified using exactly the same way as for the HPlot canvas. We will return to the HPlot3D canvas in Sect. 10.12.

3.4.3 Using a Contour Plot

The 2D functions can also be shown using a contour (density) style. This can be done with the help of the same HPlot3D canvas but adding the line c1.setContour() after the definition of the c1 object. This is shown bellow:

```
―――――――――――――― Contour plot ―――――
from java.awt import Color
from jhplot import *

c1 = HPlot3D('Canvas',600,600)
c1.setNameX('X')
c1.setNameY('Y')
c1.setContour()
c1.visible()
f1=F2D('x^2+y^2', -2.0, 2.0, -2.0, 2.0)
c1.draw(f1)
```

The execution of this script leads to the plot shown in Fig. 3.3. The color bar from the right indicates the density levels used for drawing in color. They can be redefined using several methods of this canvas.

However, it is more practical to show the F2D functions using the canvas based on the class HPlot2D. As we will discuss in Sect. 10.11, this class has significantly

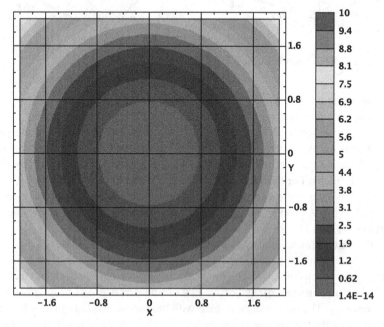

Fig. 3.3 A contour representation of the function $x^2 + y^2$ using the HPlot3D canvas

more options to display the contour and density plots since it was designed mainly
for such types of plots.

3.5 3D Functions

3.5.1 Functions in Three Dimensions

You may already have guessed that for tree-dimensional (3D) functions, jHepWork
has the F3D class. It is defined in the same way as F1D and F2D, the only difference
is it can take up to three independent variables: x, y and z.

```
>>> from jhplot import *
>>> f3=F3D('definition')
```

where 'definition' is a string that should be replaced by the actual formula. As
for the F1D and F2D classes, the function definition can be constructed from a combination of operators, mathematical functions and constants, as given in Tables 3.1,
3.2 and 3.3.
   To evaluate a 3D function at a fixed value (x, y, z), use the usual method
eval(x,y,z):

```
>>> from jhplot import *
>>> f3=F3D('sqrt(x)*sin(y)/x + z^3')
>>> f3.eval(2,0.1,4) # calculate for x=2, y=0.1, z=4
0.07159288589999416
```

   Probably, we can stop here and will not go into the drawing part of this section.
We have to be creatures leaving in four-dimensional space in order to be interested
in how to draw such functions (actually, if such a creature is reading this book
and is still interested in this option, please contact me—we may discuss how to
plot such functions). One can also project 3D functions into a 2D space by fixing
some independent variables and then using the F2D and HPlot3D to display such
projections.
   Below we will discuss more flexible classes for dealing with multidimensional
functions.

3.6 Functions in Many Dimensions

3.6.1 FND Functions

jHepWork supports functions with arbitrary number of independent variables using
the FND class. This class is more complicated and has its roots in the known JEP
Java package [1].

```
>>> from jhplot import *
>>> fn=FND('definition', 'var1,var2,...varN')
```

where the string 'definition' should be replaced by the actual mathematical formula using a combination of predefined operators, functions and constants given in Tables 3.1 and 3.3. The second argument tells which characters or strings must be declared to be independent variables in the function definition. Unlike F1D, F2D and F3D classes, variable names can be any strings (not only x, y and z). Functions of the class FND can be simplified and differentiated and, obviously, can be evaluated at fixed points. The example below shows this:

```
>>> import  jhplot  import *

>>> fn=FND('1*x^4+x^2+y+z+h','x,y,z,h')
>>> fn.simplify()
>>> print 'Simplify=',fn.toString()
Simplify= x^4.0+x^2.0+y+z+h
>>>
>>> fn.diff('x') # differentiate using x
>>> print 'Differentiate=',fn.toString()
Differentiate= 4.0*x^3.0+2.0*x
>>>
>>> fn=FND('1*x^4+x^2+y+z+h','x,y,z,h')
>>> print 'Print variables=',fn.getVars()
Print variables=array(java.lang.String,['x','y','z','h'])
>>>
>>> d=fn.eval('x=4,y=1,z=2,h=0')
>>> print 'Evaluate results=',d
Evaluate results= 275.0
```

In this example, the evaluation of the function happens after fixing all three independent variables.

3.6.2 Drawing FND Functions

The class FND for function representation is rather flexible. First of all, one can easily deal with 1D functions using any names of independent variables (remember, F1D can only accept x to define a variable). The only difference you have to keep in mind is that before drawing a FND function, you should always call the eval() method to allow for only one independent variable and friezing other variables to fixed numbers.

Let us consider an example with two independent variables. In this case, we should set the second variable to some value since we want to plot an one-dimensional function:

```
━━━━━━━━━ Drawing a FND function ━━━━━━━━━
from jhplot  import *

c1=HPlot()
f2=FND('sqrt(var1)*sin(var2)','var1,var2')
f2.eval("var1",1,100,'var2=2') # var1 in range 1-100
c1.visible()
c1.setAutoRange()
c1.draw(f2)
```

In the example above, first we evaluate the function in the range between 1 and 100 and fixing `var2` to 2 before calling the `draw(obj)` method.

Below is the example for a function with three independent variables. After fixing two variables, we plot this function in 1D:

```
━━━━━━━━━ Drawing a FND function ━━━━━━━━━
from jhplot  import *

c1 = HPlot('Example',600,400)
c1.setGTitle('FND function')
c1.visible()
c1.setAutoRange()
f2=FND('x*x+y+20+z','x,y,z')
f2.eval('x',-2,10,'y=2,z=100')
c1.draw(f2)
```

Now we have fixed two variables, y and z, and plotted the function in terms of the independent variable x between -2 and 10. It should be noted that the variables to be fixed are separated by a comma in the `eval()` statement.

3.7 Custom Functions Defined by Jython Scripts

Building a custom mathematical function using Jython scripts makes sense when the logic of the mathematical expression is so complicated that it's better to define it in a separate code block rather than to use a string and pass it during the instantiation of the function. In this way one can build infinitely complicated functions in many dimensions and evaluate them using jHepWork Java libraries.

3.7.1 Custom Functions and Their Methods

For constructing functions in this section, we will use the package `shplot` provided by jHepWork. Let us build a simple second-order polynomial function,

$y = c * x^2 + b * x + a$ using a pure-Jython approach. For an educational purpose, we will include some complication into the function definition: if $x < 0$, then we assume that the polynomial behavior will vanish and the function will be just a constant $y = a$.

Let us put the following code in a separate file called 'p2.py':

```
──────────────── File 'p2.py' ────────────────
from shplot import *

class p2(ifunc):
  def value(self, v):
    if v[0]>0:
      d=self.p[2]*(v[0]*v[0])+self.p[1]*v[0]+self.p[0]
    else:
      d=self.p[0]
    return d
```

The example features several important properties: instead of the 'jhplot' package we use the 'shplot' library, which represents a high-level Jython module based on the Java jhplot package. This module is imported automatically if one uses jHepWork. If you are using something else, one can find this module in the directory 'system/shplot'. In the latter case, we remind that in order to be able to use it, one should import the Jython module os and append the directory with the shplot module to the Jython variable to the list 'os.path'.

Secondly, our class 'p2' inherits properties of the 'ifunc' class (see the class inheritance topic in Sect. 2.11.2). This is important as it provides the necessary functionality when we will decide to draw such function.

Finally, we specify our mathematical algorithm in the function value(), which returns the results of the calculation. Obviously, self.p[] list represents our free parameters a, b, c, while v[0] corresponds to the variable x.

One can also use any mathematical function provided by the 'math' package of Jython. In this section, but we will restrict ourselves to simple examples which do not require calls to external mathematical functions.

Now, let us build a function from the class defined above. We will instantiate the function object as:

```
>>> p=function(title,dimension,paramNumber)
```

where 'title' is a string with the function title, 'dimen' is a dimension of the function and 'paramNumber' is the number of free parameters. Names for the variables and parameters will be set to default values. Alternatively, one can use custom names for the variables and parameters using this constructor:

```
>>> p=function(title,names,pars)
```

where 'name' and 'pars' are lists of strings defining the names for independent variables and parameters names (this overwrites the default names).

Let us come back to our example and instantiate the second-order polynomial function given in the module 'p2.py'. We will create a function by assigning the title 'p2', dimension of this function (1) and the number of parameters (3 parameters).

```
──────────── Building a custom function ────────────

from p2 import *

p=p2('p2',1,3)
print p.title()
print p.dimension()
print p.numberOfParameters()
print p.parameterNames()
print p.variableNames()
print p.variableName(0)
```

The execution of this script gives:

```
p2
1
3
array(java.lang.String, [u'par0', u'par1', u'par2'])
array(java.lang.String, [u'x0'])
x0
```

Let us discuss the output in more detail: We print out the title of this function with the method title() and the dimension with the method dimension(). The number of parameters is given by the method numberOfParameters(). The method parameterNames() returns the parameter names. Since we did not assign any custom names, it prints the default names, 'par0', 'par1', 'par2'. Then we print the variable name (the default is 'x0').

Now let us evaluate this function at several points. But before, let us assign some numerical values to our free parameters. One can set a single value by calling the parameter name, or setting it at once in form of list. Then we will evaluate the function at $x = 10$ and $x = -1$:

```
──────────── Setting parameters ────────────

from p2 import *

p=p2('p2',1,3)
p.setParameter('par0', 10)
p.setParameter('par1', 20)
p.setParameter('par2', 30)

# set all parameters in one go
```

```
 p.setParameters([10,20,30])

 print "Value at x=10 =", p.value([10])
 print "Value at x=-1 =", p.value([-1])
```

Note how this was done: first we pass a list to the function, and then return it with the method `value()`. As you may guess, this is necessary in case if you have several variables to pass to a function. Executing the above script gives:

```
Value at x=10 = 3210.0
Value at x=-1 = 10.0
```

Now, let us assign custom names for the parameters. This time will instantiate it using the second constructor as:

```
──────────────────────── Setting parameters ────────────────────────
 from p2 import *

 p=p2('p2',['x'],['a','b','c'])
 print p.parameterNames()
 print p.variableNames()
 p.setParameter('a', 10)
 p.setParameter('b', 20)
 p.setParameter('c', 30)

 print "Value at x=10 =", p.value([10])
 print "Value at x=-1 =", p.value([-1])
```

As you can see, we assign the names 'a', 'b', 'c'. We also set the variable name to the convenient 'x'. One can see this from the output shown below:

```
array(java.lang.String, [u'a', u'b', u'c'])
array(java.lang.String, [u'x'])
Value at x=10 = 3210.0
Value at x=-1 = 10.0
```

More advanced users can look at the methods of the object 'p'. They will find out that this object is constructed from the AIDA class IFunction and contains many other methods which have not been discussed here.

3.7.2 Using External Libraries

Previously we have shown a rather simple example illustrating how to construct a mathematical function. But we are not restricted here with constructions from

Jython: since our codding is based on Java, one can call a function or library from Java API, or Java-based external library. We can direct you, for example, to the Colt Java library [2] (included into jHepWork) which provides a large set of special functions and functions commonly used for probability densities.

To build a function, you will need first to import necessary classes from third-party libraries and replace parameters with self.p[], while variables should be replaced with v[] in the method value(). Let us show how to make a simple Jython module containing a Bessel function and Beta function:

```
––––––––––– Calling Java libraries. Module 'special.py' –––––––––––
from shplot import *

class bessel(ifunc):
   def value(self, v):
      from cern.jet.math import Bessel
      return Bessel.i0e(v[0])

class beta(ifunc):
   def value(self, v):
      from cern.jet.stat.Probability import beta
      return beta(self.p[0],self.p[1],v[0])
```

First, we import the Java package 'cern.jet.math' from the Colt library mentioned above and use this library to construct the Bessel and Beta functions. Please look at the Java API of the Colt library to learn about such functions. We will save these lines in a module called 'special.py'.

Calling both functions from the module 'special.py' is trivial. Just import this module and take care of the number of parameters you pass to such functions when calling them:

```
––––––––––––––––––– Calling special functions –––––––––––––––––––
from special import *

b1=beta('Beta function',1,2)
b2=bessel('Bessel order 0',1,0)
```

One can evaluate the functions at any allowed value 'x' after specifying values for free parameters. In case of the Bessel function, we do not have free parameters and the evaluation of such function is straightforward.

3.7.3 Plotting Custom Functions

Now we know how to build function objects from a script. Next, we will show how to manipulate with such custom functions and plot them.

First thing you probably will need to do is to convert a function to F1D or F2D
objects for plotting. Below we show how to convert IFunction to the standard
F1D, using the function Bessel defined in the previously constructed module
'special.py'.

```
——————————————— Conversion to F1D ———————————————
from special import *
from jhplot import *

p=bessel('Bessel',1,0)
f=F1D(p)
print f.eval([0.1,0.2,0.5])
```

so, it looks easy: just pass the custom function to the constructor of the F1D function. In the above example, we evaluate this function at several points using a list
of x values. The execution of this script prints the output list of y values, [0.9071,
0.8269, 0.6450].

One can return the object IFunction back as getIFunction(), a handy
method of the F1D function. We will see that the IFunction class is very important when dealing with curve fitting in Sect. 14.2.

Probably you have already realized how to plot our Bessel function. The script
below plots this function in the range [0–100]:

```
——————————————— Conversion to F1D ———————————————
from special import *
from jhplot import *

p=bessel('Bessel',1,0)
f1=F1D(p,1,100)
c1 = HPlot()
c1.visible()
c1.setAutoRange()
c1.draw(f1)
```

Plotting 2D functions is as easy as in the 1D case. First build a custom function
from the script using two variables. Then pass it to the F2D constructor exactly as in
the above example. Always pay attention to free parameters: they have to be defined
beforehand by passing the function object to the F2D constructor.

We will return to the subject of custom functions when we will discuss the class
IFunction in Sect. 14.2, which will be used for fitting experimental data. You
will also learn how to access predefined functions or create functions of the class
IFunction from a string.

3.8 Parametric Surfaces in 3D

3.8.1 FPR Functions

jHepWork has a support for drawing parametric functions or equations in 3D. This feature has its root in the initial version of the 3D Graph Explorer program [3].

We remind that a parametric function is a set of equations with a certain number of independent parameters. To build a parametric function, one should use the FPR class ("F" abbreviates the word "function" and "PR" corresponds to "parametric"). The general definition of a parametric function is:

```
>>> f1=FPR('definition')
```

where 'definition' is a string representing the equation used for the function definition. For parametric surfaces in 3D, we should define x, y and z in terms of two independent variables, 'u' and 'v', which can vary in the range [0, 1]. To construct a string representing the parametric function, one can use the predefined functions given in Table 3.1. In addition, the standard comparison operations "==", "!=", "<", ">", "<=" and ">=" can be used.

For example, the equation:

$$u = 2 * Pi * u; \qquad x = \cos(u); \qquad y = \sin(u); \qquad z = v$$

defines a cylinder. Note, that "Pi" means the predefined π value, and the multiplication sign can be replaced by a space. Each logical unit should be separated by a semicolon, and the entire expression should be passed to the FPR constructor as a string.

Let us show several other equations: A torus can be written as:

$$u = 2 * Pi * u; \qquad v = 2 * Pi * v; \qquad r = .6 + .2 * \cos(u)$$
$$z = .2 * \sin(u); \qquad x = r * \cos(v); \qquad y = r * \sin(v)$$

A cone can be written as:

$$u = 2 * Pi * u; \qquad z = 2 * (v - .5)$$
$$x = z * \cos(u) * .8 - 1; \qquad y = z * \sin(u) * .8 + .6$$

A hex cylinder can be written as:

$$u = 2 * Pi * u; \qquad x = \cos(u) * .8 + .3$$
$$y = \sin(u) * .8 - .6; \qquad z = 2 * (v - .5)$$

A sphere with a radius "r" is

$$r = 0.7; \qquad u = 2 * Pi * u; \qquad v = Pi * v; \qquad x = r * \cos(u) * \sin(v)$$
$$y = r * \sin(u) * \sin(v)$$
$$z = r * \cos(v)$$

In the above examples, we set the constants defining geometrical sizes to arbitrary values for simplicity.

As it can be seen, to vary the parameters u and v over a different range than [0,1], one should scale and shift them by a certain value. For example, if one needs to change the range of u to [5,15], use this transformation: $u = u * (15 - 5) + 5$.

The parametric functions can be shown using the HPlot3DP class (extra "P" in its name means "parametric"). This class provides a canvas for drawing parametric functions and can be used for interactive work with these functions (zooming and rotations). One can draw a surface using the usual draw(obj) method, where obj represents an object of the class FPR. One can make drawing pads as usual and navigate to the pads using the cd(i1,i2) method. One can draw several parametric functions on the same pad by repeating the draw(obj) method. This is seen from this example:

```
————————————— Drawing parametric functions ——————————

from java.awt import Color
from jhplot import *

c1 = HPlot3DP('Canvas',700,600,2,1)
c1.setGTitle('Parametric surfaces')
c1.visible()

f1=FPR('r=0.7; u=2 Pi u; v=Pi v; \
        x=r cos(u) sin(v); y=r sin(u) sin(v); z=r cos(v)')
f1.setDivisions(50,50)
f1.setLineColor(Color.blue);

f2=FPR('u=2 Pi u; v=2 Pi v; r=.7+.2cos(u); \
        z=.2 sin(u)-0.5; x=r cos(v); y=r sin(v)')
f2.setFillColor( Color(20,170,170) )

f3=FPR('ang=atan2(y,x); r2=x*x+y*y;\
        z=sin(5(ang-r2/3))*r2/3')

c1.draw(f1)
c1.draw(f2)

c1.cd(2,1)
c1.setCameraPosition(-1.0)
c1.setFog(0)
c1.draw(f3)
```

In this example, the lengthy strings which define the functions were broken to fit the page width. Figure 3.4 shows the resulting plot.

Let us discuss several graphical methods for the FPR functions. First of all, one can set colors for the lines and for the filled area, as well as the line width as:

Parametric surfaces

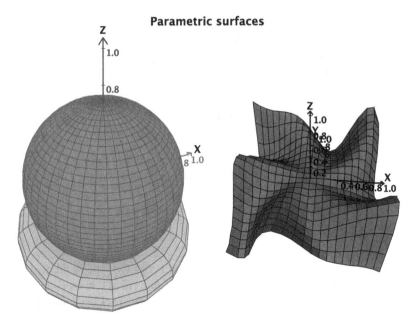

Fig. 3.4 Several parametric functions displayed using the HPlot3DP canvas

```
>>> f1.setFilled(b)
>>> f1.setFillColor(c)
>>> f1.setLineColor(c)
>>> f1.setPenWidth(i) # set line width "i"
```

where b is ether 0 (not filled area) or 1 (Java true) if the object has to be filled. 'c'
is the usual Java AWT Color class used to fill a surface. One can set the width of
the lines as setPenWidth(i), where "i" is an integer value. Finally, all above
graphical attributes can be obtained using the corresponding "getter" methods.

One can add a transparency level to the filled color by setting the so-called alpha
value for the AWT Color class. For example,

```
>>> f1.setFillColor( Color(0.1,0.9,0.7,0.5)
```

sets the transparency level for filled color to 50% (see the last argument 0.5).

The numbers of divisions for "u" and "v" independent variables are given by the
method:

```
>>> f1.setDivisions(divU, divV)
```

where divU and divV are the numbers of divisions (21 is the default value). The larger number of the divisions, the smoother surface is. However, this requires more computer resources, especially during an interactive work, such as rotations or zooming. One can also include the number of divisions during the construction of a parametric function:

```
>>> f1=FPR('definition',divU,divV)
```

We will discuss several methods associated with the class HPlot3DP in Chap. 10. One can also look at the corresponding API documentation of this class.

3.8.2 3D Mathematical Objects

In addition, jHepWork contains the 3D-XplorMathJ package [4] as a third-party library which allows to view various parametric functions in 3D. A user can specify and draw any function using the menu of this program. The package contains an impressive catalog of pre-build interesting mathematical objects, ranging from planar and space curves to polyhedra and surfaces to differential equations and fractals.

The 3D-XplorMathJ can be started using [Tools] → [3DXplorMath] from the Tool bar menu the jHepWork IDE.

3.9 Symbolic Calculations

Symbolic calculations can be performed using the SymPy library for symbolic mathematics [5]. This library is implemented in Python and included as a third-party package in the directory 'python/packages'. When working with the jHepWork IDE, this directory is automatically imported by Jython, so you do not need to worry about how to install the SymPy library.

The SymPy library can perform differentiation, simplification, Taylor-series expansions, integration. It can be used to solve algebraic and differential equations, as well as systems of equations. The complete list of features can be found on the SymPy Web page [5]. Below we will show several examples illustrating how to integrate this package with the jHepWork graphics libraries.

But, first of all, let us give several examples illustrating how to use this package for analytic functions. We will start from the example illustrating how to perform a series expansion of the expression $x^2/\cos(x)$.

```
─────────────────── Series expansion ───────────────────
from sympy import *

x = Symbol('x')
func=x**2/cos(x)
```

```
s=func.series(x, 0, 10)

print  s
print latex(s)
pprint(s)
```

The statement 'Symbol()' is rather important: in this package, symbolic variables have to be declared explicitly. The output of this example is printed in the standard ("Python") format when using the print command:

```
x**2 + x**4/2 + 5*x**6/24 + ..
```

(we have truncated the output). In addition, the example shows how to print a nicely-formatted output using the pprint() method, close to that used in papers and books. In addition, one can transform the output to the LaTeX format which is very popular within the scientific community.

We should note that, as for the jHepWork functions, all mathematical definitions given in Table 3.1 can be used to build SymPy functions as well. The functions can be in any dimensions. For example, if you have a second independent variable, y, in your mathematical formula, specify this variable using the method y=Symbol('y').

Below we show a code snippet that gives you some feeling about what mathematical operations can be performed with the SymPy library. The names of the methods shown below are self-explanatory, so we will be rather brief here:

```
————————————— SymPy examples —————————
from sympy import *

x=Symbol('x')
f=x*sin(x)
print 'Differential=',diff(f, x)
print 'Integral=',integrate(f,x)
print 'Definite integral=',integrate(f,(x,-1,1))
print 'Simplify=',S(f/x, x)
print 'Numerical simplify',nsimplify(pi,tolerance=0.01)
print 'Limit x->0=',limit(f/(x*x), x, 0)
print 'Solve=',solve(x**4 - 1, x)
y=Symbol('y')
z=Symbol('z')
print 'Combine together=',together(1/x + 1/y + 1/z)
```

The output of this script is given below:

```
Differential= x*cos(x) + sin(x)
Integral= -x*cos(x) + sin(x)
Definite integral= -2*cos(1) + 2*sin(1)
Simplify= sin(x)
Numerical simplify 22/7
Limit x->0= 1
Solve= [-1, I, 1, -I]
Combine together= (x*y + x*z + y*z)/(x*y*z)
```

The jHepWork libraries can be used for visualization of the symbolic calculations performed by the SymPy program. Use the jHepWork functions, F1D and F2D, and HPlot or HPlot3D (for 2D functions) for graphical canvases. For the conversion between Java and SymPy Python objects, you will need to use the str() method that moves SymPy objects into strings.

Below we show how to differentiate a function and show the result on the HPlot canvas, together with the original function:

```
—————————— Function differentiation ——————————
from java.awt import Color
from jhplot  import *

c1 = HPlot('Canvas')
c1.setGTitle('Example', Color.red)
c1.setName('Differential')
c1.visible()
c1.setAutoRange()

func='2*exp(-x*x/50)+sin(pi*x)/x'
f1 = F1D(func, 1.0, 10.0)

# now symbolic calculations
from sympy import *
x = Symbol('x')
d=diff(S(func), x)

f2 = F1D(str(d), 1.0, 10.0)
f2.setTitle('Differential')
f2.setColor(Color.green)
c1.draw([f1,f2])
```

The only non-trivial place of this code is when we moved the result from the differentiation into a string using the str() method. For convenience, we also used the simplify method, 'S()', which converts the string representing F1D function into the corresponding SymPy object. The rest of this code is rather transparent. The result of the differentiation will be shown with the green line.

Further discussion of the SymPy package is outside the scope of this book. Please study the original SymPy documentation.

3.10 File Input and Output

The best way to save jHepWork functions in a file is to use the Java serialization mechanism. Jython serialization, such as `pickle` or `shelve` discussed in Sect. 2.16, will not work as we are dealing with the pure-Java objects which can only be saved using the native Java serialization. However, you can use the `pickle` or `shelve` modules in case when you define a Jython function or when using a string with the function definition.

One can save a function together with its attributes (including graphical ones) using the Java serialization mechanism. For this, the `Serialized` class included to the `jhplot.io` package can be useful. It takes an Java object and saves it into a file using the method `write(obj)`. Then one can restore this object back using the method `obj=read()`. By default, all objects will be saved in a compressed form using the GZip format.

Look at the example below which shows how to save a function into a file:

```
>>> from jhplot  import *
>>> from jhplot.io import *
>>>
>>> f1=F1D('2*sin(x)')
>>> print p0.getName()
2*sin(x)
>>>
>>> Serialized.write(f1,'file.ser')
>>>
>>> # deserialize F1D from the file
>>> f2=Serialized.read('file.ser')
>>> print 'After serialization:',f2.getName()
After serialization: 2*sin(x)
```

The same can be achieved using the standard Java API, so the method `write()` from `Serialized` class is equivalent to:

```
>>> from java.io import *
>>> f=FileOutputStream("file.ser")
>>> out=ObjectOutputStream(f)
>>> out.writeObject(f1)
>>> out.close()
```

while the method read() from the Serialized class is equivalent to:

```
>>> file = File('file.ser')
>>> fin = ObjectInputStream(FileInputStream(file))
>>> f2 =fin.readObject()
>>> fin.close()
```

One can also write a list of functions, instead of a single function:

──────── Serialization of functions ────────

```
from jhplot  import *
from jhplot.io import *

f1=F1D('x*x+2')
f2=F2D("x*y+10")
# make list
a=[f1,f2]

# write list to a file
Serialized.write(a,'file.ser')

# read functions from the file:
list=Serialized.read('file.ser')

f1=list[0]
f2=list[1]
print 'After serialization:\n'
print f1.getName()
print f2.getName()
```

At this moment, the serialization mechanism is not implemented for FND. However, generally, the jHepWork functions are relatively simple entities, therefore, one can just serialize or write a string with the function definition into a file and then use it to instantiate a new object of this function.

In Chap. 11 we will discuss the input-output (I/O) issues in more detail. We will learn how to save an arbitrary mix of various jHepWork and Java objects, including lists of functions, histograms and arrays. For example, one can write a large sequence of functions persistently using the HFile class to be discussed later.

References

1. Funk, N.: A Java library for mathematical expressions. URL http://sourceforge.net/projects/jep/
2. The Colt Development Team: The COLT Project. URL http://acs.lbl.gov/~hoschek/colt/
3. Bose, A.: 3D Graph Explorer.
4. T.D.-X. Consortium: The 3D-XPLORMATHJ project. URL http://3d-xplormath.org/
5. SymPy Development Team: SYMPY: Python library for symbolic mathematics. URL http://www.sympy.org

Chapter 4
One-dimensional Data

4.1 One Dimensional Arrays

Numerical computations based on repetitive tasks with values are not too efficient in Jython, since loops over objects are slow. This has already been illustrated using Jython arrays in Sect. 2.14.2. In most cases, what we really what is to manipulate with primitive data types, such as floats or integers, rather than with immutable objects used for representation of numbers in Jython. Therefore, our strategy for this book will be the following:

> Jython will be viewed as "interface" type of language, i.e. a language designed to link and manipulate with high-level Java classes that implement repetitive operations with primitive types.

In the following chapters, we will discuss objects used for data storage and manipulation—building blocks from which a typical Java or Jython program can be constructed. Unlike Jython classes, the objects to be discussed below will be derived from pure Java classes and imported from the Java libraries of jHepWork.

In this chapter, we will discuss one-dimensional data, i.e. a data set represented by a sequence of double-precision floating-point real values, $N_1, N_2, \ldots, N_{max}$. This is a rather common case for almost any data analysis. Each number can represent, for example, a value obtained from a single measurement. Below we will discuss how to build objects which can be used for: (1) generation of one-dimensional data sets; (2) storing numerical values in containers; (3) writing to or reading from files; (4) performing numerical analyzes and producing statistical summaries; (5) finally, for visualization and comparisons with other data sets.

As we are progressing deeper into analysis of one-dimensional (1D) data, you may be unsatisfied with the simplicity of JythonShell when typing a short code snippets. We will remind that, in this case, one should write code in files with the extension '.py' and execute such macros using the key [F8] or the [run] button as described in Sect. 1.4.

S.V. Chekanov, *Scientific Data Analysis using Jython Scripting and Java,*
Advanced Information and Knowledge Processing,
DOI 10.1007/978-1-84996-287-2_5, © Springer-Verlag London Limited 2010

4.2 P0D Data Container

The P0D class is among the simplest classes of jHepWork: it does not have any graphical attributes. This class is designed to keep a sequence of real numbers. It is similar to the Java ArrayList class, but keeps only double-precision real values represented with 64 bit accuracy. In some extent, the class P0D is also similar to the Jython list which keeps a sequence of objects.

For integer values, use the class P0I which has exactly the same methods as P0D. For the P0I arrays, you may benefit from a lower memory usage and smaller file size when writing this array into files. Below we will discuss only the P0D class.

Let us construct a P0D object to keep an one-dimensional data set. It is advisable to annotate it, so we can easily obtain its attribute later:

```
>>> from jhplot import *
>>> p0=P0D('measurement')
```

where 'measurement' is a string representing the dataset title.

Let us remind again how we can learn about all methods associated with jHep-Work objects, like the p0 object described above. Here is the list of various methods to learn about this object:

- If you are using the JythonShell prompt, type 'p0.' and press [Ctrl]-[Space];
- If you are using JythonShell or jHepWork source-code editor, use the method dir(p) to print its methods;
- If you are using the jHepWork source-code editor, type 'p0.' and press [F4]. You will get a detailed description of all methods in a pop-up frame. One can select a necessary method and check its description using the mouse. In addition, one can copy the selected method to the editor area as described in Sect. 1.4.10;
- If you are using Java instead of Jython and working with Eclipse or NetBeans IDE, use the code assist of these IDEs. The description of these IDEs is beyond the scope of this book.

In all cases, you might be impressed by the significant number of methods associated with the P0D class. But let us go slowly and first discuss its most important methods.

One can obtain the title by calling the method getTitle():

```
>>> print p0.getTitle()
measurement
```

One can reassign a new title with the method:

```
>>> p0=P0D()
>>> p0.setTitle('new measurement')
```

Once the p0 object is created, one can add numbers to this container. In practice, one-dimensional data can be filled from a file or some external sources (this will be discussed later). Here we show how can add a value to the container:

```
>>> p0.add(1)       # add an integer
>>> p0.add(-4.0)    # add a float
>>> p0.add(-4E12)   # add a float
>>> p0.set(2,200)   # insert the value at the 2nd position
```

One can obtain the value at a specific index 'i' using the method get(i).

It should be noted that all values are converted into "double" representation inside the POD. The size of the container can be found using the method size(). One can remove all elements from the container using the method clear():

```
>>> print p0.size()    # size of the data
3
>>> p0.clear()         # clean up the container
>>> print p0.size()    # again print the size
0
```

Data from the p0 instance can be obtained in a form of the Jython arrays discussed in Sect. 2.14:

```
>>> array=p0.getArray()     # array with double  numbers
>>> array=p0.getArrayInt()  # array with integer numbers
```

If you want to return a list, instead of arrays, you can call the method getArrayList() method. One can convert the array into a list using the method tolist().

A POD container can be initialized from a Jython list as:

```
>>> p0.setArray([1,2,3,4])
```

It should be noted again that, for scripting with Jython, one should use the setArray() and getArray() methods which are optimized for the speed. They are much faster than pieces of codes with the methods add() or set() called inside Jython loops.

One can fill the current POD from a file (see below), generate a sequence of numbers or fill the POD with random numbers. For example, to fill a POD container with a sequence of real numbers between 0 and 100, use:

```
>>> from jhplot  import *
>>> p0=POD('sequence')
>>> p0.fill(101,0,100) # numbers from 0-100 (step 1)
```

A POD container can be instantiated from the Jython list:

```
>>> p0=POD('title',[1,2,3,4]) # with title
>>> p0.POD([1,2,3,4])         # without title
```

POD objects can also be filled with random numbers using a few built-in methods. For example, uniformly-distributed numbers can be filled as:

```
>>> p0= POD('Uniform distribution')
>>> p0.randomUniform(1000,0.0,1.0)
```

The first argument is the number of values to be filled, while the second and third specify the range for random numbers. One can also generate an array with random numbers in accordance with a Gaussian (normal) distribution with the mean 0 and the standard deviation equals one as:

```
>>> p0= POD('Normal distribution')
>>> p0.randomNormal(1000, 0.0, 1.0)
```

As before, 1000 is the total number of entries.

We will consider how to fill the POD class with many other random distributions in Sect. 9.5. In particular, we will explain how to fill a POD with random numbers using various functional forms of probability distributions in Sect. 9.6.

To print a POD on the screen, use the method toString() that converts the POD object into a string representation:

```
>>> p0=POD('measurement')
>>> p0.add(1); p0.add(2); p0.add(3);
>>> print p0.toString()
1.0
2.0
3.0
```

In some cases, it is convenient to show data as a table in a separate frame, so one can sort and search for a particular value. In this case, use

```
>>> p0.toTable()
```

This line brings up a table with filled values. Finally, one can use the print() method for printing using Java System.out.

4.2.1 POD Transformations

One can add, subtract, multiply or divide two POD objects. Having created two POD objects, say p0 and p1, one can apply several mathematical transformations resulting to new POD objects:

```
>>> p0=p0.oper(p1,'NewTitle','+') # add
>>> p0=p0.oper(p1,'NewTitle','-') # subtract
>>> p0=p0.oper(p1,'NewTitle','*') # multiply
>>> p0=p0.oper(p1,'NewTitle','/') # divide
```

To sort values or reverse their oder, use

```
>>> p0.sort()    # sort in the natural order
>>> p0.reverse() # reverse
```

Other useful methods are given below:

```
>>> m=p0.search(val) # first occurrence of 'val'
```

Here are some other useful methods:

```
>>> p0=p0.merge(p1)     # merge with p0 and p1
>>> p0.range(min,max)   # get range between min and max
```

All values inside the POD container can be transformed into a new set using an analytical function and a F1D function discussed in Sect. 3.2. Functions can either be defined from a string or using Jython functions as shown in Sect. 3.7. We remind that, to define a function, a user should use the variable x and the functions listed in Table 3.1. As usual, +, -, * or / and parenthesis () can be used in the definition.

Let us give a simple example. We will create a POD, initialize it with the list [1,2,3,4] and transform this list using the function $x^2 + 1$:

```
─────────────────── Transforming POD ───────────────────
from jhplot import *

f1=F1D('x^2+1')
p0=POD('numbers',[1,2,3,4,5])
print p0.func('squared+shift', f1)
```

The output of this code is:

```
POD squared+shift
2.0
```

```
5.0
10.0
17.0
26.0
```

Note that, in all cases, the objects remain to be the same, only the values will be modified. If one needs a copy of a POD object, use the method copy().

One can obtain a new POD object with all elements smaller, larger or equal of a specified value. Assuming that p0 is an object with filled values, this can be done using the following method:

```
>>> p1=p0.get(d,str)
```

The method returns a new array with values smaller, larger or equal to an input value d. The type of operation is specified using a string (shown as str in the above example), which can be equal to "<", ">" or "=", respectively.

4.2.2 Analyzing POD and Summary Statistics

One can obtain several useful characteristics of the POD class. First of all, let us consider the most simple methods, which return double values with certain characteristics of a p0 object of the class POD:

```
>>> m=p0.size()          # size of POD
>>> m=p0.getMin()        # min value
>>> m=p0.getMax()        # max value
>>> m=p0.getMinIndex()   # index of min value
>>> m=p0.getMaxIndex()   # index of max value
>>> m=p0.mean()          # mean value.
>>> m=p0.correlation(p1) # correlation coefficient p1
>>> m=p0.covariance(p1)  # covariance
>>> m=p0.getSum()        # sum of all values
```

In the above examples, p1 is another POD object used to find correlations between p0 and p1 arrays.

One can also perform a search for a specific value inside the POD arrays. The method contains(val) returns true if a value val is found. One can also find an index of the first occurrence of the specified element inside a POD array using the method find(d).

The second set of methods is more elaborate, but it requires execution of the method getStat(). This method returns a string charactering the entire data set. Such a summary contains a very comprehensive statistical characteristics of one-dimensional data set:

```
>>> from jhplot import *
>>> p0=P0D([1,2,3,3])
>>> m=p0.getStat()   # evaluates statistics
>>> print  m
```

The execution of the line with getStat() prints a rather long list with the summary statistics:

```
Size: 4
Sum: 10.0
SumOfSquares: 30.0
Min: 1.0
Max: 4.0
Mean: 2.5
RMS: 2.7386127875258306
Variance: 1.6666666666666667
Standard deviation: 1.2909944487358056
Standard error: 0.6454972243679028
Geometric mean: 2.213363839400643
Product: 23.99999999999999
Harmonic mean: 1.9200000000000004
Sum of inversions: 2.083333333333333
Skew: 0.0
Kurtosis: -2.0774999999999997
Sum of powers(3): 100.0
Sum of powers(4): 354.0
Sum of powers(5): 1300.0
Sum of powers(6): 4890.0
Moment(0,0): 1.0
Moment(1,0): 2.5
Moment(2,0): 7.5
Moment(3,0): 25.0
Moment(4,0): 88.5
Moment(5,0): 325.0
Moment(6,0): 1222.5
Moment(0,mean()): 1.0
Moment(1,mean()): 0.0
Moment(2,mean()): 1.25
Moment(3,mean()): 0.0
Moment(4,mean()): 2.5625
Moment(5,mean()): 0.0
Moment(6,mean()): 5.703125
25%, 50%, 75% Quantiles: 1.75, 2.5, 3.25
quantileInverse(median): 0.625
Distinct elements: [1.0, 2.0, 3.0, 4.0]
Frequencies: [1, 1, 1, 1]
```

Once the method getStat() is called, one can access the following characteristics:

```
>>> m=p0.variance()          # variance
>>> m=p0.stddeviation()      # standard deviation
>>> m=p0.standardError()     # standard error
>>> m=p0.kurtosis()          # kurtosis
>>> m=p0.skew()              # skewness
>>> m=p0.median()            # median
>>> m=p0.moment(k,c)         # k-th order moment
```

The above methods are quite self-explanatory. The last method returns k-th order moment of the distribution defined as $\sum_{i=0}((x[i] - \text{mean})^k)/size()$.

4.2.3 Displaying P0D Data

There is only one way to show a POD array: project it into a *histogram*. This topic is extensively covered in Sect. 8.1. Here we will briefly point out that a histogram is a chart of rectangles drawn on the x-axis whose areas are proportional to the frequency of a range of variables.

One can build a histogram from a POD using the two methods:

```
>>> h=p0.getH1D(bins)
>>> h=p0.getH1D(bins, min, max)
```

The first method creates a histogram with a given number of bins (bins is an integer number). The minimum and the maximum values of the X range are determined automatically. In the second case, one can explicitly specify the number of bins and the minimum (min) and maximum (max) values.

One question you may ask is this: assume we have two POD objects. One object represents x values, the second represents y values. We already know how to check correlations between these object - use the method correlation() as shown in Sect. 4.2.2. But how one can display pairs (x, y) on X–Y plots? Below we show how to do this:

```
──────────────────── Plotting P0D ────────────────

from jhplot   import *

c1 = HPlot('Canvas',600,400)
c1.setGTitle('X Y plot')
c1.visible()
c1.setAutoRange()
```

```
p1=POD('numbers',[1,2,3,4,5])
p2=p1.copy()

f1=F1D('x^2')
p2.func('squared', f1)
pp=P1D(p1,p2)
c1.draw(pp)
```

You may notice that we made an extra step by creating an object P1D from two POD arrays. We will discuss this object in Sect. 5.1.

Below is an example which shows how to generate random numbers and transform them using the F1D class. The output is shown as a histogram on the HPlot canvas:

```
————————————————————— Plotting POD —————————————
from java.awt import Color
from jhplot  import *

# build a canvas
c1 = HPlot('Canvas',600,400)
c1.setGTitle('Example of POD data array', Color.blue)
c1.visible()
c1.setAutoRange()
p0= POD('Random normal distribution')
p0.randomNormal(1000, 0.0, 1.0)

# make a copy and transform to a function
f1='x*cos(x)+2'
p01=p0.copy(f1)
p01.func(F1D(f1))

f2='exp(x)-2' # make a new copy and transform
p02=p0.copy(f2)
p02.func(F1D(f2))

h1=p0.getH1D(20)    # histogram with 20 bins
c1.draw(h1)

h1=p01.getH1D(100) # histogram with 100 bins
c1.draw(h1)

h1=p02.getH1D(200) # show again
c1.draw(h1)
```

Run this script and try to make sense of it. If it is not easy, skip this section since we will return to a very detailed discussion of the histograms in Sect. 8.1.

4.3 Reading and Writing P0D Files

To fill a P0D object from a ASCII file, assuming that each number is on a new line, use the method read(). For such files, use # or * for comments at the beginning of each line.

```
>>> p0=P0D('data from ASCII file')
>>> p0.read('FileName')
```

where 'FileName' is a string with the file name (the full path should be included). Data can also be read from a compressed ("zipped") file as:

```
>>> p0=P0D('data from a ZIP file')
>>> p0.readZip('FileName')
```

or from a gzipped format:

```
>>> p0=P0D('data from a GZIP file')
>>> p0.readGZip('FileName')
```

In all these cases, the methods read(), readZip() and readGZip() return zero in case of success. Error code 1–2 means that the file not found and 3 indicates a parse error.

To build a P0D from an ASCII file, use this constructor:

```
>>> p0=P0D('measurement', 'FileName')
```

(in this case, GZIP and ZIP form is not supported).

To write a p0 back to an ASCII file, use:

```
>>> p0.toFile('FileName')
```

Data can be written to a binary file using the so-called big endian format as:

```
>>> p0.writeBinary('FileName')
```

To read data from an existing binary file, use

```
>>> p0.readBinary('FileName')
```

In this case, the old content of the p0 object will be erased.

4.3.1 Serialization

One can save and restore a POD object containing data and other attributes using the Java object serialization. jHepWork has a short command for this:

```
>>> p0.writeSerialized('FileName')
```

The object p0 will be saved in a file with the name 'FileName' using a compressed format. The method returns zero in case of success. One can restore the object from the file as:

```
>>> p1=p0.readSerialized('FileName')
```

where p1 is a new object restored from the input file.

Now let us show an example of how to write a POD object into an external file in a serialized form and then how to restore the object back using a handy static method write(obj) of the class Serialized. To read a POD object from the file, use the method read() of the same class. The example below shows this:

```
                         Serialized I/O
from jhplot   import *
from jhplot.io import *

p0=POD('test')
p0.add(20)
p0.add(12)
print p0.toString()

# write to a file
Serialized.write(p0,'file.ser')

# deserialize POD from the file
p0s=Serialized.read('file.ser')
print 'After serialization:',p0s.toString()
```

The method write(obj) in this example writes a compressed object obj into a file. In Sect. 11.2, we will show how to deal with the case when no compression is used. Often, a program can benefit from the use of uncompressed objects since, in this case, the CPU time wasted for uncompressing files is avoided.

4.3.2 XML Format

In some cases, it is convenient to write an object to a human-readable XML file, so one can open it using any editor and look at its structure as well as written data. For the XML format, one should use the following methods from the class `Serialized` of the package `jhplot.io`:

`writeXML(obj,'FileName')`   writes a Java object (obj) into a file with the name `FileName`;

`readXML('FileName')`   reads a file with the name `FileName` and returns a stored object.

Try to replace the corresponding lines in the above examples and check how the output was written.

It should be noted that the XML-style is not recommended if there is a lot of data to be stored, since XML tags increase the file size dramatically. In addition, this approach may not work for all jHepWork classes.

For your convenience, the package `'jhplot.io'` provides two other important methods. An object can easily be converted into a XML string using the methods `toXML(obj)` and `fromXML(str)` of the `Serialized` class. Below we illustrate how to print an object in a XML form and then read it back:

```
>>> from jhplot   import *
>>> from jhplot.io.Serialized import *
>>>
>>> p0=P0D('data')
>>> str=toXML(p0)
>>> print str
>>> p0=fromXML(str)
```

To make the code shorter, we imported all methods from the class `Serialized`.

It should be noted that the serialization can be done using the native Java API. Below we rewrite the example discussed above using the standard Java serialization class:

```
————————————— Java serialized I/O ————————————
from jhplot   import *
from java.io import *

p0=P0D("test")
p0.add(20)
p0.add(12)
print p0.toString()

# serialize P0D into a file
f=FileOutputStream("filename.ser")
out =ObjectOutputStream(f)
out.writeObject(p0)
out.close()

# Deserialize P0D from the file
file = File('filename.ser')
```

```
fin = ObjectInputStream(FileInputStream(file))
p0s =fin.readObject();
fin.close();
print 'After serialization:\n',p0s.toString()
```

4.3.3 Dealing with Object Collections

The next question is how to write many POD objects to a single file. This can easily be done using Jython lists. The example below writes two POD objects into a single file:

——————————————— Writing multiple POD objects ———————————————

```
from jhplot   import *
from jhplot.io import *

p1=POD('p1')
p1.add(10)
p1.add(12)

p2=POD('p2')
p2.add(1000)
p2.add(2000)

# make list
a=[p1,p2]

# write to a file
Serialized.write(a,'file.ser')
print 'Ready!'
```

We can restore all objects from the file as:

——————————————— Reading multiple POD objects ———————————————

```
from jhplot   import *
from jhplot.io import *

# deserialize list from the file
f=Serialized.read('file.ser')

p1=f[0]
p2=f[1]
print "After serialization:\n"
print p1.toString()
print p2.toString()
```

Similarly, one can write objects and read them back using the methods `writeXML(obj)` and `readXML()` discussed before. Finally, one can use Jython dictionaries for convenient access to the data using the keys.

The serialization can be used for almost any jHepWork class. In Sect. 5.4 we will show how to use Jython dictionaries to store various objects in a serialized file and fetch them later using the keys.

To write huge sequences of `POD` arrays, the best approach would be to use the class `HFile` which will be discussed in Chap. 11 dedicated to input and output of Java objects.

Chapter 5
Two-dimensional Data

In the previous chapter we have considered the simplest possible data holder that keeps information about an one-dimensional array of numbers. More frequently, two-dimensional (or bivariate) data are necessary to consider, when each data point on the (X, Y) plane is represented by two variables, x and y. Such bivariate arrays are ideal to show a relationship between two sets of measurements.

5.1 Two Dimensional Data Structures

The P1D class is among the central classes designed for data manipulations in two dimensions (2D). The name of the P1D class is similar to that of the class F1D used to construct functions.

The declaration of a P1D object is rather similar to that considered for the POD class. However, this time, the container should be filled with at least two numbers using the method add(x,y):

```
>>> from jhplot import *
>>> p1=P1D('x-y points')
>>> p1.add(10,20)
>>> p1.add(20.,40.)
```

In this short example we declare the object P1D with the title 'x-y points' and then fill it with two points using the method add(x,y). Each point is represented by two numbers (either integer or float).

As it has been stated before, the method add(x,y) is not the most optimal, especially when it is used inside Jython loops. To avoid performance penalty, use high-level methods of this class for passing containers with numbers, rather than numbers themselves. In the example shown below, we fill a P1D from two POD arrays which, in turn, can also be filled using high-level methods as discussed in the previous chapter:

S.V. Chekanov, *Scientific Data Analysis using Jython Scripting and Java*,
Advanced Information and Knowledge Processing,
DOI 10.1007/978-1-84996-287-2_6, © Springer-Verlag London Limited 2010

```
>>> from jhplot import *
>>> p1=P1D('x-y points')
>>> p01=POD('px'); p02=POD('py')
>>> ax=p01.randomNormal(100,1.0,0.5)
>>> ay=p02.randomNormal(100,10.0,1.0)
>>> p1.fill(ax,ay)
```

In the above example, we fill two POD arrays with random numbers in accordance with the normal distributions and use these arrays to fill a P1D object. The code above can be shortened, since the POD arrays can also be passed to the P1D constructor as:

```
>>> p1=P1D('x-y points',p01,p02)
```

Of course, make sure that the sizes of the input arrays are the same.

Having filled the array P1D, one can replace data points using the method set(i,x,y), where 'i' denotes a position (index) of the (x, y) point inside the array. The array can be cleaned up with the method clear(), while the array size can be found using the method size().

Finally, one can extract data points using the methods getX(i) and getY(i) $(0 \leq i < \text{size}())$. Or, one can get arrays for x and y values without looping over all elements inside the container as:

```
>>> from jhplot import *
>>> p1= P1D('x-y points')
>>> p1.add(10,20)
>>> p1.add(20,50)
>>> ax=p1.getArrayX() # get all X values as array
>>> ay=p1.getArrayY() # get all Y values as array
```

One can also initialize a P1D from an external file as:

```
>>> p1= P1D('data from file','data.d')
```

The format of the input file is rather simple: each pair of numbers should be on a new line. To include comments, use '#' at the beginning of each line. Later we will discuss the I/O topics in more detail.

5.2 Two Dimensional Data with Errors

A P1D object is not as simple as it first may look like—it can also hold information on errors on x and y values.

Fig. 5.1 An illustration of a P1D array characterized by ten numbers: two numbers represent positions of a data point on the (x, y) plane, while other eight represent 1st and 2nd level uncertainties (see the text). The figure displays several methods to access the information on a single data point defined by an index i

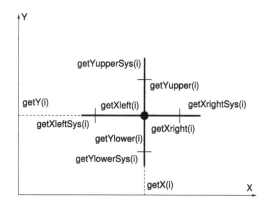

There are two different types of errors: one type comes from the fact that there is always inherent statistical uncertainty in counting random events. Usually, such uncertainties are called statistical or random errors. Also, we will call such errors "1st-level errors" when discussing various technical aspects.

The second type of uncertainties are called "systematical errors", which mainly originate from instrumental mistakes. For the P1D, we will call such uncertainties "2nd-level errors".

Thus, a single point at a position (x, y) is characterized by ten double-precision numbers: two numbers give the central position in (x, y) and other eight numbers specify its errors. A single point with 1st and 2nd errors is illustrated in Fig. 5.1, together with the "getter" methods used to access all characteristics of a P1D object.

Why do we need this complication when dealing with 2D data? Is it not enough to represent a relationship between two sets of data with two arrays? For example, a P1D can contain information on a set of points representing positions of particles in two dimensions: Each particle will be characterized by the position on the (x, y) surface, while the total number of particles will be given by the size of the P1D array. Well, in this case, you do not need indeed extra information to be stored in this data holder. Probably, you would not need even such a data holder, since one can use the usual Jython or Java arrays or the PND class to be discussed in the following chapters. However, there are many situations when each data point represents many measurements. In this case, one should store possible uncertainties associated with the measurements, which can have either statistical or systematical nature (or both). And this is why P1D becomes really handy, since this class is optimized exactly for such tasks.

Let us give a simple example. Assume we measure the average number of cars parked per week day on some parking spot. If the measurements are done during one month, there should be seven measurements for each day of the week. We calculate the average for each week day, and represent the entire measurement with seven numbers. Each number has a statistical uncertainty, which is assumed to be symmetric. The measurements can be represented by a single P1D container which, for example, can be filled for two week days as:

```
>>> from jhplot import *
>>> p1= P1D('average number of cars with errors')
>>> p1.add(1,av1,err1) # average for Monday
>>> p1.add(2,av2,err2) # average for Tuesday
```

where av1 and av2 are the averages for each week day, and err1 and err2 represent their statistical uncertainties.

Let us continue with our hypothetical example and estimate systematical uncertainties which reflect inaccuracies of our apparatus. Of course, for this particular example, we do not have any apparatus, but our measurements still may suffer from the uncertainties related to inaccuracies of our observations. We again assume that the systematical uncertainties are symmetric. In case of the 1st and 2nd level errors, we can fill the P1D as:

```
>>> from jhplot import *
>>> p1= P1D('average number of cars with errors')
>>> p1.add(1,av1,0,0,err1,err1,0,0,sys_err1,sys_err1)
>>> p1.add(2,av2,0,0,err2,err2,0,0,sys_err2,sys_err2)
```

where av1 and av2 are the averages of two different measurements, 'err1' and 'err2' their 1st-level errors (i.e. statistical errors). The 2nd-level errors (systematical errors) are added using the 'sys_err1' and 'sys_err2' values.

You may wonder, why err and sys_err have been passed to the add() twice? The reason is simple: we have used a rather general method, which can also be applied for adding asymmetrical errors. In addition, we had to type "0" which tells that there are no errors attributed for the X-axis. The last point can be clear if we will give the most general form of the add() method:

```
>>> p1.add(x,y,xLeft,xRight,       # 1st errors on X
           yUpper,yLower,          # 1st errors on Y
           xLeftSys, xRightSys,    # 2nd errors on X
           yUpperSys,yLowerSys)    # 2nd errors on Y
```

where xLeft (xRight) represents a lower (upper) 1st level error on the X-axis. For the $(x - y)$ plots, this is represented by a line started at the central position and extended to the left (right) from the central point (see Fig. 5.1). Analogously, yUpper (yLower) is used to indicate the upper (lower) uncertainty for the Y-axis. Next, four other numbers are used to show the 2nd level uncertainties for the X and Y axes. As discussed before, such uncertainties are usually due to some systematic effects reflecting instrumental uncertainties not related to statistical nature.

In many cases, we do not care about systematical uncertainties, so one can use several shortcuts. We have already shown that if one needs to specify only symmetrical statistical uncertainties on the Y-axis then one can use this method:

```
p1.add(x,y,err) # fills X,Y and symmetric error on Y
```

where `err` is a statistical error on the y value, assuming that it is equal to yUpper=yLower. All other errors are set to zero. If the error on Y is asymmetric, use this method:

```
p1.add(x,y,err_up, err_down)
```

where `err_up` and `err_down` are symmetric upper and lower error on y. If there are only 1st-level errors, then one can fill a P1D as:

```
p1.add(x,y,xLeft,xRight,yUpper,yLower)
```

while the 2nd level errors are set to zero.

Table 5.1 lists the main "setter" methods associated with the P1D class.

Occasionally, it is convenient to reset all errors to zero. For this, use the method:

```
>>> p1.setErrToZero(a)      # set 1st level errors to 0
>>> p1.setErrAllToZero(a) # set 1st and 2nd error to 0
>>> p1.setErrSysToZero(a) # set 2nd level error to 0
```

where a=0 for x values and a=1 for y values.

In addition, you may need to generate new errors from the numbers of counted events, when statistical uncertainty for each y value is the squared root of counted numbers [1]. If y represents the counted number of events, its statistical error is $\sqrt{y}$ (upper and lower). One can build a new P1D by assigning the 1st level errors for x and y separately:

```
>>> p1.setErrSqrt(a)       # set 1st level errors
```

Errors are set on x (y) when a=0 (a=1).

Table 5.2 gives "getter" methods used to access characteristics of this object.

To get index with either minimum or maximum value, use:

```
>>> p1=p1.getMinIndex(a) #  index with min
>>> p1=p1.getMaxIndex(a) #  index with max
```

where a=0 for x values and a=1 for y values.

Finally, one can use the method `integral(i1,i2)` that returns the integral between two indexes, `i1` and `i2` of a P1D data. For this operation, the integration is a sum over all y values, so it is rather similar to the integration of one-dimensional functions.

Table 5.1 Some methods used to fill a P1D container. In the methods shown in this table, "b" indicates a boolean value (1 for true and 0 for false), while "i" is an integer parameter

Methods	Definitions
add(x,y)	add (X, Y) (all errors are 0)
add(x,y,err)	add (X, Y) and 1st level error on Y
add(x,y,up,down)	add (X, Y) and asymmetric 1st level errors on Y
add(x,y,left,right,up,down)	add (X, Y) and asymmetric 1st level errors on X and Y
add(x,y,left,right,upper,lower,leftSys, rightSys,upperSys,lowerSys)	add (X, Y) and asymmetric 1st and 2nd level errors on X and Y insert (X, Y) at position i (all errors are 0)
set(i,x,y,err)	insert (X, Y) and 1st level error on Y
set(i,x,y,up,down)	add (X, Y) and asymmetric 1st level errors on Y
set(i,x,y,left,right,up,down)	insert (X, Y) and asymmetric 1st level errors on X and Y
set(i,x,y,left,right,upper,lower,leftSys, rightSys,upperSys,lowerSys)	insert (X, Y) and asymmetric 1st and 2nd level errors on X and Y
replace(i,x,y)	replace (X, Y) at position i (all errors are 0)
replace(i,x,y,err)	replace (X, Y) and 1st level error on Y
replace(i,x,y,up,down)	replace(X, Y) and asymmetric 1st level errors on Y
replace(i,x,y,left,right,up,down)	replace (X, Y) and asymmetric 1st level errors on X and Y
replace(i,x,y,left,right,upper,lower,leftSys, rightSys,upperSys,lowerSys)	replace (X, Y) and asymmetric 1st and 2nd level errors on X and Y
fill(arrayX,arrayY)	fills from 2 arrays
fill(arrayX,arrayY,arrayE)	fills from 3 arrays with errors
fill(arrayX,arrayY,arrayUP,arrayDOWN)	fills from 4 arrays with errors
fill(arrayX,arrayY ..+ 8 arrays with errors)	fills from 10 arrays with full sets of errors
fill(P0D,P0D)	fills from 2 P0D arrays
fill(P0D,P0D,P0D)	fills from 3 P0D arrays with errors
setTitle('text')	set a title

5.2.1 Viewing P1D Data

In addition to the Java serialization mechanism used to store the P1D containers, which will be discussed in detail later, one can store entries from a P1D in a human-readable format. For example, one can write data in ASCII files using the method:

```
>>> p1.toFile('newfile.d')
```

Table 5.2 Some methods for accessing information about a P1D object. In the methods shown in this table, "b" indicates a boolean value ("1" is false, "0" is true), while "a" is an integer parameter indicating the axis (a=0 means X-axis, a=1 means Y-axis). The notation "d" denotes a float value, while 'text' represents a string

Methods	Definitions
copy()	copy to new P1D
size()	size of the data
getMin(a)	min value for axis
getMax(a)	max value for axis
getMaxIndex(a)	get index of max value for axis
getMinIndex(a)	get index of min value for axis
mean()	mean value
integral(i1,i2)	sum up all Y-values between i1 and i2
updateSummary()	update summaries after adding a value
getArrayX()	get array with all X values
getArrayY()	get array with all Y values
getArrayXleft()	get array with all left X-errors
getArrayXright()	get array with all right X-errors
getArrayXleft()	get array with all left X-errors
getArrayXrightSys()	get array with all right X-2nd level errors
getArrayXleftSys()	get array with all left X-2nd level errors
getArrayYlower()	get array with all lower Y-errors
getArrayYupper()	get array with all upper Y-errors
getArrayYlowerSys()	get array with all lower 2nd-level Y-errors
getArrayYupperSys()	get array with all upper 2nd-level Y-errors
getTitle()	get title
getX(i)	X value at index i
getY(i)	Y value at index i
getXleft(i)	left error on X at index i
getXright(i)	right error on X at index i
getXleftSys(i)	left 2nd error on X at index i
getXrightSys(i)	right 2nd error on X at index i
getYupper(i)	upper error on Y at index i
getYlower(i)	lower error on y at index i
getYupperSys(i)	upper 2nd error on Y at index i
getYlowerSys(i)	lower 2nd error on Y at index i

where 'newfile.d' is the name of the output file. One can also export data into a LaTeX table as

```
>>> from java.text import DecimalFormat
>>> format=DecimalFormat("##.####E00")
>>> p1.toFileAsLatex("Output.tex",xformat,xformat)
```

One should specify an appropriate format for the numbers to be stored the LaTeX table. We remind that the class DecimalFormat is used to format decimal numbers in the Java platform. The pound sign ('#') denotes a digit, and the dot is a placeholder for the decimal separator. Please refer to the Java API documentations of this class.

Also, one can print a P1D container on the console as

```
>>> p1.print()
```

To print the stored values on the JythoShell prompt, convert the p1 object into a string and print it using the standard print method:

```
>>> print p1.toString()
```

In some cases, it is convenient to show data as a table in a separate frame, so one can sort and search for a particular value. In this case, use the statement:

```
>>> p1.toTable()
```

which brings up a frame showing the data inside a table. The method calls the class HTable to be discussed in Sect. 12.4. Analogously, data can be exported to a spreadsheet as discussed in Sect. 12.4.

5.2.2 Plotting P1D Data

In oder to display a P1D representing a set of bivariate data, the usual HPlot canvas discussed in Sect. 3.3 can be used. To plot data, follow the same steps as for drawing F1D functions: first, create a canvas and then use the draw(obj) method to display a P1D object on a *scatter* plot:

```
>>> from jhplot import *
>>> c1=HPlot('Canvas')
>>> c1.visible()
>>> c1.setAutoRange()
>>> p1=P1D('x-y points', 'data.d')
>>> c1.draw(p1)
```

Table 5.3 The most important methods for graphical representation of a P1D. "b" indicates a boolean value (1 for true and 0 for false), while "i" is an integer parameter. The notation "d" indicates a float value. The attributes "c" and "f" correspond to the Color and Font classes of Java AWT, while 'text' represents a Jython string

Methods	Definitions
setStyle('text')	set as symbols ("p") or line ("l")
setSymbolSize(i)	symbol size
setSymbol(i)	symbol type i=0-12:
	0: not filled circle
	1: not filled square
	2: not filed diamond
	3: not filled triangle
	4: filled circle
	5: filled square
	6: filed diamond
	7: filled triangle
	8: plus (+)
	9: cross as (x)
	10: star (*)
	11: small dot (.)
	12: bold plus (+)
setColor(c)	set the line color
setPenWidh(i)	width of the line
setPenDash(i)	dashed style with "i" being the length
setLegend(b)	set (b=1) or not (b=0) the legend
setTitle('text')	set a title

In this example, we first create a P1D object from an ASCII input file, and then we display data as a collection of points on the (x, y) plane.

Table 5.3 shows the most important graphical attributes associated with the P1D class.

It should be noted that one can edit the plot using a pop-up menu Edit which allows to change some attributes. Click on the right mouse button to access a GUI-driven dialog with the menu.

There are many methods which come together with the P1D data holder to display error bars, assuming that error values have been filled before. To display errors, use the methods:

```
>>> p1.setErr(1)     # show  1st level errors
>>> p1.setErrSys(1)  # show  2nd level errors
```

which should be set before drawing the p1 object on the canvas.

Table 5.4 The most important methods for a graphical representation of the P1D errors. "b" indicates a boolean value (1 for true and 0 for false), while "i" is an integer parameter. The notation "d" indicates a float value. The attributes "c" and "f" correspond to the Color and Font classes of Java AWT, while 'text' represents a Jython string

Methods	Definitions
setErrAll(b)	set all errors (1st and 2nd level)
setErrX(b)	set error on X or not
setErrY(b)	ser error on Y or not
setErrColorX(c)	color used for X error bars
setErrColorY(c)	color used for Y error bars
setPenWidthErr(i)	line width for 1st level errors
setPenWidthErrSys(i)	line width for 2nd level errors
setErrSysX(b)	set or not 2nd level error on X
setErrSysY(b)	set or not 2nd level error on Y
setErrFill(b)	fill or nor the area covering errors
setErrFillColor(c)	fill color
setErrFillColor(c,d)	fill color + transparency level "d"
setErrSysFill(b)	fill or not 2nd level errors
setErrSysFillColor(c)	fill color
setErrSysColor(c,d)	as before + transparency level "d"

Table 5.4 shows various methods for controlling the attributes of the errors bars. As you can see, there are two separate methods to modify the horizontal and vertical error bars. The "getter" methods are similar, but start with the "get" string instead of "set".

5.2.3 Contour Plots

The HPlot canvas can be used to display P1D data in a form of a contour plot. For such type of plots, we draw colored regions which show the density population, instead of showing separate particles. It is required to bin a (x, y) plane in x and y: the smaller the bin size is, the more chances to resolve a fine structure for the plotted density distribution.

To set up the canvas HPlot for showing contour plots, use the method setContour(1). Table 5.5 shows some methods to setup a contour plot.

It should be noted that there is a special canvas, called HPlot2D, which is designed to show the contour plots and has more options for manipulations. Please read Sect. 10.11 for more details.

Table 5.5 HPlot methods for displaying contour plots. "b" indicates a boolean value (1 for true and 0 for false), while "i" is an integer parameter

Methods	Definitions
setContour(b)	sets (or not) the contour style
setContourLevels(i)	the number of color levels
setContourBins(iX,iY)	the number of bins in X and Y
setContourBar(b)	set (or not) a color line showing levels
setContourGray(b)	set (or not) white-black style

5.3 Manipulations with P1D Data

The P1D containers are designed from the ground to support numerous mathematical operations. The operations do not create new objects, but just modify the original containers. To create a new P1D object, use the method copy(). For example:

```
... p1 is created above...
>>> p2=p1.copy() # now p2 is different object
>>> c1.draw(p1)  # draw two different objects
>>  c1.draw(p2)
```

One can merge two P1D containers into one using the method merge(). One can also add, divide, subtract and multiply the P1D objects. Let us read two P1D containers from files and perform such operations:

```
>>> p1=P1D('first','data1.d')
>>> p2=P1D('second','data2.d')
>>> p3=p1.merge(p2)              # merge 2 P1D's into one
>>> p1.oper(p2,'NewTitle','+') # add   p1 and p2
>>> p1.oper(p2,'NewTitle','-') # subtract p2 from p1
>>> p1.oper(p2,'NewTitle','*') # multiply p1 by p2
>>> p1.oper(p2,'NewTitle','/') # divide   p1 by p2
```

The execution speed of these operations is significantly faster compare to equivalent Jython codes based on loops, since all such methods are implemented in the form of Java libraries. One can skip the string with a new title if you want to keep the same title as for the original P1D. For example, in this case, the additive operation will be p1.oper(p2,'+'). All graphical attributes are preserved during such data manipulations.

For the above operations, the errors on p3 will be propagated accordingly assuming that all 1st and 2nd-level errors associated with p1 and p2 are independent of each other. For the error propagation, we use a rather standard prescription [1].

To scale a P1D with a number, use the statement:

```
>>> p1.operScale(a, scaleFactor)
```

where `scaleFactor` is a double or an integer number. If a=0 (Java boolean 'false'), the scaling is applied for *x*, if a=1, the scaling is applied for *y*. The title is optional for this operation. It is important to know that the factor 'scaleFactor' scales the 1st and 2nd levels uncertainties as well. If you need to scale only errors, use:

```
>>> p1.operScaleErr(a, scaleFactor)
```

which scales only the 1st-level errors for either *X* (a=0) or *Y* (a=1) axis. If one needs to scale also the 2nd-level errors, use:

```
>>> p1.operScaleErrSys(a, scaleFactor)
```

which works exactly as the method `operScaleErr()`, but this time it is applied for the 2nd-level errors.

Finally, to extract a range of P1D points from the original data container, use

```
>>> p1=p1.range(min,max)
```

where `min` and `max` are integer numbers denoting the range.

5.3.1 Advanced P1D Operations

5.3.1.1 Operations with Correlations

The operations considered above assume that there are no correlations between two data holders. In reality, data from different measurements can correlate, so do P1D containers corresponding to such measurements. In this case, one can also specify a correlation coefficient and use it for mathematical operations. A correlation coefficient should be represented with an additional P1D container used for the actual mathematical manipulations. The correlation coefficients should be added using the `add()` method and included at the positions of errors (statistical or systematical). Look at the example below where we assume that there is a 50% correlation between two data sets:

```
―――――――――― Adding data with correlations ――――――――――
from jhplot import *
```

```
# fill data set 1 points with 2 points
p1= P1D('data1')
# use only 1st-level errors on Y
p1.add(10,100,5,5)
p1.add(20,50,5,5)

# fill data set 2 with 2 points
p2= P1D('data2')
# use only statistical errors on Y
p2.add(10,40,5,5)
p2.add(20,40,5,5)

# add with 50% correlations
corr=P1D('correlation coefficients')
corr.add(0,0,0.5,0.5)
corr.add(0,0,0.5,0.5)
# add them. Do not do anything with X
p50=p1.copy()
p0=p1.copy()
p50.oper(p2,'added with 50% corr.','+','Y',corr)
print p50.toString()
p0.oper(p2,'added with 0% corr.','+')
print p0.toString()
```

The output of this script shows that the (x, y) values remain to be the same, but the statistical errors are different for the case with 50% correlations and without any correlations.

Analogously, one can include correlations for the 2nd-level errors (see the API description for this method). The same feature is supported for any operation, such as subtraction, multiplication and division.

5.3.1.2 Functional Transformation

One can also transform a P1D data using a mathematical function. The error propagation is done for x or y components (or for both). The following functions are supported: "inverse" ($1/y$), square ($y * y$) "sqrt" (square root), "exp" (exponential), "log" ($\log_{10}(y)$) and all trigonometrical functions. The example given below illustrates a generic usage of this functional transformation:

```
>>> p1=p1.move('function', 'a' )
```

where 'function' is a string defining a function used for the transformation, and 'a' is a string indicating the axis, which can either be "X" (apply the transformation for x values), "Y" (apply the transformation for y values) or "XY" (transform both x and y). For example:

```
>>> p1=p1.move('log', 'Y' )
```

transforms all y values to $\log_{10}(y)$. Errors for the p1 container will be transformed appropriately.

5.3.1.3 Smoothing

For some situations, you may be interested in smoothing P1D values. This can be done by averaging over a moving window of a size specified by the method parameter: if the value of the parameter is "k" then the width of the window is $2 * k + 1$. If the window runs off the end of the P1D only those values which intersect are taken into account. The smoothing may optionally be weighted to favor the central value using a "triangular" weighting. For example, for a value of "k" equal to 2 the central bin would have weight $1/3$, the adjacent bins $2/9$, and the next adjacent bins $1/9$. Errors are kept to be the same.

All of this can be achieved using the command:

```
>>> p2=p1.operSmooth(a,b,k)
```

where "a" defines the axis to which the smoothing is applied, i.e. it can be either a=0 (for the "X" axis) or a=1 (for the "Y" values). When b=1 (boolean "true") then x or y values are weighted using a triangular weighting scheme favoring bins near the central bin, and "k" is the smoothing parameter which must be non-negative. If zero, the original P1D object will be returned with no smoothing applied.

One can also convert a P1D into a Gaussian smoothed container in which each band of the original P1D is smoothed by discrete convolution with a kernel approximating a Gaussian impulse response with the specified standard deviation. This can be done using the command:

```
>>> p2=p1.operSmoothGauss(a,sDev)
```

where "sDev" is the standard deviation of the Gaussian smoothing kernel (must be non-negative).

5.3.2 Weighted Average and Systematical Uncertainties

When there are several different measurements with different values x_i and known errors σ_i (or variance, σ_i^2), then it is reasonable to combine the measurements using the so-called weighted average method. In this case, the value x for best estimate and its variance are given by

$$x = \sum (x_i / \sigma_i^2) / \sum (1/\sigma_i^2)$$
$$\sigma^2 = 1 / \sum (1/\sigma_i^2)$$

Such calculation can be done in one line as:

```
>>> p1.addAndAverage(P1D[])
```

Let us explain this method. Assume there is a measurement represented by a p1 and additional measurements represented by an array of P1D objects. Then, after calling the above method, p1 will represent a weighted average with the corresponding 1st-level errors.

The class P1D is also very useful when one needs to evaluate systematical uncertainties of many measurements. If we have several measurements with different resulting outcomes, and each measurement is represented by a P1D data holder, one can obtain a P1D representing the final measurement using the method getSys() which returns the final P1D object with systematical uncertainties. Let us illustrate this using generic example:

```
>>> p0= P1D('default') # original measurement
>>> pp=[]
>>> p1= P1D('data1')    # other measurements
>>> pp.append(p1)
>>> p2= P1D('data2')
>>> pp.append(p2)
>>> p3= P1D('data3')
>>> pp.append(p3)
>>> ....
>>> psys=p0.getSys(pp) # build uncertainties
>>> psys.setErr(1)
>>> c1.draw(psys)
```

It should be noted that the systematical uncertainties are added in quadrature, thus they are assumed to be independent of each other.

One can also face with the situation like this: there are three P1D objects, the first contains an array with the central values and two others represent lower and upper deviations. One can build a new P1D with the 2nd level errors that represent the differences between the central P1D values and the upper and the lower P1D using the method operErrSys('title',a, p1,p2), where 'a' represents the axis (0 for X, 1 for Y), and p1 and p2 are objects for the lower and upper errors. The example below illustrates this for two data points:

```
——————————— Errors from two input P1D ———————————
from java.awt import Color
from jhplot  import *
```

```
c1 = HPlot('Canvas')
c1.visible()
c1.setRange(5,25,70,120)
c1.setGTitle('Uncertainties',Color.blue)

p1= P1D('Central')
p1.setColor(Color.blue)
p1.add(12,100); p1.add(22,80)

p2= P1D('Lower')
p2.add(10,90); p2.add(20,75)
c1.draw(p2)

p3= P1D('Upper')
p3.add(10,110); p3.add(20,96)
c1.draw(p3)

p0=p1.operErrSys('Data',1,p2,p3)
p0.setErrSys(1)
p0.setPenWidthErrSys(2)
c1.draw(p0)
```

This script plots a p0 object with new 2nd-level errors given by the difference between the input P1D containers. The output plot is shown in Fig. 5.2. The container with the central values can contain statistical errors, while statistical errors in the input P1D containers are ignored.

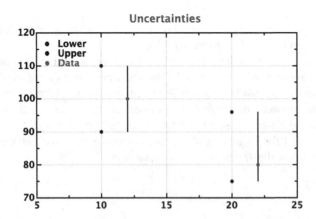

Fig. 5.2 Displaying a new P1D object with the 2nd-level errors given by two other P1D containers

5.4 Reading and Writing P1D Data

5.4.1 Dealing with a Single P1D Container

The P1D containers can be saved into a file and restored later. As for the POD class, one can fill a P1D container from a text ASCII file using the method read(). We remind that each (x, y) pair should be on a separate line, and x and y should be separated by a space. One can use the characters # or * at the beginning of each line for comments:

```
>>> p0=P1D('data from ASCII file')
>>> p0.read('FileName')
```

In this example, 'FileName' is a string with the file name (the full path should be included). One can use a shorter version of the code above by passing a file name directly to the constructor:

```
>>> p0=P1D('data from ASCII file', 'FileName')
```

as was illustrated before. Data can also be read from a compressed (Zip) file:

```
>>> p0=P1D('data from a ZIP file')
>>> p0.readZip('FileName')
```

or from a file in the Gzip format:

```
>>> p0=P1D('data from a GZIP file')
>>> p0.readGZip('FileName')
```

In all cases, the methods read(), readZip() and readGZip() return zero in case of success. The error code 1 or 2 tells that the file not found. If a parse error occurs, the reaGZip() method returns 3.

To write a P1D data into an ASCII file, use the following method:

```
>>> p0.toFile('FileName')
```

There is another handy method: one can store data in the LaTeX format as:

```
>>> from java.text import DecimalFormat
>>> format=DecimalFormat('##.####E00')
>>> p0.toFileAsLatex('FileName', format, format)
```

where 'format' is an instance of the DecimalFormat class from the Java package java.text. We passed the object 'format' twice, one for *x* values and one for *y* values.

It should be pointed out that, for the above examples, we could only write (read) data. All graphical attributes are completely lost after saving the data in ASCII files. But there is another more elegant way to save a P1D object: As for any jHepWork object, one can save and restore P1D data and other attributes (including those used for graphical representation) using the Java serialization mechanism.

To serialize the entire P1D object into a file, use the method below:

```
>>> p0.writeSerialized('FileName')
```

which writes the object p0 including their graphical attributes into a file with the name 'FileName'. The method returns zero in case of no I/O problem occurs. One can restore the object from this file as:

```
>>> p1=p0.readSerialized('FileName')
```

where p1 is a new object from the file.

The example below shows how to write a P1D into an external file in a serialized form and then restore the object back:

```
──────────────── P1D serialization ────────────────
from jhplot  import *
from jhplot.io import *

p1=P1D('x-y data')
p1.add(10,20)
p1.add(12,40)
print p1.toString()

# write to a file
Serialized.write(p1,'file.ser')

# deserialize P1D from the file
p1s=Serialized.read('file.ser')

print 'After serialization:',p1s.toString()
```

The serialization to the XML format can be achieved using the writeXML() and readXML() methods of the same Serialization class. This class also allows to convert the container into XML strings using the toXML() method.

5.4.2 Reading and Writing Collections

One can write any number of P1D objects into a single file using the serialization mechanism. The idea is simple: put all P1D objects into a list and then serialize the entire list in one step. Of course, one can use Jython tuples or dictionaries instead of using lists.

Below we show how to save two different objects, P0D and P1D, into one file and then restore them later:

```
────────────── Serialization of multiple containers ──────────────
from jhplot  import *
from jhplot.io import *

p1=P1D('p1 data')
p1.add(10,20);  p1.add(12,40)
print p1.toString()

p2=P0D('p2 data')
p2.add(1000);  p2.add(2000)
print p2.toString()

a=[p1,p2] # make a list

# write to a file
Serialized.write(a,'file.ser')

# deserialize list from the file
list=Serialized.read('file.ser')

p1,p2 = list[0], list[1]
print 'After serialization:\n'
print p1.toString()
print p2.toString()
```

We should note again that one can also use the standard Java for the serialization mechanism as shown in Sect. 4.3.

It is more convenient to use Jython dictionaries to store different objects, since there will be no need for memorizing the order of objects in the list or tuple holding other objects. We remind that, in case of dictionaries, we have one-to-one relationship between the keys and the corresponding values. Below we write three objects into a file, P1D, P0D and F2D, using the keys for easy retrieval:

```
────────────── Serialization using dictionaries ──────────────
from jhplot  import *
from jhplot.io import *

hold = {} # create empty dictionary
```

```
p1=P1D('p1dobject')
p1.add(10,20);    p1.add(12,40)
hold[p1.getTitle()] = p1

p2=P0D('p0dobject')
p2.add(1000); p2.add(2000)
hold[p2.getTitle()] = p2

f1=F2D('2*x*sqrt(2*y)')
hold['f2dobject'] = f1

# write the dictionary to a file
Serialized.write(hold,'file.ser')

# deserialize the dictionary
newhold=Serialized.read('file.ser')

print newhold.keys() # print all keys

# fetching objects using keys and print them
print 'After serialization:\n'
print 'P1D = ',newhold['p1dobject'].toString()
print 'P2D = ',newhold['p0dobject'].toString()
print 'F2D = ',newhold['f2dobject'].getName()
```

In this example, we create a dictionary hold to store different objects using the keys (strings). We used the title strings as the keys for the P1D and P0D objects. Then we write this dictionary into a file. In the second part of this code, we read the dictionary from the file and restore all the objects back using the corresponding keys. For example, the object newhold['p1dobject'] gives an access to the P1D object stored in the dictionary. It should be noted that the keys can be any Jython objects, not only strings.

Finally, one can store and retrieve data using the HFile class which is designed to work with large sequences of arbitrary Java objects, including P1D (see Chap. 11).

5.5 Real-life Example I: Henon Attractor

We will illustrate a typical program based on the P1D class. For our example, we will consider a Henon map (or Henon attractor) [2, 3]. The Henon map can be written as the coupled equations:

$$x_{n+1} = y_n + 1 - a * x_n^2$$
$$y_{n+1} = b * x_n$$

The parameters are usually set to the canonical values, $a = 1.4$ and $b = 0.3$. This simple equation is known to exhibit properties of an attractor with a fractal structure of its trajectories (the so-called strange attractor).

Let us show how to program this attractor using the P1D class. To visualize the attractor on the (x, y) plane, the SPlot canvas will be used as a light-weight alternative to the HPlot class considered before. The SPlot class will be discussed in detail in Sect. 10.9. Here we will note that this canvas has a low memory footprint, and can be used for easy zooming in to rectangle.

The code snippet that implements the Henon map with 10000 iterations are shown below:

```
                              Henon attractor

from jhplot   import *

a = 1.4;   b = 0.3
p=P1D('Henon attractor')
p.setSymbol(11)
x=0
y=0
for i in range(10000):
    x1=x
    x=1+y-a*x*x
    y=b*x1
    p.add(x,y)

c1 = SPlot()
c1.setGTitle('Henon attractor')
c1.visible()
c1.setAutoRange()
c1.draw(p)
```

Figure 5.3 shows the resulting image for $a = 1.4$ and $b = 0.3$. One can examine the fine structures of this attractor by zooming into a specific rectangular area of the plot by clicking and dragging the mouse to draw a rectangle where desired. One can also replace the method setAutoRange() with the method setRange(xmin,xmax,ymin,ymax) and re-running the script (the arguments of this method define the zooming rectangle).

In Sect. 10.9, we will rewrite this example using another approach in which we will directly populate the canvas with (x, y) points without using the intermediate step based on the P1D class.

5.6 Real-life Example II. Weighted Average

In this section we will consider how to find a weighted average of several measurements and plot it together with the original measurements. The weighted average

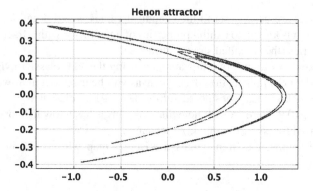

Fig. 5.3 The Henon attractor. Use the mouse for zooming in order discover its fine structure

of a list with P1D objects was already considered in Sect. 5.3.2. In addition, this example shows somewhat technical issue: how to mix Java with Jython classes.

For an educational purpose, we will diverge from our original concept which stated that all CPU intensive calculations should be managed inside Java libraries. For this example, we create a custom Jython class "measurement" mixing Jython with Java class P1D. The latter will be used mainly for graphical representation.

Let us create a file "measurement.py" with the lines:

```
                      ────── "measurement.py" module ──────
from jhplot import *

class measurement:
  def __init__(self, number, value, error):
    "A single measurement"
    self.number = number
    self.v  = value
    self.err =  error
    self.p1=P1D(str(self.number))
    self.p1.setSymbolSize(8)
    self.p1.setSymbol(4)
    self.p1.setPenWidthErr(4)
    self.p1.setPenWidthErrSys(2)
    self.p1.add(self.v,self.number,self.err,self.err,0,0)
  def echo(self):
    print  self.number,self.v,self.err
  def getValue(self):
    return self.v
  def getError(self):
    return self.err
  def getNumber(self):
    return self.number
  def getPoint(self):
    return self.p1
```

An object of this class keeps information about a single measurement characterized by an integer number (which defines the type of measurement), measured value (can be accessed with the method getValue()), its statistical error (accessed as getError()). Finally, we will return the measurement represented in the form of a P1D object. Note the way how to fill this object: unlike the previous example, we assign errors in x direction, rather than for the y axis. This is mainly done for better representation of the final result.

Let us test this module. We will assume that two measurements have been performed. We fill a list with these two measurements and then print the filled values. Assuming a "counting" experiment in which the statistical error is the square root of the counted numbers of events, our Jython module to add the measurements can look as:

─────────── Adding data ───────────

```
from measurement import *
import math

data=[]
data.append( measurement(1,100, math.sqrt(100)))
data.append( measurement(2,120, math.sqrt(120)))

for m in data:
    m.echo()
```

Now let us calculate a weighted average of an arbitrary number of measurements with errors. The weighted-average and its error for two measurements with a common error σ_C are:

$$v3 = \frac{v_1/(\sigma_1^2 - \sigma_c^2) + v_2/(\sigma_2^2 - \sigma_C^2)}{1/(\sigma_1^2 - \sigma_c^2) + 1/(\sigma_2^2 - \sigma_C^2)}$$

$$\sigma_3^2 = \frac{1}{1/(\sigma_1^2 - \sigma_c^2) + 1/(\sigma_2^2 - \sigma_c^2)} + \sigma_C^2$$

where v_1 and v_2 are measured values, σ_1 and σ_2 are their statistical errors, respectively. We have introduced the common error for generality; in most cases, the common error $\sigma_C = 0$. The module which calculates the weighted average from a list of measurements and a common error can be written as:

─────────── module "average.py" ───────────

```
from measurement import *
import math

def average(meas,c):
    "Calculate weigthted average"
    s1,s2=0,0
```

```
      for m in meas:
        e=m.getError()*m.getError()-c*c
        w1= 1.0/e
        w2= m.getValue()*w1
        s1=s1+w1
        s2=s2+w2
      err=math.sqrt((1.0/s1)+c*c)
      return measurement(len(meas)+1,s2/s1,err)
```

Put these lines into a file with the name "`average.py`".

Now let us use this module and plot the original measurements as well as the weighted average (for which we used a different symbol). This time, we will use the class `HPlotJa` which has very similar methods as for the `HPlot` canvas, but can be used for an interactive drawing with the mouse, as discussed in Sect. 10.10.

```
───────────────── A weighted average ─────────────
from measurement import *
from average import *
import math

c1 = JaPlot('Canvas')
c1.removeAxes()
c1.showAxis(0)
c1.setGridAll(0,0)
c1.setGridAll(1,0)
c1.setRange(50,150,0,5)
c1.setNameX('Measurements')
c1.visible()

data=[]
data.append( measurement(1,100, math.sqrt(100)))
data.append( measurement(2,120, math.sqrt(120)))

p=average(data,1.)
for m in data:
    c1.draw(m.getPoint())

p1=p.getPoint()
p1.setSymbol(5)
c1.draw(p1)
```

Figure 5.4 presents the result. Two filled circles show the original measurements, while the third point shows their weighted average calculated by the module "`average.py`".

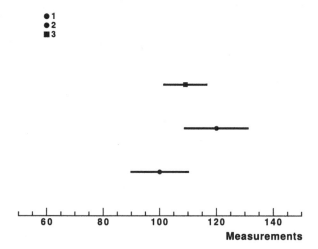

Fig. 5.4 A weighted average (marked by the filled square and labeled as "3") of two independent measurements (indicated using the keys "1" and "2")

References

1. Taylor, J.: An Introduction to Error Analysis: The Study of Uncertainties in Physical Measurements. University Science Books, Herdon (1997)
2. Henon, M.: Numerical study of quadratic area-preserving mappings. Q. Appl. Math. **27**, 291–312 (1969)
3. Gleick, J.: Chaos: Making a New Science. Penguin Books, New York (1988)

Chapter 6
Multi-dimensional Data

6.1 P2D Data Container

Now let us discuss how to deal with data in dimensions larger than two. A natural extension of the P1D class for 3-dimensional data is the class P2D. It is rather similar to the P1D, the only difference is that it keeps data in 3D phase space (x, y, z). Also, it has less options for drawing and, in addition, statistical and systematical errors are not supported. In Jython, one can add values to this container in the same way as for P1D, only this time the method add(x,y,z) takes three arguments. In the example below we create a P2D object and append a single point with the components $(1, 2, 3)$:

```
>>> from jhplot import *
>>> p2= P2D('x-y-z points')
>>> p1.add(1,2,3)
```

Table 6.1 shows the main methods of this class. We will not discuss in detail since the P2D arrays are similar to P1D, and one can always look at the Java API documentation to learn about this data holder.

6.1.1 Drawing P2D and HPlot3D Canvas

To draw a P2D, one should use the 3D canvas based on the HPlot3D class. This class was discussed in Sect. 3.4.2 and used to draw F2D functions. In the example below we draw two data sets shown in blue and red colors:

```
———————————— Drawing P2D data in 3D ————————————
from jhplot import *
from java.awt import Color

c1 = HPlot3D('Canvas')
```

Table 6.1 The main P2D methods. In this table, "b" denotes a boolean value (1 for Java "true" and 0 for "false"), while "i" is an integer parameter. "a" indicates the axis (a=0,1,2)

Methods	Definitions
add(x,y,z)	add (x, y, z)
set(i,x,y,z)	insert (x, y, z) at position i
fill(arrayX, arrayY, arrayZ)	fills from 3 arrays
setTitle("text")	set title
getTitle()	get title
size()	get size
clear()	clean from data
copy()	copy to a new P2D
getArrayX()	get arrays with X
getArrayY()	get array with Y
getArrayZ()	get array with Z
getX(i),getY(i),getZ(i)	get points at index i
getMax(a)	get max for axis
getMin(a)	get min for axis
mean(a)	get mean for axis
toTable()	show in a table
merge(p2d)	merge with another P2D

```
c1.visible()
c1.setNameX('X axis')
c1.setNameY('Y axis')
c1.setRange(-5,10,-5,5,-10,20)

h1= P2D('blue data')
h1.setSymbolSize(6)
h1.setSymbolColor(Color.blue)
h1.add(1,2,3)
h1.add(4,4,5)
h1.add(3,2,0)
c1.draw(h1)

h1= P2D('red data')
h1.setSymbolSize(6)
h1.setSymbolColor(Color.red)
for i in range(10):
   h1.add(0.1*i, 0.2*i, 0.5*i)
c1.draw(h1)
```

Table 6.2 lists several most important methods used to draw P2D data on the 3D canvas. The number of methods is not very large, since many drawing methods belong to the actual 3D canvas, rather then attributed to the P2D object itself. We will discuss the HPlot3D methods in Sect. 10.12.

Table 6.2 Graphical methods for displaying P2D data. In the methods shown in this table, "c" denotes the Java Color class, while "i" is used to indicate an integer parameter

Methods	Definitions
setSymbolColor(c)	Set symbol color
setSymbolSize(i)	Set symbol size
getSymbolColor()	Get symbol color
getSymbolSize()	Get symbol size

We will remind that the HPlot3D canvas is a rather similar to HPlot: One can display several plots on the same canvas and change the plotted regions using the cd(i1,i2) method:

```
>>> from jhplot import *
>>> c1=HPlot3D('Canvas',600,400,2,2)  #  2x2 pads
>>> c1.visible()          # set visible
>>> c1.cd(1,1)            # go to the 1st pad
... draw some object ..
>>> c1.cd(1,2)            # go to the 2nd pad
```

As for the HPlot canvas, first two integers in the constructor HPlot3D define the size of the canvas (600 × 400 pixels for this example), while two other integers define how many drawing regions (pads) should be shown (2 regions in X and 2 regions in Y).

Let us give a more concrete example of how to work with the P2D:

```
———————————— Graphical options for drawing P2D ————————————
from java.util import Random
from java.awt import Color
from jhplot import *

c1 = HPlot3D('Canvas')
c1.setGTitle('Interactive 3D')
c1.setNameX('X')
c1.setNameY('Y')
c1.visible()
c1.setRange(-5,10,-5,5,-10,30)

p1= P2D('3D Gaussian 1')
p1.setSymbolSize(6)
p1.setSymbolColor(Color.blue)

rand = Random()
for i in range(200):
    x=1+rand.nextGaussian()
    y=1+0.5*rand.nextGaussian()
    z=10+4.5*rand.nextGaussian()
    p1.add(x,y,z)
```

Fig. 6.1 Two P2D objects
displayed using the
HPlot3D canvas

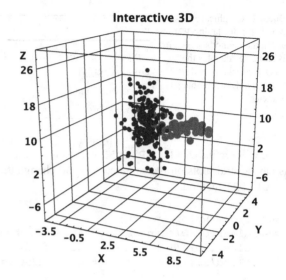

```
p2= P2D('3D Gaussian 2')
p2.setSymbolSize(10)
p2.setSymbolColor(Color.red)

for i in range(50):
    x=2+2*rand.nextGaussian()
    y=4+0.5*rand.nextGaussian()
    z=6+1.5*rand.nextGaussian()
    p2.add(x,y,z)

c1.draw(p1)
c1.draw(p2)
```

In this script, we fill two P2D containers with 3D Gaussian numbers and plot them on the same canvas. The result of this script is shown in Fig. 6.1. Note that the most efficient way to fill the containers is to fill POD with Gaussian numbers using the methods of the POD class, and use three POD containers for the input of the P2D constructor.

6.2 P3D Data Container

You may wonder, what could be shown with the object called P3D, since by analogy, it must contain points in a 4-dimensional space. The P3D container, by design, still can be used to show 3-dimensional data, but this time, points can have some extension in 3D space. All methods of the P3D are very similar to that of P2D, the only difference is that each data point in x, y, z, has an additional parameter representing an extension of the point in the corresponding direction. As before, the canvas HPlot3D should be used for drawing such objects. To fill a P3D container, one should use the method add(x,dx,y,dy,z,dz) which takes exactly 6 arguments, with dx, dy and dz being the extensions in corresponding direction.

Here we will stop the discussion of this class since it has a limited use for mathematical manipulations. Rather, we will show a simple example of how to use the P3D to draw various shapes (lines, cubes and surfaces):

```
———————————————— Working with P3D ————————————————
from jhplot import *
from java.util import Random
from java.awt import Color

c1 = HPlot3D('Canvas')
c1.setGTitle('3D plot with P3D objects')
c1.setNameX('X')
c1.setNameY('Y')
c1.visible()

c1.setRange(-5,10,-4,10,0,20)
# build P3D shape
h1 = P3D('3D form in blue')
h1.setPenColor(Color.blue)
#   build a 3D cube
h1.add(4.0,1.0,8.0,2.0,3.0,4.0)

# build 2D panel (Z extension is 0)
h1.add(5.0,2.0,3.0,1.0,8.0, 0.0)

# build 1D lines
h1.add(5.8,0.0,3.0,0.0,10.0, 3.0)
h1.add(-1.2,4.0,-2.0,0.0,10.0, 0.0)
h1.add(-1.2,0.0,-2.0,2.0,10.0, 0.0)

# build a 3D cube
h2 = P3D('3D form in red')
h2.setPenColor(Color.red)
h2.add(-0.5,3.0,-1.0,2.0,6.0,2.0)

c1.draw(h1)
c1.draw(h2)
```

Figure 6.2 shows the resulting plot.

We should note that the P3D class was not designed for complicated drawings in 3D. You may find more appropriate to visualize geometrical shapes using the HView3D class discussed in Sect. 10.13.2 or the class HPlot3DP discussed in Sect. 10.12.

Fig. 6.2 P3D objects
displayed using the
HPlot3D canvas

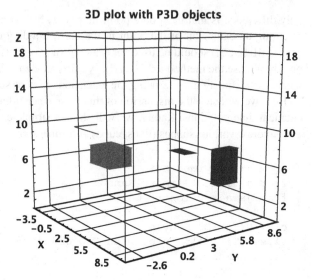

3D plot with P3D objects

6.3 PND Data Container

The PND class can be used to store data in many dimensions (this explains the
appearance of the "N" in its name). As usual, "D" in the class name means that the
object stores double values. Use the class PNI to store integer values.

Unlike other classes discussed so far, this class does not have any graphical attributes, since it is mainly designed for data manipulations. To be able to visualize
multidimensional data, one should always project such data into a lower dimension,
or use the P1D or P2D classes for drawing.

Below we will discuss the class PND, since its clone for integer values, PNI, has
exactly the same methods.

As usual, one should first initialize a PND object and then fill it. This can be done
by appending lists with numbers using the method add(list). The lists can have
any size. The example below shows how to create and fill a PND:

```
>>> from jhplot import *
>>>
>>> p0=PND('example')     # build a PND object
>>> p0.add([1,2,3])       # append some values
>>> p0.add([2,3,4])
>>> p0.add([2,3,4,3,4])   # append more columns
>>> p0.add([2,3])         # append less columns
>>> print p0.getDimension()
2
>>> print p0.toString()
PND: example
1.0 2.0 3.0
2.0 3.0 4.0
```

```
2.0 3.0 4.0 3.0 4.0
2.0 3.0
```

It should be noted that the dimension for each row is not fixed, i.e. one can add an array with arbitrary length. The method getDimension() returns the dimension of the last appended array. If the dimension is different for each row, one could expect problems for methods based on two-dimensional arrays with fixed number of rows and columns. So, try to avoid the use of rows with different length to avoid problems in future.

As for any other class, one can build a PND from an ASCII file. Once the object is created, one can easily obtain an array from a certain column or row in the form of POD using the following methods:

```
>>> p0=PND('PND from file', 'FileName')
>>> p0.getRow(i)     # get a row at index "i" as POD
>>> p0.getColumn(i)  # get a column at index "i" as POD
```

Table 6.3 shows the most important methods of the class PND.

To draw a PND object, create a histogram object first. This collects all values stored in the array and project them in one dimension. Let us give a simple example of how to display a PND values:

──────────── Showing PND as a histogram ────────────

```
from jhplot import *

c1 = HPlot('Canvas')
c1.visible()
c1.setAutoRange()

pnd=PND('array')
pnd.add([1,2,3,4])
pnd.add([5,6,7,8])
h1=pnd.getH1D(10)
c1.draw(h1)
```

We should note that the minimum and maximum values of the histogram are determined automatically. We have only specified the number of bins for data projection. Alternatively, one can make a histogram from a given column of the PND array.

6.3.1 Operations with PND Data

One can perform various operations with the PND data. Below we will discuss the most important methods. Please refer to the Java API documentation of this class.

To scale all data by a constant factor, use this method:

Table 6.3 Some methods of the PND class. "i" and "j" denote integer indexes

Methods	Definitions
setTitle()	set title
clear()	clean from entries
get(i)	returns row "i" as array
get(i,j)	returns value at row "i" and column "j"
toString()	convert data to a string
getColumn(i)	get POD from column at index "i"
getRaw(i)	get POD from row at index "i"
getDimension()	get dimension (last added entry)
getArrayList()	get data in form of array list
getArray()	get double array
getH1D(bins)	get H1D histogram with "bins"
add(POD)	add POD as row
add(array[])	add array
set(i,POD)	set POD at index i
set(i,array[])	set array at index i
setArray(array[][])	set from double array
remove(i)	remove a row at index i
getMin()	get minimum value
getMax()	get minimum value
size()	get the size
copy('text')	new copy with a title 'text'
standardize()	convert each column to $(x_i - \bar{x})/\sigma$
rescale(i)	rescale each column to $[0,1]$ (i=0) or $[-1,1]$ (i=1)

```
>>> p0.operScale(scale)
```

where 'scale' is a number used to scale all elements.

A PND can be rescaled to the range $[-1, 1]$ or $[0, 1]$ with the rescale() method. Another operation is the so-called standardize(), which is useful for the neural-network studies to be discussed in Sect. 15.1.1.

One can add, subtract, multiply and divide two PND arrays. Assuming p0 and p1 are objects of the PND class, this can be done as:

```
>>> p2=p0.oper(p1,'NewTitle',operation)
```

where 'operation' is a string which can either be "+", "/", "*" or "/" (they are self-explanatory). Finally, 'NewTitle' is an optimal title which can be dropped.

6.4 Input and Output

All the containers discussed before, P2D, P3D and PND, can be serialized and re-stored back in exactly the same way as discussed in Sect. 5.4.

In addition, one can write and read ASCII files with the instances pn of the class PND using the methods:

```
>>> pn.toFile('FileName') # write to a file
>>> pn.write('FileName')
>>> pn.read('FileName')    # read from files:
```

In case if a multidimensional collection of data must be stored in a single file, the most convenient way would be to populate a Jython list or a dictionary with the PND objects and serialize it to a file as shown in Sects. 4.3.3 and 5.4.

We will come back to the I/O methods in Chap. 11.

Chapter 7
Arrays, Matrices and Linear Algebra

jHepWork contains many types of arrays that can be used to hold data in the form of primitive numbers. Typically, they come from third-party Java libraries.

One should not consider such external arrays as being completely independent of the native jHepWork containers which come with the package jhplot. There are many ways to convert the third-party arrays into the POD, P1D and PND objects for manipulation and graphical representation.

7.1 Jaida Data Containers

First, we will start from the Jaida library which is included to the jHepWork package by default.

To keep data points with errors, one can use the Measurement class. It keeps information on a single measurement with its errors. There are several constructors for this class:

```
>>> from hep.aida.ref.histogram import *
>>> m=Measurement(d)
>>> m=Measurement(d,err)
>>> m=Measurement(d,errMinus, errPlus)
```

where 'd' is a central measurement represented by a double value, 'err' is its error. 'errMinus' and 'errPlus' can be used to add asymmetric uncertainties (i.e. a lower and an upper error).

To retrieve the measurement value and its errors, one should call the following methods:

value()   to obtain the measurement value
errorMinus()   to obtain a lower error
errorPlus()   to obtain upper error

S.V. Chekanov, *Scientific Data Analysis using Jython Scripting and Java*,
Advanced Information and Knowledge Processing,
DOI 10.1007/978-1-84996-287-2_8, © Springer-Verlag London Limited 2010

One should note that this class is conceptually rather similar to the P1D container.

A measurement can also be kept in a more general container, called Data-Point. It has the following constructors:

```
>>> from hep.aida.ref.histogram import *
>>> DataPoint(d[])
>>> DataPoint(d[], err[])
>>> DataPoint(d[], errMinus[], errPlus[])
```

In contrast to the class Measurement, the DataPoint object can hold values and their errors in many dimensions. One can access the values as:

dimension()   get the data dimension
lowerExtent(i)   the lower value at "i"
upperExtent(i)   the upper value at "i"

Jaida has a special container to hold data represented by the class DataPoint, called DataPointSet. It was designed for holding a set of DataPoint objects. Below we give two most popular constructors:

```
>>> from hep.aida.ref.histogram import *
>>> DataPointSet(name, title, int dim)
>>> DataPointSet(name, title, int dim, int capacity)
```

The strings 'name' and 'title' are self-explanatory, dim is the dimension (integer) and capacity is the default capacity (integer). Once the DataPointSet object is created, one can add a DataPoint using the method addPoint(p), where 'p' is a data point represented by a DataPoint object.

A DataPointSet object can be translated into the usual P1D object discussed in Sect. 5.1 and used for visualization using the jHepWork canvas. What one needs to do is to build a P1D object passing a DataPointSet as argument:

```
>>> from hep.aida.ref.histogram import *
>>> from jhplot import *
>>> #.. create dp using DataPointSet class
>>> p1=P1D(dp) # convert to P1D
```

In this operation, only 1st-level errors of the P1D array will be filled. Table 7.1 lists some methods of this container.

7.1.1 Jaida Clouds

Jaida cloud is a set of data points. These objects are rather similar to P0D, P2D etc. containers of the jHepWork library.

Table 7.1 Some methods of the `DataPointSet` from the Jaida library

Methods	Definitions
addPoint(p)	add a DataPoint
clear()	clear the array
dimension()	returns the dimension
point(i)	returns DataPoint at index 'i'
removePoint(i)	removes point at index 'i'
scale(d)	scales by a factor 'd'
scaleValues(d)	scales values by a factor 'd'
scaleErrors(d)	scales errors by a factor 'd'
upperExtent(i)	get a lower extent at index 'i'
lowerExtent(i)	get an upper extend at index 'i'

To create a cloud in 1D, the following constructor should be used:

```
>>> from hep.aida.ref.histogram import *
>>> c=Cloud1D()
```

Once the cloud object has been initialized, one can fill it using two methods:

```
>>> c.fill(d)
>>> c.fill(d,w)
```

The second method fills a cloud with a value 'd' which has a weight 'w' (both numbers are double). The notion of the weight is rather similar to that to be discussed in the section devoted to histograms (see the following chapter). In simple words, the number 'w' represents importance of a data point in the data set. By default, the first method assumes w=1. To display a cloud, one needs to convert it into a histogram. Table 7.2 shows some most important methods of the clouds:

A cloud can be created in 2D using the `Cloud2D` constructor:

```
>>> from hep.aida.ref.histogram import *
>>> c=Cloud2D()
```

which can be filled with two numbers (say, x, y):

```
>>> c.fill(x,y)
>>> c.fill(x,y,w)
```

As before, one can specify a weight 'w' for each 2D point (x, y) (the default weight is 1). The methods for this cloud are the same as those shown in Table 7.2, but this time there are more methods since we have two values, instead of one. Therefore,

	Methods	Definitions
Table 7.2 Some methods of the class Cloud1D from the Jaida library. 'd' indicates a double value and 'i' is an integer index	fill(d)	fill with data point 'd'
	fill(d,w)	fills with 'd' and a weight 'w'
	histogram()	create a histogram
	lowerEdge()	cloud's lower edge
	mean()	the mean value
	entries()	the number of entries
	rms()	the standard deviation
	scale(d)	scale with a factor 'd'
	upperEdge()	get the upper edge
	value(i)	get a value at the index 'i'
	weight(i)	get a weight at the index 'i'

each method characterizing the Coud1D class has extra "X" or "Y" string at the end of each method name. For example, meanX() denotes the mean value in x, meanY() is the mean value in y.

Similarly, one can build a cloud in 3D using the constructor Cloud3D(). We will leave the reader here as it is rather easy to guess how to fill and how to access values of such cloud, reminding that one single point now is represented by three numbers, (x, y, z).

7.2 jMathTools Arrays and Operations

Another package which is incorporated into jHepWork is jMathTool[1]. It contains a collection of Java classes designed for engineering and general scientific computing needs.

7.2.1 1D Arrays and Operations

Let us first create an one-dimensional array with double numbers as:

```
>>> from jhplot.math.DoubleArray import *
>>> a=one(N,d)
```

This example instantiates a 1D array with 'N' numbers, all of which have the same double value 'd'. Check the type which corresponds to the object 'a' as type(a). You will see that the object 'a' is <type 'array.array'>. One can find all the methods which could be used for manipulation with this object using the code assist.

One can also build an array with numbers incremented by some value:

```
>>> a=increment(N,begin,val)
```

again, 'N' is the total number of values, 'begin' is an initial value, and 'val' represents a double value used to increment it, x[0]=begin, x[n]= x[n-1]+val. Below we describe some important methods for the 1D arrays (denoted by 'a'):

copy(a)   returns a copy
min(a)   the minimum value for array
max(a)   the maximum values for array
minIndex(a)   get index of the minimum value
maxIndex(a)   get index of the maximum value
sum(a)   sums of all values, $\sum_{i=0}^{N} x_i$
cumSum(a)   array with the cumulative sum, $b_k = \sum_{i=0}^{k} x_i$
product(a)   product of all values, $\prod_{i=0}^{N} x_i$
cumProduct(a)   array with the cumulative product, $b_k = \prod_{i=0}^{k} x_i$
sort(a)   sorts array
insert(a1,i,a2)   inserts an array a2 to array a1 starting from index "i"
increment(N,d,p)   initializes an array with size N, $x[i] = d + i * p$
deleteRange(a,i,j)   deletes range from 'i' to 'j'
delete(a,i)   deletes the range starting from 'i'
random(i)   creates an array of the size 'i' with random numbers
random(i,min,max)   creates an array of size 'i' with random numbers between min and max

One can print the 1D array using the io.ArrayString method. Here is an complete example of how to create an array, print it and sum-up its values:

```
>>> from  jhplot.math.DoubleArray import *
>>> from  jhplot.math.io.ArrayString import *
>>> a=increment(10,0,1)
>>> print 'Array='+printDoubleArray(a)
Array=0.0 1.0 2.0 3.0 4.0 5.0 6.0 7.0 8.0 9.0
>>> sum(a)
45
```

We remind that one can learn more about this type of arrays using the method dir(obj). For the DoubleArray static Java class, call this lines:

```
>>> from  jhplot
>>> dir(jhplot.math.DoubleArray)
```

7.2.2 2D Arrays

DoubleArray class can be used to build 2D arrays almost in the same way as it for the 1D case above. Such arrays are ideal for storing matrix values. To initialize a 2D array, the following class can be used:

```
>>> from jhplot.math.DoubleArray import *
>>> a=one(i1,i2,d)
```

where 'd' is a number to be assigned for the entire matrix with the number 'i1' of rows and the number 'i2' of columns.

```
>>> from  jhplot.math.DoubleArray import *
>>> from  jhplot.math.io.ArrayString import *
>>> a=one(2,5,1)
>>> print printDoubleArray(a)
1.0 1.0 1.0 1.0 1.0
1.0 1.0 1.0 1.0 1.0
```

Below we give a list of the most important methods of this class, where 'a' indicates an input array.

copy(a)	returns exact copy
min(a)	minimum value for array
max(a)	maximum values for array
minIndex(a)	an index of the minimum value
maxIndex(a)	an index of the maximum value
sum(a)	sums of all values, $\sum_{i,j=0}^{N} a_{i,j}$
cumSum(a)	array with cumulative sum, $b_k = \sum_{i,j=0}^{k} a_{i,j}$
product(a)	product of all values, $\prod_{i,j=0}^{N} a_{i,j}$
cumProduct(a)	array with cumulative product, $b_k = \prod_{i,j=0}^{k} a_{i,j}$
increment(i,j, b[], p[])	initialize array $i x j$ as $a[i][j] = b[j] + i * p[j]$
getSubMatrixRangeCopy(a,i1,i2,j1,j2)	get submatrix using ranges i1–i2 (rows) and j1–j2 (columns)
getColumnsRangeCopy(a,i1,i2)	get columns between 'i1–i2'
getColumnCopy(a,i)	obtains a column 'i'
getColumnsRangeCopy(a,i1,i2)	obtains columns between 'i1–i2'
getRowsRangeCopy(a,i1,i2)	obtains rows between 'i1–i2'
getRowCopy(a,i)	obtains a row at 'i'
getColumnDimension(a,i)	the dimension of column at 'i'
insertColumns(a1, a2, i)	inserts a2[][] to a1[][] at column 'i'
insertColumn(a1, a2, i)	inserts a2[] to a1[][] at column 'i'

`insertRows(a1, a2, i)`	inserts a2[][] to a1[][] at row 'i'
`insertRow(a1, a2, i)`	inserts a2[] to a1[][] at row 'i'
`deleteColumnsRange(a,i1,i2)`	deletes columns between 'i1' and 'i2'
`deleteRowRange(a,i1,i2)`	deletes rows between 'i1' and 'i2'
`sort(a)`	sorts the array
`random(i)`	creates a 2D array of size 'i' with random numbers
`random(i,min,max)`	creates a 2D array of the size 'i' with random numbers between min and max

7.3 Colt Data Containers

The Colt package [2] provides several enhanced data containers to store primitive values and to perform manipulations with them. The arrays support quick access to its elements which is achieved by non-bounds-checking methods. There are several classes used to build the Colt arrays:

- `IntArrayList`
- `DoubleArrayList`
- `LongArrayList`

We will consider the `DoubleArrayList` container for further examples since all these classes are very similar.

```
>>> from cern.colt.list import *
>>> a=DoubleArrayList()     # empty list
>>> a=DoubleArrayList(d[]) # with elements d[]
>>> a=DoubleArrayList(initialCapacity)
```

The constructors above illustrate various ways to initialize the arrays. Try to look at the methods of this class with the code assist. Table 7.3 lists some of the most important methods of the `DoubleArrayList` class.

The data stored in the `DoubleArrayList` array can be analyzed using the `DynamicBin1D` class from the same package:

```
>>> from cern.colt.list import *
>>> from hep.aida.bin   import *
>>> a=DoubleArrayList()
>>> bin=DynamicBin1D()
>>> bin.addAllOf(a);
>>> print bin.toString()
```

This example prints a comprehensive statistical summary of the array. One should also note that one can build a POD from the `DoubleArrayList` object as:

Table 7.3 Methods for the DoubleArrayList. (IntegerArrayList and LongArrayList have the same methods)

Methods	Definitions
add(d)	add a double value
size()	get the size
copy()	returns a DoubleArrayList copy
elements()	returns elements as list
get(i)	get double value at index 'i'
getQuick(i)	get double at index 'i' without checking
set(i,d)	set double 'd' at index 'i'
reverse()	reverses the order
shuffleFromTo(i1,i2)	randomly permutes from index 'i1' to 'i2'
contains(d)	returns true (1) if "d" exists
delete(d)	deletes first element 'd'
addAllOf(a)	appends an array 'a'
indexOf(d)	returns index of the first occurrence
quickSortFromTo(i1,i2)	Sorts the range [i1–i2] into ascending numerical order
toList()	returns ArrayList
clear()	clear the array

```
>>> from cern.colt.list import *
>>> from jhplot  import *
>>> a=DoubleArrayList()
>>> p0=P0D(a)
>>> print p0.getStat()
```

File-based input/output can be achieved through the standard Java built-in serialization mechanism.

The Colt package also includes the class `ObjectArrayList`, which is similar to the Java list discussed in the previous sections.

7.4 Statistical Analysis Using Jython

In addition to the Java libraries provided by jHepWork, one can also use third-party libraries implemented using the Python language. The Python/Jython third-party packages are located in the directory `python/packages`. There are several advantages in using programs implemented in Jython: (1) one can directly access the code with the implemented numerical algorithms. (2) One can reuse the libraries in CPython programs.

We remind that the directory `'python/packages'` is imported automatically when using the jHepWork IDE, so there is no need to worry about appending this directory location to the `'sys.path'` variable. If one uses another editor, make sure that Jython looks at the directory `'python/packages'` to import the third-party packages. For example, you can do it as:

```
>>> import sys
>>> sys.path.append(SystemDir+'/python/packages')
```

where `SystemDir` is the directory with the installed jHepWork. Below we assume that the reader uses the jHepWork IDE and does not need to specify this directory.

Here we will consider the module `'stats'` which provides basic statistical functions for Jython collections. It allows calculations of simple characteristics from a Jython list, such as the mean value and the standard deviation:

```
>>> from statlib.stats import *
>>> a = [1,2,3,4,5,6,7,8,10 ]
>>> print 'Mean=',mean(a)
Mean= 5.1111
>>> print 'Standard deviation=',stdev(a)
Standard deviation= 2.93446
```

Analogously, one can calculate other statistical characteristics, such as the median, variation, skewness, kurtosis and moments. Let us give their definitions: A moment of an order n for a list with N elements is given as

$$M_n = \frac{1}{N} \sum_i^N (\text{list}(i) - \text{mean})^n$$

A skewness is defined via the moments, $M_3/(M_2)^{1.5}$, while the kurtosis is $M_4/(M_2)^2$. A variation is given as the ratio of the biased standard deviation to the mean. All such statistical characteristics can be assessed in one line of the code using the `describe(list)` function:

```
>>> a = [1,2,3,4,5,6,7,8,10]
>>> stat=describe(a)
>>> print 'N=',stat[0]
N= 9
>>> print 'tuple=',stat[1]
tuple= (1, 10)
>>> print 'mean=',stat[2]
mean= 5.1111
>>> print 'standard deviation=',stat[3]
standard deviation= 2.93
>>> print 'skewness=',stat[4]
```

```
skewness= 0.199
>>> print 'kurtosis=',stat[5]
kurtosis= 2.00
```

The output values have been truncated to fit the page width.

In the next example, we will extend this code snippet by including calculations of various characteristics of a list with 100 numbers. Moments about the mean for this list will be calculated in a loop (up to the order nine):

```
—————————— Statistical analysis of lists ——————————
from statlib.stats import *

a=range(100)
print 'geometric mean=',geometricmean(a)
print 'median=',median(a)
print 'variation=',variation(a)
print 'skew=',skew(a)
print 'kurtosis=',kurtosis(a)
for n in range(2,10):
    print 'moment order '+str(n)+':',moment(a,n)
```

The output of this code snippet is given below:

```
geometric mean= 0.0
median= 49.005
variation= 58.315
skew= 0.0
kurtosis= 1.799
moment order 2: 833
moment order 3: 0.0
moment order 4: 1249
moment order 5: 0.0
moment order 6: 2230
moment order 7: 0.0
moment order 8: 4.335+12
moment order 9: 0.0075
```

Again, we have reduced the precision of the output numbers to fit the page width.

The module allows calculations of correlation coefficients between two lists. Let us generate two correlated lists using a Gaussian distribution and estimate their correlations using several tests:

```
—————————— Correlation coefficients ——————————
from statlib.stats import *

from random import *
ran=Random()
```

```
mu,sigma=2.0,3.0

x,y=[],[]
for i in range(100):
  t=ran.gauss(mu,sigma)
  x.append(t)
  y.append(t*2+ran.gauss(mu,sigma))

print stdev(x), '+/-',sterr(x)
print mean(y),'+/-',sem(y)
print 'Covariance=',lcov(x,y)
print 'Pearson (correlation coeff. prob)=',lpearsonr(x,y)
print 'Spearman rank-order correlation=',lspearmanr(x,y)
```

The output is shown below:

```
2.917 +/- 0.2917
6.610 +/- 0.6750
Covariance= 17.126
Pearson (correlation coefficient,prob)=(0.869,8.945e-32)
Spearman rank-order correlation=(0.848,8.361-29)
```

For clarity and in order to fit the output to the page width, we truncated several numbers to three decimal digits.

The module 'stats' has many statistical tests (including the Anova test) and functions, please refer to the package description and the source code.

7.5 Matrix Packages

The construction of matrices and their manipulation can be performed with several third-party Java libraries [3, 4]. They allow for basic linear algebra calculations based on two-dimensional matrices of double-precision floating-point numbers. Here is a short summary of features of such packages:

- Creation and manipulation with matrices;
- Calculation of derived quantities, such as condition numbers, determinants, ranks etc.;
- Elementary operations, such as addition, subtraction, multiplication, scalar multiplication etc.;
- Various decompositions.

We will start our discussion with the Jama package [3] implemented in Java. Below we show how to build a 2×2 matrix and print its values:

```
>>> from Jama import *
>>> m=Matrix([[1.,2.],[3.,4.]])
>>> print m.toString(3,2)
     1.00 2.00
     3.00 4.00
```

The example is rather simple: first we initialize the matrix from a Jython list, and
then print it out using the toString() method which converts the matrix into a
string. This method takes two integer parameters: the column width and the number
of digits after the decimal point. One can also pass the DecimalFormat instance
from Java API for nice printing, but this requires more typing.

Let us construct a "null" matrix or a matrix holding a constant value, say 100:

```
>>> from Jama import *
>>> m=Matrix(2,3)
>>> print m.toString(3,1)
     0.0  0.0  0.0
     0.0  0.0  0.0
>>> m=Matrix(2,3,100)
>>> print m.toString(3,1)
     100.0 100.0 100.0
     100.0 100.0 100.0
```

The constructor Matrix takes 2 arguments: the number of rows and the number of
columns. One can obtain a single value using the matrix induces:

```
>>> m=Matrix(2,3,100)
>>> m.set(1,1,200)
>>> print m.get(1,1)
200.0
```

It is relatively easy to construct a matrix from any array supported by jHepWork.
For example, one can build a matrix from two POD classes discussed in Sect. 4.1
using the method getArray().

```
>>> from jhplot import *
>>> from Jama import *
>>> p1=POD([2,3])
>>> p2=POD([4,5])
>>> m=Matrix([p1,p2])
```

Finally, one can fill a matrix with the method m.random(n,m) using uniformly
distributed random elements. In this method, n is the number of rows and m is the
number of columns.

One can insert a sub-matrix of a matrix with the method:

```
>>> m.setMatrix(int[] r,  int[] c, X)
```

where r is an array of row indexes and c is an array of column indexes. X is the actual sub-matrix defined as A(r(:),c(:)).

Now let us consider the question of extraction of the information about a matrix. First, one can access 2D array of a matrix using the method getArray(). To get a single value, use the usual method get(i1,i2), where i1 is the row number and i2 is the column number. One can return the entire matrix object or only a sub-matrix using getMatrix(i1,i2,j1,j2), where i1 is the initial row index, i2 is the final row index, j1 is the initial column index and j2 is the final column index. One can also learn about other methods by looking at the corresponding API documentation.

A matrix or a group of matrices can be saved into dictionaries or lists and serialized into a file as any jHepWork object discussed in Chap. 11. This is possible because the class Matrix implements the Java Serializable interface.

7.5.1 Basic Matrix Arithmetic

Assuming that you have two matrices, "A" and "B", one can perform the following operations:

B.minus(A)	subtract matrix A from B
A.timesEquals(d)	multiply a matrix by a scalar "d", $A = d * A$
B.times(A)	Linear algebraic matrix multiplication, $A * B$
B.plus(A)	$C = A + B$
A.plusEquals(B)	$A = A + B$
A.minusEquals(B)	$A = A - B$
A.minus(B)	$C = A - B$
A.arrayRightDivide(B)	Element-by-element right division, $C = A./B$
A.arrayRightDivideEquals(B)	Element-by-element right division in place, $A = A./B$
A.arrayLeftDivide(B)	Element-by-element left division, $C = A.$ divide B
A.arrayLeftDivideEquals(B)	Element-by-element left division in place, $A = A.$ divide B
A.arrayTimes(B)	Element-by-element multiplication, $C = A. * B$

`A.arrayTimeEquals(B)`	Element-by-element multiplication, $A = A. * B$
`A.uminus()`	unary minus, i.e. $-A$

Finally, one can use the `inverse()` method to inverse all elements. One can obtain the basic normalization methods, such as:

`norm1()`	the maximum column sum, i.e. summing up absolute values of all column numbers
`normInf()`	the maximum row sum, i.e. summing up absolute values of all row numbers

7.5.2 Elements of Linear Algebra

The Jama package provides more advanced operations with matrices than those discussed above. The determinant of a matrix can be obtained with the method `det()`, while one can access its rank as `rank()`.

There are several basic matrix decompositions:

- Cholesky Decomposition of symmetric, positive definite matrices;
- QR Decomposition of rectangular matrices;
- LU Decomposition of rectangular matrices;
- Singular Value Decomposition of rectangular matrices;
- Eigenvalue Decomposition of both symmetric and non-symmetric square matrices.

The decompositions are accessed by the class `Matrix` to compute solutions of simultaneous linear equations, determinants and other matrix functions.

Let us give a small example illustrating the power of the Jama package:

```
───────────────── Solving a linear system ─────────────────
from Jama import *

A=Matrix([[2.,2.,3],[4.,5.,6.],[7.,8.,4.]])
print 'Determinant',A.det()
B=Matrix([[7.,8.,1],[1.,7.,0.],[4.,1.,9.]])
X = A.solve(B)
print X.toString(1,2)
R = A.times(X).minus(B)
print 'normInf=',R.normInf()

print 'EigenvalueDecomposition:'
Eig=A.eig()
D=Eig.getD()
V=Eig.getV()
print 'D=',D.toString(1,3), 'V=',V.toString(1,3)
```

The output is shown below:

```
Determinant -13.0
 14.77 8.85 4.23
 -13.00 -9.00 -2.00
 1.15 2.77 -1.15
normInf= 1.50990331349e-14
EigenvalueDecomposition:
D=
 13.765 0.000 0.000
 0.000 0.307 0.000
 0.000 0.000 -3.072
V=
 0.291 0.761 -0.319
 0.627 -0.675 -0.481
 0.722 0.021 0.860
```

7.5.3 Jampack Matrix Computations and Complex Matrices

The package Jampack (Java Matrix PACKage) [4] is a complementary package. Unlike the previous package, it supports complex matrices. In fact, it supports only complex matrices, since the design proceeded from a more general case to the less general. The package has all basic linear algebra operations and many decomposition methods:

- The Pivoted LU decomposition;
- The Cholesky decomposition;
- The QR decomposition;
- The eigendecompostition of a symmetric matrix;
- The singular value decomposition;
- Hessenberg form;
- The Schur Decomposition;
- The eigendecompostition of a general matrix.

In this package, a complex number is represented by the class Z(a,b), where a represents a real part while b is an imaginary part of a complex number. One can perform several simple arithmetical operations as shown below:

```
>>> from Jampack import *
>>> z1=Z(10,6)
>>> z2=Z(8,1)
>>> z1.Minus(z2)
>>> print 'Real part=',z1.re, ' Imaginary part=',z1.im
```

The `print` statement is used to print real and imaginary parts of the complex number. Below we will give an example to help to get started with the complex matrices:

```
――――――――――― Complex matrices ――――――――――――

from Jampack import *

# initializes real and imaginary parts
A=Zmat([[1,2],[3,4]], [[3,4],[5,6]])
print Print.toString(A,4,3)


# Frobenius normalization
nrm = Norm.fro(A)
# then product
A=Times.o(H.o(A), A)
print Print.toString(A,4,3)
```

One should look at the API documentation of this package or use the jHepWork help assist to learn about the package.

7.5.4 Jython Vector and Matrix Operations

In addition to the Java-based packages for matrix operations, jHepWork also contains pure Jython packages for vector, matrix, and table math operations. These operations are based on the Jython package `'statlib'` which is located in the `'python/packages'`. We remind that this directory is imported by the jHep-Work IDE during the startup. More details on how to import Jython libraries have been discussed in Sect. 1.4.2.

The `'statlib'` package provides many basic vector operations. Below we will show an example of how to call this library and perform a few basic operations with two vectors, `'a'` and `'b'`:

```
――――――――――― Vector operations ―――――――――――

from statlib.matfunc import Vec

a = Vec( [ 4, -2,  5 ] )
b = Vec( [ 3, 10, -6 ] )
print a.dot(b)        # the dot product of 'a' and 'b'
print a.norm()        # length of the vector
print a.normalize()   # length 1
print a.cross(b)      # cross product of 'a' and 'b
```

One can learn more about the methods of this package using the code assist or by calling the statement `dir(obj)` (`obj` represents an object of the class `Vec`). Another option is to look at the module directly. Just open the Jython module `matfunc.py` in the directory `'python/packages/statlib'`.

Let us consider several examples of how to work with Jython-implemented matrices. A matrix can be defined from a list of lists as in this example:

```
>>> from statlib.matfunc import *
>>> m=Mat( [[1,2,3], [4,5,1,], [7,8,9]] )
>>> print m
1    2    3
4    5    1
7    8    9
>>> print 'eigs:', m.eigs()
eigs: 2.692799    -0.846943    13.154144
>>> print 'det(m):', m.det()
det(m): 30.0
```

The class Mat has many useful methods, such as tr() (for transpose), star() (for Hermetian adjoints), diag(), trace(), and augment(). All such methods have their usual mathematical meaning. Matrix multiplications are accomplished by the method mmul(), and matrix division by solve(b).

Let us give a more detailed example:

```
––––––––––––––––––––––– Vector operations –––––––––––––
from statlib.matfunc import *

C = Mat( [[1,2,3], [4,5,1,], [7,8,9]] )
print C.mmul( C.tr())
print C ** 5
print C + C.tr()
A = C.tr().augment( Mat([[10,11,13]]).tr() ).tr()
q, r = A.qr()
print 'q:\n', q, '\nr:\n',r
q.mmul(r) == A
print '\nQ.tr()&Q:\n', q.tr().mmul(q)
print '\nQ*R\n', q.mmul(r)
b = Vec([50, 100, 220, 321])
x = A.solve(b)
print 'x:  ', x
print 'b:  ', b
print 'Ax: ', A.mmul(x)
inv = C.inverse()
print   'inverse=\n',inv
print 'C * inv(C):\n', C.mmul(inv)
```

Because of the lengthy output, we will not show it in this book. Try to make sense of all these operations. They are well documented in the file 'matfunc.py' inside the directory 'python/packages/statlib'.

7.5.5 *Matrix Operations in SymPy*

Another package designed for matrix operations, including operations with symbols, is SymPy [5]. It was already discussed in Sect. 3.9. We remind that this library is implemented in Python and included as a third-party package in the directory 'python/packages'.

Let us build a matrix using the SymPy package. Our matrix can contain numbers as well as symbols which have to be declared with the statement Symbol():

```
>>> from sympy import *
>>> Matrix([[1,2], [3,4]])
[1, 2]
[3, 4]
>>> x = Symbol('x')
>>> Matrix([[1,x], [2,3]])
[1, x]
[2, 3]
```

There are several ways to construct predefined matrices based on the statements eye (the identity matrix), zeros and ones:

```
>>> from sympy import *
>>> eye(2)
[1, 0]
[0, 1]
>>> zeros((2, 3))
[0, 0, 0]
[0, 0, 0]
```

One can use the standard mathematical operations, like *, /, -, + as shown in the example:

```
>>> from sympy import *
>>> M1=Matrix(([1,2,3],[4,5,6]))
>>> M2=Matrix(([1,1,1],[2,2,2]))
>>> M3=M1-M2
>>> M3
[0, 1, 2]
[2, 3, 4]
```

A few linear-algebra operations are shown below:

```
>>> from sympy import *
>>> M1 = eye(3)
>>> M1.det()
```

```
1
>>> M1.inv()
[1, 0, 0]
[0, 1, 0]
[0, 0, 1]
```

There are many decomposition methods associated with the class `Matrix` of this package. Check the SymPy web site [5] or look at the package `'sympy/matrices'` inside the directory 'python/packages'.

7.6 Lorentz Vector and Particle Representations

Containers that hold a group of numbers (arrays, vectors, matrices) are useful abstraction for keeping specific quantities in a structural form. In this section we will discuss a concrete physics implementation—a representation of a particle using four-momentum which is typically used in relativistic calculations. Such representation is particular useful for various transformations involving simulated (or real) relativistic particles.

7.6.1 Three-vector and Lorentz Vector

Before going into the depth of four-vectors, let us first take a look at the simplest case when a vector is represented by three coordinates, $\mathbf{p} = (p_x, p_y, p_z)$. This situation is represented by a `Hep3Vector` class which can be instantiated as:

```
>>> from hephysics.vec import *
>>> v=Hep3Vector(x,y,z)
```

where x, y, z are coordinates of this vector. This class is useful for usual three-vector transformations as illustrated below:

```
>>> from hephysics.vec import *
>>> v1=Hep3Vector(1,1,1)
>>> v2=Hep3Vector(2,2,2)
>>> v1.add(v2)              # add vector v2
>>> v1.mult(10)            # multiply by  10 (scale)
>>> v1.sub(v2)             # subtract v2
>>> print v1.dot(v2)       # dot product
168.0
>>> print v1.toString()   # print
[28.0, 28.0, 28.0]
```

Next, a four-momentum of a particle can be represented by four numbers: three-momentum $\mathbf{p} = (p_x, p_y, p_z)$ and energy e. Similarly, a position of particle in space and time can be represented with four coordinates, (x, y, z, t), where t is time. This means that either position, in the momentum space or space-time, can be described by four numbers. A class which can be used for such description is HepLorentzVector:

```
>>> from hephysics.vec import *
>>> hp=HepLorentzVector(px,py,pz,e)
```

or, in the case of space-time:

```
>>> hp=HepLorentzVector(x,y,z,t)
```

There is a number of useful methods implemented for this class. First of all, one can perform the standard arithmetic operations as with the usual three-vector. One can access angles and perform transformations as with any other vector representation. Let us give a few examples:

```
>>> from hephysics.vec import *
>>> # (px,py,pz,energy)=(10,20,30,100)
>>> hp=HepLorentzVector(10,20,30,100)
>>> print hp.mag()    # magnitude of 3-vector
>>> print hp.phi()    # azimuthal angle
>>> print hp.perp()   # transverse momenta.
>>> print hp.m()      # invariant mass
>>> print hp.theta()  # polar angle
>>> h.add(HepLorentzVector(1,2,3,10)) # add a new vector
```

The transverse momentum is calculated as $\sqrt{px^2 + py^2}$, while the invariant mass is $\sqrt{e^2 - px^2 - py^2 - pz^2}$. Please refer to the corresponding API to find more methods of this class.

Now let us show how to visualize an object of the HepLorentzVector class. One natural way to do this is to use the 3D canvas HPlot3D, and show a Lorentz vector as a point with a symbol size proportional to the forth component (either energy or time) as show below:

```
———————————— Visualizing a Lorenz vector ————————————
from jhplot  import *
from  hephysics.vec import *

hp=HepLorentzVector(10,20,30,10)

c1 = HPlot3D("Canvas",600,400)
c1.setGTitle("Lorenz Particle")
```

```
c1.setRange(0,100,0,100,0,100)
c1.setNameX('pX'); c1.setNameY('pY'); c1.setNameZ('pZ')
c1.visible()

p= P2D('LorenzParticle')
p.setSymbolSize(int(hp.e()))
p.add(hp.px(),hp.y(),hp.z())
c1.draw(p)
```

7.6.2 Classes Representing Particles

A particle in the package hephysics can be represented by the two classes, LParticle (a "light-weight" class which is a direct extension of the class HepLorentzVector) and HEParticle. The latter contains not only the basic Lorentz-type coordinates, but also can be characterized by a spin, parity and other characteristics which are typically used in particle physics. For simplicity, below we will concentrate on the LParticle class.

The class LParticle can hold information about four-momentum, particle name and even another LParticle object representing a parent particle. Let us create a particle with the name "proton" and with a known mass (in MeV units):

```
>>> from hephysics.particle import *
>>> p1=LParticle('proton',939.5)
>>> p1.setPxPyPzE(10,20,30,300)
```

The second line of this code sets the three-momentum and energy. One can also set charge of this particle via the method setCharge(c). One can access the information about this particle using various "getter" methods which can be found either using the code assist or the class API.

As for any other Jython/ or Java object, one can create a list of particles and store them in a file. In case of Java, use the ArrayList class, in case of Jython, one can use a Jython list:

```
>>> from hephysics.particle import *
>>> p1=LParticle('proton',939.5)
>>> p1.setPxPyPzE(10,20,30,300)
>>> p2=LParticle('photon',0.0)
>>> p2.setPxPyPzE(1,2,3,30)
>>> list=[p1,p2] # keep them in a list
```

The object list (as well as the objects of type LParticle) can be serialized into a file as any other Java object to be discussed in Chap. 11. Particles can be visualized in exactly the same way as shown for the `HepLorentzVector` class, since the class `LParticle` is an extension of the class `HepLorentzVector`.

We will give detailed examples of how to work with the `LParticle` class in Sect. 17.5 where we will discuss a simple Monte Carlo model for simulation of particle decays.

References

1. JMATHTOOLS Java Libraries. URL http://jmathtools.berlios.de
2. The Colt Development Team: The COLT Project. URL http://acs.lbl.gov/~hoschek/colt/
3. Hicklin, J., et al.: Jama, a Java Matrix Package. URL http://newcenturycomputers.net/projects/dif.html
4. Stewart, G.: Jampack, a Java package for matrix computations. URL ftp://math.nist.gov/pub/Jampack/Jampack/
5. SymPy Development Team: SYMPY: Python library for symbolic mathematics. URL http://www.sympy.org

Chapter 8
Histograms

A histogram is an elegant tool to project multidimensional data to lower dimensions, a tool which is designed for graphical representation and visual inspection.

> A histogram is a summary graph showing counts of data points falling into various ranges, thus it gives an approximation of the frequency distribution of data.

The histogram shows data in the form of a bar graph in which the bar heights display event frequencies. Events are measured on the horizontal axis, X, which has to be binned. The larger number of bins, the higher chances that a fine structure of data can be resolved. Obviously, the binning destroys the fine-grain information of original data, and only the histogram content can be used for analysis.

In this respect, the histogram representation is useful if one needs to create a statistical snapshot of a large data sample in a compact form. Let us illustrate this: Assume we have N numbers, each representing a single measurement. One can store such data in a form of Java or Jython arrays. Thus, one needs to store $8 \times N$ bytes (assuming 8 bytes to keep one number). In case of a large number measurements, we need to be ready to store a very big output file, as the size of this file is proportional to the number of events. Instead, one can keep for future use only the most important statistical summary of data, such as the shape of the frequency distribution and the total numbers of events. The information which needs to be stored is proportional to the number of bins, thus the file storage has nothing to do with the size of the original data.

8.1 One-dimensional Histogram

To create a histogram in one dimension (1D), one needs to define the number of bins, Nbins, and the minimum (Min) and the maximum (Max) value for a certain variable. The bin width is given by (Max-Min)/Nbins. If the bins are too wide

S.V. Chekanov, *Scientific Data Analysis using Jython Scripting and Java,*
Advanced Information and Knowledge Processing,
DOI 10.1007/978-1-84996-287-2_9, © Springer-Verlag London Limited 2010

(Nbins is small), important information might get lost. On the other hand, if the bins are too narrow (Nbins is large), what may appear to be meaningful information really may be due to statistical variations of data entries in bins. To determine whether the bin width is set to an appropriate size, different bin sizes should be tried.

jHepWork histograms are designed on the bases of the JAIDA FreeHEP library [1]. For an one-dimensional histogram, use the class H1D. To initialize an empty histogram, the following constructor can be used:

```
>>> from jhplot import *
>>> h1=H1D('data', 100, 0, 20)
```

This creates a 1D histogram with the title 'data', the number of bins Nbins= 100, and the range of axis X to be binned, which is defined by the minimum and the maximum values, Min=0 and Max=20. Thus, the bin width of this histogram is fixed to 0.2. The bin size and the number of bins are given by the following methods:

```
>>> d=h1.getBinSize()
>>> i=h1.getBins()
```

We should note that a fixed-size binning is used. In the following sections, we will consider a more general case when the histogram bin size is not fixed to a single value.

The method fill(d) fills a histogram with a single value, where 'd' is a double number. The histograms can be displayed on the HPlot canvas using the standard draw(h1) method.

We will discuss the main methods of the histogram class in the following section. Here, to illustrate the methods discussed before, we give a complete example of how to fill three histograms with Gaussian random numbers and then display them on different pads.

```
———————————— Plotting histograms ————————————
from java.awt import Color
from java.util import Random
from jhplot import *

c1 = HPlot('Canvas',600,400,2,1)
c1.visible()

h1 = H1D('First',20,-2.0,2.0)
h1.setFill(1)
h1.setFillColor(Color.green)

h2 = H1D('Second',100,-2.5,2.5)
```

```
r=Random()
for i in range(500):
  h1.fill(r.nextGaussian())
  h2.fill(r.nextGaussian())

h3 = H1D('Third',20,0.0,10.0)
h3.setFill(1)
for i in range(50000):
  h3.fill(2+r.nextGaussian())

c1.cd(1,1)
c1.setAutoRange()
c1.draw([h1,h2])

c1.cd(2,1)
c1.setAutoRange()
c1.draw(h3)
```

After execution of this script, you should see the plots shown in Fig. 8.1. By default, the lines on the histogram bars are drawn to indicate the size of statistical uncertainty in each bin using the Gaussian estimation of uncertainties for counting experiments, i.e. $\mathrm{Err} = \sqrt{N}$, where N is the number of events in each bin.

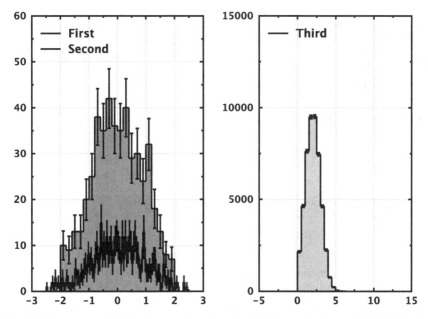

Fig. 8.1 Histograms with Gaussian numbers plotted with different numbers of bins (*left*) and a larger number of events (*right*)

In the above script, we have used two new methods of the H1D class. Due to their importance, we will discuss them here:

setFill(b)   fill a histogram area when b=1 (b=0 means the area is not filled)
setFillColor(c)   color for filling (Java AWT class)

As before, 'c' denotes the Java Color class discussed in details in Sect. 3.3.1.

The example above illustrates the following important features:

- The height of bins depends on the bin size. Even when the number of entries is the same, histograms are difficult to compare in shape when the histogram bins are different, see Fig. 8.1 (left).
- Relative size of errors decreases with increasing the number of entries.

Below we will show a few basic manipulations useful for examining the shapes of histograms, assuming that the underlaying mechanism for occurrence of events reveals itself in shapes of event distributions, rather than in the overall statistics or chosen bin size. The shape of the distributions is very important as it conveys information about the probability distribution of event samples.

First of all, let us get rid of the bin dependence. To do this, we will divide each bin height by the bin size. Assuming that 'h1' represents a histogram, this can be done as:

```
>>> width=h1.getBinSize()
>>> h1.scale(1/width)
```

After this operation, all histogram entries (including statistical uncertainties) will be divided by the bin width. You may still want to keep a copy of the original histogram using the method h2=h1.copy(), so one can come back to the original histogram if needed.

Different histograms can contain different normalization. In case if we are interested in the shapes only, one can divide each bin height by the total number of histogram entries. This involves another scaling:

```
>>> entries=h1.allEntries()
>>> h1.scale(1/entries)
```

Obviously, both operations above can be done using one line:

```
>>> h1.scale(1/(h1.getBinSize()*h1.allEntries()))
```

The second step in comparing our histograms would be to shift the bins of the third histogram. Normally, we do not know the exact shift (what should be don in this case will be considered later). At this moment, for the sake of simplicity, we assume that the shift is known and equals to -2. There is a special histogram operation which does such shift:

```
>>> h2.shift(-2) # shift all bins by -2
```

Now we are ready to modify all the histograms and to compare them. Look at the example below:

——————————————— Histogram operations ———————————————
```
from java.util import Random
from jhplot import *

c1 = HPlot('Canvas',600,400)
c1.visible()
c1.setRange(0,-2,2)

h1 = H1D('First',20,  -2.0, 2.0)
h2 = H1D('Second',100,-2.5, 2.5)
r=Random()
for i in range(500):
 h1.fill(r.nextGaussian())
 h2.fill(r.nextGaussian())

h3 = H1D('Third',20, 0.0, 10.0)
for i in range(50000):
  h3.fill(2+r.nextGaussian())


h1.scale(1/(h1.getBinSize()*h1.allEntries()))
h2.scale(1/(h2.getBinSize()*h2.allEntries()))
h3.shift(-2)
h3.scale(1/(h3.getBinSize()*h3.allEntries()))

c1.draw(h1)
c1.draw(h2)
c1.draw(h3)
```

After execution of this script, you will find three overlaid histograms. The shapes of all histograms will be totally consistent with each other, i.e. all bin heights will agree within their statistical uncertainties.

The problem of the histogram comparison discussed above is not a theoretical one. One can find many situations in which you may be interested in how well histogram shapes agree to each other. For example, let us assume that each histogram represents the number of days with rainfall measured during one year for one state. If the distributions are shown as histograms, it is obvious that bigger states have a larger number of days with rainfalls compare to small states. This means that all histograms are shifted (roughly by a factor proportional to the area of states, ignoring other geographical differences). The measurements could be done by different weather stations and the bin widths could be rather different, assuming that there is no agreement between the weather stations about how the histograms should be defined. Moreover, the measurements could be done during different time intervals, therefore, the histograms could have rather different numbers of entries. How one can compare the results from different weather stations, if we are only interested in some regularities in the rainfall distributions? The answer to this question is in the example above: all histograms have to be: (1) normalized; (2) shifted; (3) a bin dependence should be removed.

The only unclear question is how to find the horizontal shifts, since the normalization issue is rather obvious and can be done with the method discussed above. This problem will be addressed in the following chapters when we will discuss a statistical test that evaluates the "fit" of a hypothesis to a sample.

8.1.1 Probability Distribution and Probability Density

The examples above tell that there are several quantities which can be derived from a histogram. One can extract a *probability distribution* by dividing histogram entries by the total number of entries. The second important quantity is a *probability density*, when the probability distribution is divided by the bin width, so that the total area of the rectangles sums to one (which is, in fact, the definition of the probability density).

Both the probability distribution and the density distribution can be obtained after devisions of histogram entries as discussed above. However, these two characteristics can be obtained easier by calling the following methods:

```
>>> h2=h1.getProbability()
>>> h2=h1.getDensity()
```

which return two new H1D objects: the first represents the probability distribution and the second returns the probability density. In addition to the obvious simplicity, such methods are very useful for variable-bin-size histograms, since this case is taken into account automatically during the devisions by bin widths.

Also note the following: one can save computation time in case of the calculation of the probability distributions if you know the total number of events (or entries) N_{tot} beforehand. In this case, one can obtain the probability distribution by using the weight $w1 = 1.0/N_{tot}$ in the method fill(x,w1), without subsequent call to the method getProbability(). After the end of the fill, the histogram will represent the probability distribution normalized to unity by definition. In addition, one can remove the bin dependence by specifying the weight as $w2 = 1.0/BinSize$. Finally, the density distribution can be obtained using the weight $w3 = w1 * w2$.

8.1.2 Histogram Characteristics

In this subsection we will continue to discuss the most important characteristics of the H1D histogram class.

   The most popular characteristics of a histogram are the median and the standard
deviations (RMS). Assuming that `h1` represents a `H1D` histogram, both (double)
values can be obtained as:

```
>>> d=h1.mean()
>>> d=h1.rms()
```

   We already know that one can obtain the number of entries with the method
`allEntries()`. However, some values could fall outside of the selected range
during the `fill()` method. Luckily, the histogram class has the following list of
methods to access the number of entries:

```
>>> i=h1.allEntries()    # all entries
>>> i=h1.entries()       # number entries in the range
>>> i=h1.extraEntries()  # under and overflow entries
>>> i=h1.getUnderflow()  # underflow entries
>>> i=h1.getOverflow()   # overflow entries
```

All the methods above return integer numbers.
   Another useful characteristics is the histogram entropy. It is defined to be the
negation of the sum of the products of the probability associated with each bin with
the base-2 log of the probability. One can get the value of the entropy with the
method:

```
>>> print 'Entropy=',h1.getEntropy()
```

8.1.3 Histogram Initialization and Filling Methods

In this subsection we will consider the major histogram methods used for histogram
initialization and filling. Previously, it has been shown how to initialize a histogram
with fixed bin sizes. One can also create a histogram using a simpler constructor,
and then using a sequence of methods to set histogram characteristics:

```
>>> h1=H1D('Title')
>>> h1.setMin(min)
>>> h1.setMax(max)
>>> h1.setBins(bins)
```

which are used to set the minimum, maximum and the number of bins. These methods can also be useful to redefine these histogram characteristics after the histogram
was created using the usual approach.

One can also build a variable bin-size histogram by passing a list with the bin edges as shown in this example:

```
>>> bins=[0,10,100,1000]
>>> h1=H1D('Title',bins)
```

This creates a histogram with three bins. The bin edges are given by the input list. This constructor is handy when a data is represented by a bell-shaped or falling distribution; in this case it is important to increase the bin size in certain regions (tails) in order to reduce statistical fluctuations.

As we already know, to fill a histogram with numbers, use the method fill(d). More generally, one can assign a weight "w" to each value as

```
>>> h1.fill(d, w)
```

where 'w' is any arbitrary number representing a weight for a value 'd'. The original method fill(d) assumes that all weights are 1.

But why do we need weights? We have already discussed in Sect. 8.1.1 that the weights are useful to reduce the computational time when the expected final answer should be either a probability distribution or density distribution. There are also other cases when the weights are useful. We should note again that a histogram object stores the sum of all weights in each bin. This sum runs over the number of entries in a bin only when the weights are set to 1. Events may have smaller weights if they are relatively unimportant compared to other events. It is up to you to make this decision, this depends on a concrete situation.

The method fill(d) is slow in Jython when used inside loops, therefore, it is more efficient to fill a histogram at once using the method fill(list), where list is an array with the numbers passed from another program or file. As before, fill(list, wlist) can be used to fill a histogram from two lists. Each number in list has an appropriate weight given by the second argument.

Instead of Jython (or Java) lists, one can pass a POD array discussed in Sect. 4.2.3:

```
>>> h1.fill(p0d)
```

where p0d represents an object of the POD class.

Analogously, one can fill a histogram by passing a PND multidimensional array discussed in Sect. 6.3. This can be done again with the method fill(pnd), where pnd is an array with any size or dimension. One can specify also weights in the form of an additional PND object passed as a second argument to the method fill(pnd, w).

Histograms can be filled with weights which are inversely proportional to the bin size—as it was shown in the previous section, removing the bin size dependence is one of the most common operations:

```
>>> h1.fillInvBinSizeWeight(d)
```

It should be noted that this method works even when histograms have irregular binning.

Finally, one can set the bin contents (bin heights and their errors) from an external source as shown below:

```
>>> h1.setContents(values, errors)
>>> h1.setMeanAndRms(mean,rms)
```

where `values` and `errors` are input arrays. Together with the settings for the bin content, the second line of the above example shows how to set the global histogram characteristics, such as the mean and the standard deviation. There are more methods dealing with external arrays; advanced users can find appropriate methods in the API documentation of the class `H1D` or using the code assist.

8.1.4 Accessing Histogram Values

One-dimensional histograms based on the `H1D` class can easily be viewed using the following convenient methods designed for visual inspection:

`toString()`   convert a `H1D` histogram into a string
`print()`   print a histogram
`toTable()`   show a histogram as a table

Once we know that a histogram is initialized and filled, the next question is to access the histogram values. We will be brief in this section, since most methods are obvious. Table 8.1 shows the most important `H1D` methods.

It should be noted that the bin heights and the numbers of entries are the same when histogram weights used to fill histogram are set to one, i.e. when the method `fill(d)` is used.

Finally, one can view the `H1D` histograms using already known `toTable()` method. This method passes all histogram attributes to a pop-up table for easy visual inspection.

8.1.5 Integration

Histogram integration is similar to the `F1D` functions considered in the previous chapters: We simply sum up all bin heights. This can be done using the method `integral()`. More often, however, it is necessary to sum up heights in a certain bin region, say between a bin 'i1' and 'i2'. Then use this method:

Table 8.1 Several methods designed to access information about the H1D class. The following notations are used: "i" denotes an integer, "d" is a double value, "a" corresponds to a 1D array and "aa" is a 2D array

Methods	Returns	Definitions
allEntries()	i	number of all entries
entries()	i	number of entries in the range
getTitle()	text	get histogram title
mean()	d	mean of the histogram
rms()	d	RMS of histogram
sumAllBinHeights()	d	sum of all bin heights
getEntropy()	d	entropy of histogram
integral(i1, i2)	d	integrate between 'i1' and 'i2' bins
integralRegion(x1, x2)	d	integrate region [x1,x2]
extraEntries()	i	number of entries outside the range
getUnderflow()	i	underflow entries
getUnderflowHeight()	d	underflow heights
getOverflow()	i	overflow entries
getOverflowlowHeight()	d	overflow heights
getMin()	d	min value for bins
getMax()	d	max value for bins
getValues(i)	a	arrays with bins, heights and errors
		i=0 when bin means, i=1 for bin centers
getProbability()	H1D	probability distribution
getDensity()	H1D	probability density
maxBinHeight()	d	maximum bin height
minBinHeight()	d	minimum bin height
binCenter(i)	d	center of ith bin
binCenters()	a	bin centers
binEntries(i)	i	entries in i^{th} bin
binEntries()	a	entries in all bins
binError(i)	d	errors on entries in i^{th} bin
binErrors()	a	errors on all entries
binHeight(i)	d	height of ith bin
binHeights()	a	heights for all entries
binLowerEdge(i)	d	low edge of ith bin
binLowerEdges()	a	low edges for all entries
binUpperEdge(i)	d	upper edge of ith bin
binUpperEdges()	d	upper edges of all bins
binMean(i)	d	mean value in i^{th} bin
binRms(i)	a	RMS value in i^{th} bin
findBin(d)	i	find a bin number from a coordinate

```
>>> sum=h1.integral(i1,i2)
```

We should note that the integral is not just the number of events between these two bins: the summation is performed using the bin heights. However, if the weights for the method `fill()` are all set to one, then the integral is equivalent to the summation of numbers of events.

The integration shown above does include multiplication by a bin width. If one needs to calculate an integral over all bins by multiplying the bin content by the bin width (which can be either fixed or variable), use the method:

```
>>> sum=h1.integral(i1,i2,1)
```

where the last parameter should be set to 1 (or to 'true' in case of Java codding).

The next question is how to integrate a region in X by translating X-coordinates into the bin indexes. This can be done by calling the method `findBin(x)`, which returns an index of the bin corresponding to a coordinate X. One can call this method every time when you need to identify the bin index before calling the method `integrate()`. Alternatively, this can be done in one line as:

```
>>> sum=integralRegion(xmin,xmax,b)
```

The method returns a value of the integral between two coordinates, `xmin` and `xmax`. The bin content will be multiplied by the bin width if the boolean value `b` is set to 1 (boolean 'true' in Java).

8.1.6 Histogram Operations

Histograms can be added, subtracted, multiplied and divided. Assuming that we have filled two histograms, `h1` and `h2`, all operations can be done using the following generic method:

```
>>> h1.oper(h2,'NewTitle','operation')
```

where `'operation'` is a string which takes the following values: "+" (add), "−" (subtract), "*" (multiply) and "/" (divide). The operations are applied to the histogram `h1` using the histogram `h2` as an input. One can skip the string with a new title if one has to keep the same title as for the original histogram. In this case, the additive operation will look as `h1.oper(h1,'+')`

To create an exact copy of a histogram, use the method `copy()`. Previously, we have already discussed the `scale(d)` and `shift(d)` operations.

A histogram can be smoothed using the method:

```
>>> h1=h1.operSmooth(b,k)
```

This is done by averaging over a moving window. If $'b=1'$ then the bins will be weighted using a triangular weighting scheme favoring bins near the central bin ($'b=0'$ for the ordinary smoothing). One should specify the integer parameter $'k'$ which defines the window as '$2*k + 1$'. The smoothing may be weighted to favor the central value using a "triangular" weighting. For instance, for '$k=2$', the central bin would have weight $1/3$, the adjacent bins $2/9$, and the next adjacent bins $1/9$. For all these operations, errors are kept to be the same as for the original (non-smoothed) histogram.

One can also create a Gaussian smoothed version of a H1D histogram. Each band of the histogram is smoothed by a discrete convolution with a kernel approximating a Gaussian impulse response with the specified standard deviation.

```
>>> h2=h1.operSmoothGauss(rms)
```

where rms is a double value representing a standard deviation of the Gaussian smoothing kernel (must be non-negative).

One useful technique is histogram re-binning, i.e. when groups of bins are joined together. This approach could be used if statistics in bins is low; in this case, it makes sense to make bins larger in order to reduce relative statistical uncertainty for entries inside bins (we remind that in case of counting experiments, such uncertainty is $\sqrt{N}$, where N is a number of entries). The method which implements this operation is called rebin(group), where group defines how many bins should merged together. This method returns a new histogram with a smaller number of bins. However, there is one restriction: the method rebin cannot be used for histograms with non-constant bin sizes.

8.1.7 Accessing Low-level Jaida Classes

The H1D class is based on the two classes, IAxis and Histogram1D of the Jaida FreeHep library. Assuming h1 represents a H1D object, these two Jaida classes can be obtained as:

```
>>> a=h1.getAxis() # get IAxis object
>>> h=h1.get()     # get Histogram1D class
```

Both objects are rather useful. Although they do not contain graphical attributes, they have many methods for histogram manipulations, which are not present for the higher-level H1D class. The description of these Jaida classes is beyond the scope of this book. Please look at the Java documentation of these classes or use the code assist.

8.1.8 Graphical Attributes

Sometimes one has to spend a lot of typing and playing with various graphical options to present analysis results in an attractive and clear form. This is especially important when one needs to show several histograms inside a single canvas. jHep-Work provides many methods designed to draw histograms using different styles.

First of all, histograms can be shown either by lines (default) or by using symbols. For the default option (lines), one can consider either to fill histogram area or keep this area empty. The following methods below can be useful:

```
>>> h1.setFill(b)
>>> h1.setFillColor(c)
```

For the first method, Jython boolean 'b=1' means to fill the histogram, while 'b=0' (false) keeps the histogram empty. If the histogram area has to be filled, you may consider to select an appropriate color by specifying Java AWT Color object 'c'. How to find appropriate color has been discussed in Sect. 3.3.1.

Histograms can be shown using symbols as:

```
>>> h1.setStyle('p')
```

The style can be set back to the default style (histogram bars). This can be done by passing the string 'h' instead of 'p'. One can also use symbols connected by lines; in this case, use the character 'l' (draw lines) or the string 'lp' (draw lines and symbols).

Table 8.2 lists the most important graphical attributes of the H1D class. The graphical attributes can be retrieved back using similar methods after substituting the string "set" by "get" in the method names.

8.2 Histogram in 2D

A histogram in two dimensions (2D) is a direct extension of the 1D histogram discussed in the previous section. To initialize such histograms, one should define ranges and bins in X and Y. A 2D histogram can be visualized using the 3D canvas discussed before.

The 2D histograms are implemented using the H2D class, which is based on the JAIDA FreeHEP Histogram2D class [1]. To create a 2D histogram, one needs to

Table 8.2 Methods for graphical representation of the H1D histograms. The following notations are used: "i" means an integer, "d" means a double value, "b" corresponds to a boolean ("b=1" means Java true and "b=0" means false), "c" is Java AWT Color class

Methods	Definitions
setStyle("text")	"p"—show symbols, "l"—show lines
	"lp"—lines and symbols, "h"—as histogram bars
setPenWidthErr(i)	line width
setPenDash(i)	dashed line style,
	"i" is the length of dashed line
setColor(c)	set color for drawing
setFill(b)	b=1—fill histogram area (b=0 not fill)
setFillColor(c)	set AWT Java color for fill area
setFillColorTransparency(d)	set the transparency ($0 \leq d \leq 1$)
setErrX(b)	show or not errors on X
setErrY(b)	show or not errors on Y
setErrColorX(c)	set Java color for errors on X
setErrColorY(c)	set Java color for errors on Y
setSymbol(i)	symbol type: 0: circle;
	1: square; 2: diamond;
	3: triangle; 4: filled circle;
	5: filled square; 6: filed diamond;
	7: filled triangle; 8: plus (+);
	9: cross; 10: star; 11: small dot;
	12: bold plus;
setSymbolSize(i)	set symbol size "i"

define the number of bins for X and Y axis, and the minimum and the maximum values for each axis. Then the histogram can be initialized as:

```
>>> from jhplot import *
>>> h2= H2D('Title',binsX,minX,maxX,binsY,minY,maxY)
```

where binsX(binsY) is the number of bins for X (Y), minX(minY) and maxX(maxY) are the minimum and the maximum values for the X (Y) axis, respectively.

In addition to the fixed-bin-size case, one can create histograms with variable bin sizes. One should call the constructor below which allows to pass lists with the bin edges:

```
>>> from jhplot import *
>>> h2= H2D('Title',[1,2,3],[1,2,4,8])
```

This constructor shows how to define the bin edges in X (the first input list) and Y (the second input list).

To fill a 2D histogram, the method `fill(x,y)` should be used, where `'x'` and `'y'` values for the X and Y axis. It should be noted that the bin heights and the numbers of entries are the same when weights used to fill the histogram are set to one. Non-unity weights `'w'` can be specified in the method `fill(x,y,w)`.

Table 8.3 shows the main methods of the H2D histogram class. Unlike the H1D histogram, the H2D class has the setter and getter methods for each axis, X and Y. For example, `getMeanX()` returns the mean value in X, while `getMeanY()` returns the mean value in Y.

You may find that the methods given above are not enough for some complicated operations. We should remind that the H2D class is based on the Jaida classes, IAxis and Histogram2D, which are located in the Jaida package `hep.aida.ref.histogram.*`. As a consequence, one can build H2D histograms by creating the object of the class IAxis , which represents the axis in one dimension, and then pass it to the H2D constructor.

```
>>> from hep.aida.ref.histogram import *
>>> from jhplot import *
>>> xAx=FixedAxis(10,0.0,1.0)
>>> yAy=FixedAxis(20,0.0,1.0)
>>> h2=H2D('Title',xAx,yAy)
```

Again, as for the one-dimensional case, both Jaida classes can be obtained as:

```
>>> aX=h2.getAxisX() # get IAxis object for axis X
>>> aY=h2.getAxisY() # get IAxis object for axis Y
>>> h=h2.get()       # get Histogram2D class
```

assuming that h2 represents a H2D object. We will return to the Jaida histogram package in Sect. 8.3.

8.2.1 Histogram Operations

All histogram operations are exactly the same as for the H1D class discussed in Sect. 8.1.6. The H2D histograms can be added, subtracted, multiplied and divided using the generic method:

```
>>> h1=h1.oper(h2,'NewTitle','operation')
```

where `'operation'` is a string which can have the following values: "+" (add), "−" (subtract), "*" (multiply) and "/" (divide) histograms, and h1 and h2 are objects of the class H2D. The histograms can be scaled using the method `scale(d)`.

Table 8.3 Some methods of the H2D class. The table uses the following notations: "i" indicates an integer value, "d" means a double value, "a" corresponds to a 1D array, "aa" denotes a 2D array

Methods	Returns	Definitions
fill methods		
fill(x,y)	–	fill x and y
fill(x,y,w)	–	fill x and y with weight w
clear()	–	clean from all entries
getter methods		
copy()	H2D	exact copy
getMeanX()	d	get the mean value in X
getMeanY()	d	get the mean value in Y
getRmsX()	d	get RMS in X
getRmsY()	d	get RMS in Y
getMinX()	d	get min value in X
getMaxX()	d	get max value in Y
getMinY()	d	get min value in Y
getMaxY()	d	get max value in Y
getBinsX()	i	number of bins in X
getBinsY()	i	number of bins in Y
allEntries()	i	number of all entries
entries()	i	number of entries in the range
getTitle()	text	get histogram title
sumAllBinHeights()	d	sum of all bin heights
extraEntries()	i	number of entries outside the range
getUnderflowX()	i	underflow entries in x
getUnderflowHeightX()	d	underflow heights in x
getUnderflowY()	i	underflow entries in y
getUnderflowHeightY()	d	underflow heights in y
getOverflowX()	i	overflow entries in x
getOverflowlowHeightX()	d	overflow heights in x
getOverflowY()	i	overflow entries in y
getOverflowlowHeightY()	d	overflow heights in y
binEntries(ix,iy)	i	entries in bins (ix,iy) bin
binError(ix,iy)	d	errors on a (ix,iy) bin
binHeight(ix,iy)	d	height of a (ix,iy) bin
getDensity()	H2D	density distribution
getProbability()	H2D	probability distribution
integral(i1,i2,j1,j2)	d	integral in the range
integralRegion(d1,d2,d1,d2)	d	integral in the range (coordinates)

Table 8.3 (*Continued*)

Methods	Returns	Definitions
`setter` methods		
setContents(d[][], e[][])		set heights and errors (double arrays)
setBinError(ix,iy,d)		set the bin error for (ix,iy) bin
setMeanX(d)		set the mean value for X
setMeanY(d)		set the mean value for Y
setRmsX(d)		set RMS for X
setRmsY(d)		set RMS for Y

The probability and density distributions can be obtained as discussed in Sect. 8.1.1:

```
>>> h2=h1.getProbability()
>>> h2=h1.getDensity()
```

Finally, one can obtain the sum of the bin contents in the range defined in terms of the bin indexes or coordinate values using the methods shown in Sect. 8.1.5.

Finally, one can calculate an integral of histogram entries using the `integral()` method. A histogram region can be integrated by passing bin indexes.

8.2.2 Graphical Representation

The 2D histograms should be plotted using the three-dimensional canvas `HPlot3D` which has been used before to plot `F2D` functions. One can find more details in Sect. 3.4. The example below shows how to fill and display a `H2D` histogram:

```
———————————— Filling 2D histogram ————————————
from jhplot import *
from java.util import Random
c1 = HPlot3D('Canvas')
c1.visible()
c1.setNameX('X axis')
c1.setNameY('Y axis')
h1=H2D('2D Test',30,-3.0, 3.0, 30, -3.0, 3.0)
r=Random()
for i in range(1000):
    h1.fill(r.nextGaussian(),r.nextGaussian())
c1.draw(h1)
```

Here we fill a 2D histogram with Gaussian random numbers and then plot it using exactly the same `draw(obj)` method (with `obj` being an object for drawing) method as for the `F2D` functions.

Below we show a more complicated example. We will fill several histograms using Gaussian numbers after shifting the means for the sake of better illustration. In addition, we show how to: (1) plot two histograms on the same plot; (2) plot a histogram and a 2D function on the same plot. In both cases, we use the method draw(obj1,obj2), where obj1 and obj2 could be either H2D or F2D object.

```
———————————— Multiple 2D histograms ————————————
from java.awt import Color
from java.util import Random
from jhplot import *

c1  = HPlot3D('Canvas',600,700, 2,2)
c1.visible()
c1.setGTitle('H2D drawing options')

h1=H2D('H2D Test1',30,-4.5,4.5,30,-4.0, 4.0)
h2=H2D('H2D Test 2',30,-3.0, 3.0, 30, -3.0, 3.0)
f1=F2D('8*(x*x+y*y)', -3.0, 3.0, -3.0, 5.0)

r=Random()
for i in range(10000):
   h1.fill(r.nextGaussian(),0.5*r.nextGaussian())
   h2.fill(1+0.5*r.nextGaussian(),-2+0.5*r.nextGaussian())

c1.cd(1,1)
c1.setScaling(8)
c1.setRotationAngle(30)
c1.draw(h1)

c1.cd(1,2)
c1.setScaling(8)
c1.setColorMode(2)
c1.setRotationAngle(30)
c1.draw(h1,h2)

c1.cd(2,1)
c1.setColorMode(4)
c1.setLabelFontColor(Color.red)
c1.setScaling(8)
c1.setRotationAngle(40)
c1.draw(f1,h2)

c1.cd(2,2)
c1.setColorMode(1)
c1.setScaling(8)
c1.setElevationAngle(30)
c1.setRotationAngle(35)
c1.draw(h1)
```

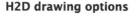

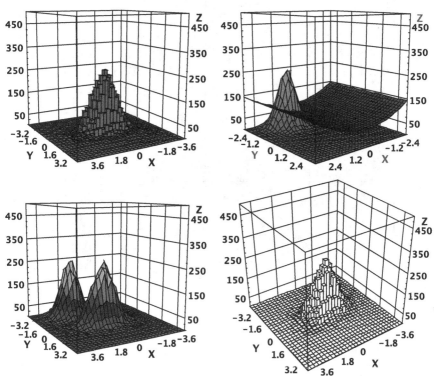

Fig. 8.2 H2D and F2D objects shown on the HPlot3D canvas using different styles

The resulting plots are shown in Fig. 8.2. One can see that the default drawing option for the H2D is histogram bars. The 2D functions are usually shown using a surface-type representation.

We should note that, at the moment when this book was written, there was no support for drawing 2D histograms with variable bin sizes.

The 2D histograms (as well as 2D functions) can be shown as a contour (or density) plot, an alternative graphical method when each region in X and Y is represented by different color, depending on the density population of each area with data points. To show such plot, one can use the method setContour(). The small code snippet below illustrates this:

```
―――――――――――――――― 2D Contour histogram ――――――――
from jhplot   import *
from java.util import Random

c1 = HPlot3D('Canvas',600,600)
c1.setNameX('X')
c1.setNameY('Y')
```

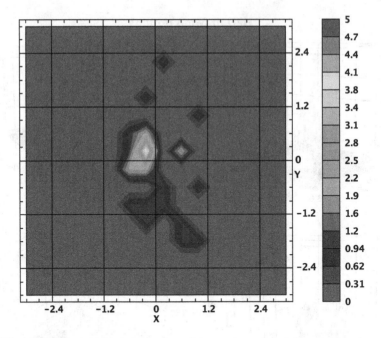

Fig. 8.3 A H2D histogram shown as a contour plot

```
c1.setContour()
c1.visible()

h1 = H2D('Contour',30,-3.0,3.0,30,-3.0,3.0)
rand = Random()
for i in range(100):
  h1.fill(0.5*rand.nextGaussian(),rand.nextGaussian())
c1.draw(h1)
```

The execution of this script brings up a window with the contour plot shown in Fig. 8.3.

As for the F2D and P1D objects, one can use the canvas HPlot2D for showing H2D histograms. This canvas class was specifically designed to show the density and contour plots as will be discussed in Sect. 10.11.

8.3 Histograms in Jaida

We have already mentioned that there is a way to access the so-called Jaida histograms from the FreeHep scientific library. Essentially, every histogram class in jHepWork is a derivative of the corresponding Jaida class.

The Jaida histograms can be created after importing the Java package hep.aida.ref.histogram. Generally, one can use the so-called Java factories to create the Jaida histograms. In this section, however, we will concentrate on a more basic histogram construction using Jaida.

Before building a Jaida histogram, first you have to define the "axis" object, which can either contain fixed or variable size bins. The example below shows how to create an one-dimensional Jaida histogram from the class `Histogram1D`:

```
>>> from hep.aida.ref.histogram import *
>>> ax=FixedAxis(bins,min,max)
>>> h1=Histogram1D('name','title',ax)
```

where 'bins' represents the number of bins, 'min' and 'max' is the minimum and the maximum value for the X range. The code above builds a 1D Jaida histogram using fixed-size bins. One should also specify the histogram name and its title. They may not need to be the same. The passed name is used for the internal Jaida purpose, so one can just set it to the histogram title.

Using the code assist, try to check the available methods of the object h1. You may notice that they are rather similar to those of the H1D class. The only difference is that the `Histogram1D` does not have any graphical attributes.

Analogously, one can build a Jaida histogram using a variable bin size. The code is essentially the same, with the only one difference: the axis should be replaced by the line with `axis=VariableAxis(edges)` with edges being an array with the bin edges.

A two-dimensional Jaida histogram, `Histogram2D` can be constructed in a similar way:

```
>>> from hep.aida.ref.histogram import *
>>> ax=FixedAxis(binsX,minX,maxX)
>>> ay=FixedAxis(binsY,minY,maxY)
>>> h1=Histogram2D('name','title',ax,ay)
```

Again, to make a 2D histogram with variable bin sizes, use the `VariableAxis()` instead of `FixedAxis`.

To add graphical attributes to these Jaida classes, we have to move them to the full-featured jHepWork histograms, H1D or H3D. One can do this rather easily by using these constructors:

```
>>> from hep.aida.ref.histogram import *
>>> from jhplot import *
>>> h1d=H1D(h1) # build H1D from Jaida histogram
>>> h2d=H2D(h2) # build H2D from Jaida histogram
```

One can retrieve the Jaida histograms back as:

```
>>> h1=h1d.get() # get Histogram1D object
>>> h2=h2d.get() # get Histogram2D object
```

But why do we need to use the Jaida histograms, if the jHepWork histograms are direct derivatives of the corresponding Jaida classes? The answer is simple: jHepWork does not map every single method of the Jaida histogram classes. In most cases, jHepWork histograms inherent only the most common methods (and add extra methods not present in Jaida). Therefore, if you find that jHepWork histograms do not have necessary methods, try to access the IAxes, Histogram1D and Histogram2D objects that may have the methods you need for your work.

8.4 Histogram in 3D

Histograms in three dimensions are a bit tricky. One cannot use them for a graphical representation, therefore, jHepWork does not add extra features compared to those present in the Jaida Histogram3D class.

In jHepWork, 3D histograms are implemented using the H3D class. In case of a fixed-size binning, these histograms can be defined by building three axes in X, Y and Z, and passing them to the histogram constructor:

```
>>> from jhplot import *
>>> from hep.aida.ref.histogram import *
>>> ax=FixedAxis(binsX,minX,maxX)
>>> ay=FixedAxis(binsY,minY,maxY)
>>> az=FixedAxis(binsZ,minZ,maxZ)
>>> h3=H3D('title', ax, ay, az)
```

The methods of the H3D class are rather similar to those discussed for H2D. The only difference is this: now we should take care of the additional Z axis. For example, to fill this histogram with weights, one needs to use the method fill(x,y,z). We remind again that this type of histograms does not have any graphical attributes—essentially, it is the exact mapping of the corresponding Histogram3D Jaida class.

8.5 Profile Histograms

A profile histogram is used to show the mean value in each bin of a second variable. The errors on the bin heights usually represent statistical uncertainties on the mean values or data spreads (i.e. standard deviations) of event distributions inside bins.

8.5.1 Profile Histograms in 1D

The profile histograms in one dimension are implemented in the HProf1D class. Such histograms are filled using the method fill(x,y), similar to the two-dimensional histograms. The first variable x is used for binning, while the second argument represents the variable for which the mean is calculated.

To show a profile histogram, use the method getH1D(). This method converts the profile histogram into the usual H1D discussed in Sect. 8.1, which then can be used for graphical representation. It can also accept a string to define a new title after the conversion. By default, errors on the mean values in each bin are shown by vertical lines. One can also display the mean values of y and their root-mean-square (RMS) deviations from the mean for each bin. The RMS values are shown as errors on the histogram heights when using the option 's' during the conversion with the method getH1D('title','s').

Below we calculate the mean values of a Gaussian distribution as a function of the second variable with uniform random numbers between zero and ten.

```
───────────────── Profile histogram ─────────────────
from jhplot import *
from java.util import Random

c1 = HPlot('Canvas')
c1.setGTitle('Profile histogram')
c1.setRange(0,11,1.5,2.5)
c1.setNameX("X")
c1.setNameY('Gaussian mean')
c1.visible()

h2=HProf1D('Profile1D',10,0.0, 11.0)
r=Random()
for i in range(2000):
   h2.fill(10*r.nextDouble(),r.nextGaussian()+2)

h1=h2.getH1D()
h1.setStyle('p')
c1.draw(h1)
```

The result of this script is shown in Fig. 8.4.

8.5.2 Profile Histograms in 2D

In contrast to the HProf1D histograms, the class HProf2D is designed to construct a profile histogram in two dimensions. As before, such histograms represent the mean of some distribution in each bin of two additional variables in X and Y.

Fig. 8.4 A profile histogram showing the mean values as a function of a second variable

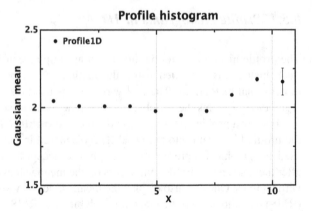

A HProf2D histogram can be created by specifying the number of bins in X and Y, as well as the minimum and maximum values for each axis. Alternatively, one can pass arrays with the bin edges, if the bin sizes should not be fixed to a constant value. The HProf2D histograms should be converted to H2D histograms (see Sect. 8.1) for graphical representation. The conversion can be done by either calling the method getH2D() (error on the mean in each bin) or getH2D('title','s') (errors correspond to RMS in each bin). Obviously, the HPlot3D canvas should be used for plotting.

Below we show a simple example of how to display the mean of a Gaussian distribution in two dimensions.

```
———————————— 2D profile histogram ————————————
from jhplot import *
from java.util import Random

c1 = HPlot3D('Canvas')
c1.setGTitle('Gaussian mean')
c1.setRange(0,10,0.0,5)
c1.setNameX('X')
c1.setNameY('Y')
c1.visible()

h2=HProf2D('Profile2D',10,0.0,10.0,10,0.0,5.0)
r=Random()
for i in range(5000):
   x=10*r.nextDouble()
   y=5*r.nextDouble()
   z=r.nextDouble()+2
   h2.fill(x,y,z)

c1.draw( h2.getH2D() )
```

8.6 Histogram Input and Output

All the histograms discussed above can be written into files and restored later. As example, one can use the method `toFile()` to write a histogram into a file. As for any jHepWork object, one can save collections of histograms using Jython lists, tuples or dictionaries. Alternatively, one can use the Java containers, such as arrays, sets or maps.

In case of Jython lists, one should always remember the order that was used to store histograms inside the lists. In case of Jython dictionaries or Java maps, one can use a human-readable description as the key for each histogram entry. Below we illustrate how to use a Jython dictionary to store several histograms (including their graphical attributes) in a serialized file: The code below writes a collection of histograms filled with random numbers into the file 'file.ser':

```
───────────────── Writing histograms ─────────────
from jhplot  import *
from jhplot.io import *
from java.awt import Color
from java.util import Random

hold = {} # define a dictionary

h1=H1D('Simple1',20,-2.0,2.0)
h1.setFill(1)
h1.setFillColor(Color.green)
h2=H2D('Simple2',20,-3.0, 3.0, 20, -3.0, 3.0)
h3=HProf1D('Profile1D',10,0.0, 11.0)

r=Random()
for i in range(1000):
  h1.fill(r.nextGaussian())
  h2.fill(r.nextDouble(),r.nextGaussian())
  h3.fill(10*r.nextDouble(),r.nextGaussian()+50)

# put all objects in a dictionary with description
hold['describe']='Collection of histograms'
hold['h1']=h1
hold['h2']=h2
hold['h3']=h3

# write the collection into a file
Serialized.write(hold,'file.ser')
```

Next, we will restore the histograms from the written file. First, we will read the dictionary and then will fetch all the histograms from the file using their keys:

```
─────────────────── Reading histograms ───────────────
from jhplot  import *
from jhplot.io import *

c1 = HPlot('Canvas')
c1.visible()
c1.setAutoRange()

# deserialize dictionary object
hold=Serialized.read('file.ser')

# print all keys
print hold.keys()
print 'Description: '+hold['describe']

# draw two 2D histograms
c1.draw(hold['h1'])
c1.draw(hold['h3'].getH1D())
```

In this example, we draw two histograms retrieved from the serialized file, H1D and HProf1D. The latter histogram has to be converted to a H1D object for visualization. We will remind that, in order to draw the H2D histogram, we will need a 3D drawing canvas, HPlot3D.

Serializations into XML files can be done using the writeXML() and readXML() methods from the same Serialization class. In addition, one can convert a histogram into a XML string using the toXML() method.

Finally, we recall that one can store and retrieve multiple objects, including jHep-Work histograms, using the class HFile. It is designed to work with a large sequence of any Java objects (see Chap. 11 for detail). In addition, one can use a GUI browser to look at all stored objects inside the files created with the class HFile and then plot them.

8.6.1 External Programs for Histograms

8.6.1.1 CFBook Package

Histograms can be filled by an external C++ or FORTRAN program. For this, use the CFBook [2] or CBook [3] packages. The first package can be used to fill histograms from a C++ or FORTRAN program, and write them into specially designed XML files (which are optimized for storing numerical data). The second package is based on compressed records constructed using the Google's Protocol Buffers. This approach will be discussed in Sect. 11.4.

The package CFBook package generates two static libraries: libcbook.a (to be linked with C++) or libfbook.a (to be linked with FORTRAN). Histograms filled by the CFBook library can be retrieved and visualized using the HBook class. Let us give one example. We assume that histograms are filled by a C++ external program and are kept in the 'cpp.xml'. One can read and retrieve histograms from this file as:

```
>>> from jhplot import *
>>>
>>> hb = HBook()          # create HBook instance
>>> hb.read('cpp.xml')    # read input XML file
>>> print hb.listH1D()    # list all histograms
>>> h1 = hb.getH1D(10)    # get H1D histogram with ID=10
>>> c1.draw(h1)           # draw it
```

In the code above, 'c1' represents an instance of the HPlot class.

The HBook class can also be used to save histograms or other objects into XML files. In this case, one should use its method write():

```
>>> from jhplot import *
>>>
>>> hb = HBook('hbook')
>>> h3=H1D('test',2,0.0,1.)
>>> h4=H2D('test',5,0.0,1.,4,0.,1.)
>>> hb.add(30,h3)         # add to HBook with ID=30
>>> hb.add(40,h3)         # add to HBook with ID=40
>>> p1=P1D('test")        # create P1D objects
>>> hb.add(10,p1)         # add to HBook with ID=10
>>> print hb.listH1D()    # list all histograms
>>> h1 = hb.getH1D(30)    # get H1D histogram with ID=30
>>> ...
>>> hb.write('out.xml')   # write to an XML file
```

This example illustrates that one can insert the histogram H1D, H2D or even the containers like P1D into the HBook holder using some identification numbers (10, 20, 30). These numbers can be used later for object retrieval. Also, we show how to write all such objects into an external XML file.

8.6.1.2 ROOT Package

jHepWork can also read histograms saved into ROOT files [4, 5]. One can read ROOT histograms using Jython scripts or using a GUI browser in order to navigate to a certain histogram object inside the ROOT files. These topics will be discussed in Sect. 11.5.1.

8.7 Real-life Example. Analyzing Histograms from Multiple Files

We will continue with the example discussed in Sect. 2.17. What we what to do now is more complicated than before: (1) We want to identify all files with the extension '.dat' and read them. The files contain numbers (one per line) and can be compressed. (2) Then we will analyze all numbers in all these files by putting them into a histogram. This histogram can be used to build statistical summaries of data, such as the number of entries, the mean value and the standard deviation.

While the described tasks are notably more complicated than those considered so far, our code will be at least a factor two smaller than that discussed[1] in Sect. 2.17. Below we show a Jython code which does the complete data analysis:

```
————————————— Reading histograms —————————

from jhplot  import *
from utils import *

c1 = HPlot('Analysis')
c1.visible()
c1.setAutoRange()

list=FileList.get('/home/','.dat$')
sum=POD('sum')
for file in list:
    p0=POD('data',file)
    sum=sum.merge(p0)
h1=H1D('test',100,0,200)
h1.fill(sum)
c1.draw(h1)
c1.drawStatBox(h1)
```

The execution of this script brings up a frame with the filled histogram. A box with calculated statistics, displayed after calling the method drawStatBox(), tells about all major statistical features of the data sample. Note that the method FileList was used to scan all files which have the extension '.dat'. Obviously, one can use rather sophisticated Java regular expressions to find a necessary patten in the file names. Also, it should be noted that we first created a POD object from ASCII files. If the input files are zipped (gzipped), use the methods readZip('name') or readGZip('name') instead. If you know that files are serialized using Java, replace the read methods with the method readSerialized('name').

In exactly the same fashion one can use various jHepWork containers, such as P1D, PND etc. To plot a particular slice of data (row or column), one can convert these objects into Java arrays, Jython lists or POD objects, which then can be passed to either H1D or H2D histograms.

---

[1]Below we will skip the part which was necessary to remove duplicate files, since this task described in Sect. 2.17 was used for an illustration only.

References

1. FreeHEP Java Libraries. URL http://java.freehep.org/
2. Chekanov, S.: CFBOOK histogram library. URL http://jwork.org/jhepwork/cbook
3. Chekanov, S.: CBOOK histogram library. URL http://jwork.org/jhepwork/cbook
4. Brun, R., Rademakers, F., Canal, P., Goto, M.: Root status and future developments. ECONF C0303241 (2003) MOJT001
5. Brun, R., Rademakers, F.: ROOT: An object oriented data analysis framework. Nucl. Instrum. Methods A **389**, 81 (1997). URL http://root.cern.ch/

Chapter 9
Random Numbers and Statistical Samples

A random number, i.e. a number chosen by chance from a specified distribution, is an essential concept for many scientific areas, especially for simulations of physical systems using Monte Carlo methods.

For a set with random numbers, no individual number can be predicted from knowledge of any other number or group of numbers. However, sequences of random numbers in a computer simulation eventually contain repeated numbers after generation of many millions of random numbers. Thus, it is only a good approximation to say that the numbers are random, and the definition "pseudo-random" is more appropriate.

Another notion which is usually associated with a sequence of random numbers is the so-called "seed" value. This is a number that controls whether the random number generator produces a new set of random numbers after the code execution or repeats a certain sequence.

For debugging of programs, it is often necessary to start generating exactly the same random number sequence every time you start the program. In this case, one should initialize a random number generator using the same seed number.

The seed must be changed for each run if you want to produce completely different sets of random numbers every time the program is executed. Usually, this can be done by generating a new seed using the current date and time, converted to an integer value.

9.1 Random Numbers in Jython

This section is going to be very short, since we have discussed this topic in Sect. 2.8. The standard Jython (as well as CPython) module that implements a random number generator is called 'random'. It must be imported using the usual statement import.

As before, it is advisable to use the standard jHepWork libraries to create arrays with random numbers where possible, instead of filling lists with random values using Jython loops. There are two reasons for this: (1) less chances that a mistake

S.V. Chekanov, *Scientific Data Analysis using Jython Scripting and Java*,
Advanced Information and Knowledge Processing,
DOI 10.1007/978-1-84996-287-2_10, © Springer-Verlag London Limited 2010

can be made; (2) programs based on the standard Java libraries are significantly faster. Look at the examples below: the first program is rather inefficient (and long), while the second code snippet is a factor five faster (and shorter). Both programs fill histograms with random numbers in the range between 0 and 100 and show them in a canvas.

```
─────────────── Jython random numbers ───────────────
from java.util import Random
from jhplot  import *
from time import clock

start=clock()
h1 = H1D('Uniform distribution',100, 0.0, 100.0)

rand = Random()
for i in range(2000000):
   h1.fill(100*rand.nextDouble())

c1 = HPlot('Canvas')
c1.visible()
c1.setAutoRange()

c1.draw(h1)
send=clock()
print 'Time elapsed = ',send-start,' seconds'
```

The program below does the same, but it is faster by a factor seven:

```
─────────────── jHepWork random numbers ───────────────
from jhplot  import *
from time import clock
from jhplot.math.StatisticSample import randUniform

start=clock()
# fill histogram with random numbers
h1 = H1D('Uniform distribution',100, 0.0, 100.0)
h1.fill( randUniform(2000000,0.0,100.0) )

c1 = HPlot('Canvas')
c1.visible()
c1.setAutoRange()
c1.draw(h1)
send=clock()
print 'Time elapsed = ', send-start,' seconds'
```

Here we recall that the method fill() of the class H1D accepts not only separate values, but also arrays of different type—in this case, we build such array on-fly using the method randUniform().

Below we will discuss in more detail how to use pre-build libraries from jHep-Work and third-party libraries. But first will discuss the most common classes to generate random numbers in Java.

9.2 Random Numbers in Java

Random numbers provided by Java API have already been used in the previous sections. Let us remind that the class Random can be used to generate a single random number. Below we check the methods of this class:

```
>>> from java.util import *
>>> r=Random()      # seed from the system time.
>>> r=Random(100L) # user defined seed=100L
>>> dir(r)
[.. 'nextDouble', 'nextFloat', 'nextGaussian',
...'nextInt', 'nextLong' ..]
```

In the first definition, the default seed comes from the computer system time. In the second example, we initiate the random sequence from an input value to obtain reproducible results for every program execution.

Below we describe the most common methods to generate random numbers: As usual, 'i' denotes an integer, 'l' represents a long integer, 'd' means a double value while 'b' corresponds to a boolean value.

i=r.nextInt(n)	random int ≥ 0 and $\leq n$
i=r.nextInt()	random int (full range)
l=r.nextLong()	random long (full range)
d=r.nextDouble()	random double ≥ 0.0 and ≤ 1.0
b=r.nextBoolean()	random boolean, true (1) or false (0)
d=r.nextGaussian()	double from a Gaussian distribution with mean 0.0 and standard deviation 1

To build a list containing random numbers, invoke a Jython loop. For example, this code typed using JythonShell builds a list with Gaussian random numbers:

```
>>> from java.util import *
>>> r=Random()
>>> g=[]
>>> for i in range(100):
>>> ... g.append(r.nextGaussian())
```

It was already discussed before that Jython loops are not particularly fast, therefore, it is recommended to use the predefined jHepWork methods to build lists. This is also not the only problem: using the predefined methods to build collections with random numbers grantees that the code is sufficiently short and free of errors.

9.3 Random Numbers from the Colt Package

The Colt package provides a comprehensive list of methods to create random numbers. The classes necessary to build random numbers come from the package cern.jet. As example, let us consider a generation of reproducible random numbers using the MersenneTwister class from the sub-package random.engine. The macro below creates an array POD with random numbers and then prints the statistical summary of a Gamma distribution:

```
————————————————— Colt random numbers ————————————————
from cern.jet.random.engine import *
from cern.jet.random import *
engine = MersenneTwister()
alpha=1
lamb=0.5
gam=Gamma(alpha,lamb,engine)
from jhplot import POD
p0=POD()
for i in range(100):
    p0.add(gam.nextDouble())
print p0.getStat()
```

Here we used the so-called "Mersenne-Twister" algorithm, which is one of the strongest uniform pseudo-random number generators. We did not specify any argument for the engine, therefore, the seed is set to a constant value and the output is totally reproducible. One can use the current system date for a seed to avoid reproducible random numbers:

```
>>> import java
>>> engine=MersenneTwister(new java.util.Date())
```

Learn about all possible methods of this package as usual:

```
>>> import cern.jet.random
>>> dir(cern.jet.random)
```

The above command prints the implemented distributions:

```
Beta, Binomial, BreitWigner, BreitWignerMeanSquare,
ChiSquare, Empirical, EmpiricalWalker, Exponential,
ExponentialPower, Gamma, Hyperbolic, HyperGeometric,
Logarithmic, NegativeBinomial, Normal, Poisson,
PoissonSlow, StudentT, Uniform, VonMises, Zeta
```

All these classes operate on a user supplied uniform random number generator.

Once you know which random number is necessary for your program, use the code assist or Java API documentation to learn more.

There is one special distribution you have to be aware of. One can generate random numbers from an array of predefined set of numbers given by some function. Such distribution is called "Empirical". The probability distribution function (PDF) should be provided as an array of positive numbers. The PDF does not need to be provided in the form of relative probabilities, absolute probabilities are also accepted. If LINEAR_INTERPOLATION constant is set, a linear interpolation within the bin is computed, resulting in a constant density within each bin. When NO_INTERPOLATION is passed, no interpolation is performed and the result is a discrete distribution.

```
——————————— Empirical PDF from the Colt library ———————
rom cern.jet.random.engine import *
from cern.jet.random import *
engine = MersenneTwister()
pdf=[1.,4.,4.,4.,3.,2.,1.,1.,1.]
enterpolation=Empirical.LINEAR_INTERPOLATION
em=Empirical(pdf,enterpolation,engine)
from jhplot import P0D
p0=P0D()
for i in range(100):
        p0.add(em.nextDouble())
print p0.getStat()
```

Look also at the class EmpiricalWalker which implements the so-called Walker's algorithm.

9.4 Random Numbers from the jhplot.math Package

jMathTool [1] classes further extend the Java random number generators. This library is included into the package 'jhplot.math'.

The example below shows how to generate random numbers using the jMathTool package:

```
>>> from jhplot.math.Random import *
>>> r=rand()              # random number between 0 and 1
>>> i=randInt(min,max) # integer in the range [min,max]
```

Below we will show other possible options:

uniform(min,max)   a random number between min and max.
dirac(d[], p[])   a random number from a discrete random variable, where d[] array with discrete values, and p[] is the probability of each value.

normal(m,s)	a random number from a Gaussian (Normal) distribution with the mean ('m') and the standard deviation ('s').
chi2(i)	a random number from a ξ^2 random variable with 'i' degrees of freedom.
logNormal(m,s)	a LogNormal random variable with the mean ('m') and standard deviation ('s').
exponential(lam)	a random number from an exponential distribution (mean = 1/lam, variance = 1/lam**2).
triangular(min,max)	a random number from a symmetric triangular distribution.
triangular(min,med,max)	a random number from a non-symmetric triangular distribution ("med" means a value of the random variable with max density).
beta(a,b)	a random number from a Beta distribution. 'a' and 'b' is the first and second parameter of the Beta random variable.
cauchy(med,s)	a random number from a Cauchy distribution (Mean = Inf, and Variance = Inf). 'med' is a median of the Weibull random variable, 's' is the second parameter of the Cauchy random variable.
weibull(lam,c)	a random number from a Weibull distribution. "lam" is the first parameter of the Weibull random variable, 'c' is the second parameter of the Weibull random variable.

Finally, one can generate a random number from an analytical function using the known rejection method. This requires building a F1D function first and then passing its parsed object to the rejection() method. Below we show how this can be done:

```
>>> from jhplot  import *
>>> from jhplot.math.Random import *
>>> f=F1D('x*exp(x)',1,2)
>>> p=f.getParse()
>>> print rejection(p,15,1,2)
1.4
```

The method rejection() takes three arguments: a parsed function, a maximum value of the function (15 in this case) and a minimum and a maximum value for the abscissa. The method returns a random number between 1 and 2, since these numbers have been specified in the rejection() method.

9.4.1 Apache Common Math Package

Random numbers can also be generated using the Apache common math package. A random generator can be initialized using the `RandomDataImpl()` class. Then, call its methods to generate a random number. The code below shows how to generate a single random number from different distributions:

```
>>> from org.apache.commons.math.random import *
>>> r=RandomDataImpl()
>>> dir(r)                      # check all methods
>>> d=r.nextUniform(0,1)        # a random number in [0,1]
>>> d=r.nextExponential(1)      # Exponential with the mean 1
>>> d=r.nextGaussian(0,1)       # Gaussian (mean=0 and sigma=1)
>>> d=r.nextPoisson(1)          # Poisson with mean 1
>>> s=r.nextHexString(10)       # hex string of length len
```

Check the corresponding API documentation for more options. One can reseed the random numbers using the `reSeed()` method (it sets the seed of the generator to the current computer time in milliseconds). One can also reseed the random number generator with the supplied seed using the method `reSeed(i)`, where `'i'` is an arbitrary integer number.

There are also the so-called "secure" methods, much are slower than those given above. A secure random sequence has the additional property that knowledge of values generated up to any point in the sequence does not make it any easier to predict subsequent values. Such values are useful for a cryptographically secure random sequence. As an example, check the method `nextSecureHexString(10)` which generates a "secure" random string of size ten.

9.5 Random Sampling

When one needs to create a large array with random numbers obeying some probability distribution, it is inconvenient (and slow!) to create data containers using Jython loops as shown at the beginning of this chapter. Instead, one should use the Java libraries implementing generations of arrays with random numbers.

We have already discussed in Sect. 4.1 how to build a `POD` and populate it with random numbers:

```
>>> from jhplot import *
>>> p0= P0D()
>>> p0.randomUniform(1000,0.0,1.0) # Uniform distribution
>>> p0.randomNormal(1000,0.0,1.0)  # Gaussian distribution
```

In both cases we will fill an array with 1000 numbers populated by either a uniform random generator (in the range [0–1]) or a Gaussian generator with the mean zero, and with the standard deviation equals unity.

Below we will consider more options that specify how to fill the POD with custom random numbers.

9.5.1 Methods for 1D Arrays from jhplot.math

Static methods of the package 'jhplot.math' can also be useful to generate arrays with random numbers. We remind that the package is based on the original class of jMathTools [1].

We will start from a simple example of how to create an array of size 10 with random integer numbers distributed between 0 and 100. After creation, we print out the array:

```
>>> from jhplot.math.StatisticSample import *
>>> a=randomInt(10,0,100)
>>> print a
array('i',[93, 19, 70, 36, 55, 43, 52, 50, 67, 38])
```

Analogously, randUniform(N,min,max) generates 'N' random numbers uniformly distributed between 'min' and 'max'.

Below we show a list of methods designed to generate arrays with random numbers:

randomDirac(N,d[],p[])	a random array from a discrete random variable, where d[] array with discrete values and p[] is the probability of each value.
randomNormal(N,m,s)	a Gaussian (Normal) random numbers with the mean ("m") and standard deviation ("s").
randomChi2(N,i)	a random array with ξ^2 random numbers with 'i' degrees of freedom.
randomLogNormal(N,m,s)	a random array with a LogNormal random numbers.
randomExponential(N,lam)	an array with an exponential random variable (mean $= 1/$lam, variance $= 1/$lam^2).
randomTriangular(N,m1,m2)	an array with symmetric triangular random variable in the range [m1,m2].
randomTriangular(N,m1,m,m2)	an array from a non-symmetric triangular random distribution in the range [m1,m2] ('m' indicates a value of the random variable with the maximal density).

`randomBeta(N,a,b)`	an array with Beta distribution, "a" and "b" are first and second parameter of the Beta random variable.
`randomCauchy(N,m,s)`	an array from a Cauchy random distribution (Mean = Inf, and Variance = Inf). 'm' is the median of the Weibull random variable, 's' is second parameter of the Cauchy random variable.
`randomWeibull(N,lam,c)`	an array with Weibull random variable. 'lam' is the first parameter of the Weibull random variable ('lam'), 'c' is the second parameter of the Weibull random variable.

In all cases, the names of the methods are exactly the same as those shown in Sect. 9.4. The only difference now is in the extra argument "N" which specifies how many random values should be generated in the output arrays.

Obviously, once a sequence of random numbers is generated, the next step would be to verify it. This can be done as following:

- Convert the random array into a `POD` object;
- Display accumulated statistics with the method `getStat()` or convert it to a `H1D` histogram for visualization.

As example, let us generate an array with log-normal distribution, print statistics and plot log-normal distribution in a form of histogram:

```
                        ──── Checking random numbers ────

from jhplot   import *
from jhplot.math.StatisticSample import *

a=randomLogNormal(1000,0,10)
p0=POD(a)
print p0.getStat()

h=H1D('LogNormal',40,-50,50)
h.fill(a)

c1 = HPlot('Show histogram')
c1.setGTitle('LogNormal')
c1.visible()
c1.setAutoRange()
c1.draw(h)
```

By running this script you will see a very detailed statistical summary of the log-normal distribution, plus a histogram with random numbers from this distribution.

9.5.2 Methods for 2D Arrays from jhplot.math

The generation of 2D arrays (i.e. matrices) is rather straightforward. We will continue with the above example used to generate a random 1D array; This time, however, we will add an extra argument representing the number of rows in the matrix. This time our code snippet creates a matrix 3×2 with random integer numbers from 0 to 100:

```
>>> from jhplot.math.StatisticSample import *
>>> a=randomInt(3,2,0,100)
>>> print a
array([I, [array('i', [79, 92]), array('i', [78, 81]),
             array('i', [92, 72])])])
```

We should point out that all methods to generate random 2D arrays have the same names and the meaning as for the 1D case. The only difference now is that all such methods have an additional argument representing the number of rows.

Now let us consider how to build arrays of random values in accordance with a functional form. Below we will give an example which: (1) shows how to generate a vector with 1000 random numbers distributed between 1 and 2 using the analytic function $x * \exp(x)$; (2) fill a histogram with such numbers and plot them together with the function on the same canvas

```
———————— Plotting random arrays with arbitrary PDF ————
from jhplot  import *
from jhplot.math.StatisticSample  import *

c1 = HPlot('Canvas',600,400)
c1.setGTitle('Title')
c1.visible()
c1.setAutoRange()

f=F1D('x*exp(x)',1,2)
c1.draw(f)

p=f.getParse()
a=randomRejection(1000,p,15,1,2)
h=H1D('x*exp(x)',100,1,2)
h.fill(a)
c1.draw(h)
```

Analogously, one can build a 2D array by adding one extra argument to the `randomRejection()` method representing the number of columns.

9.6 Sampling Using the Colt Package

We have already considered how to generate separate random numbers using the Colt random engine in Sect. 9.3. Now we will learn how to use this package to create 1D and 2D random arrays.

First, let us discuss how to fill a POD container with random numbers using the Colt package. Consider the code below:

```
———————————— Statistical summary of the Binomial PDF ————————

from cern.jet.random.engine import *
from cern.jet.random import *
from jhplot import *
engine=MersenneTwister()
n,p=10,0.5
gam=Binomial(n,p,engine)
a=POD()
a.random(100,gam)
print a.getStat() # print statistics
```

If you need to look at the actual numbers, just append the line 'print a.toString(). The method applied to populate the POD with random numbers is rather powerful since many predefined distributions are available. In addition, one can build an empirical distribution without making any assumptions about the functional form of the population distribution that the data come from. See the class Empirical() discussed in Sect. 9.3 for detail.

Analogously, one can fill the native Colt array called DoubleArrayList. This is exactly what we will do in the next example: we change the random generator engine, fix the seed value to 99 and then create an array with the random numbers:

```
———————————— Generating a Gamma PDF ————————

from cern.jet.random.engine import *
from cern.jet.random import *
from cern.colt.list import *
engine=DRand(99)
alpha,lamb = 1, 0.5
gam=Gamma(alpha,lamb,engine)
a=DoubleArrayList()
a.fillRandom(100,gam)
print a
```

Similarly, one can populate the array IntArrayList with integer random numbers.

References

1. jMathTools Java Libraries. URL http://jmathtools.berlios.de

Chapter 10
Graphical Canvases

In this chapter we will step back from the numerical computations and discuss somewhat technical issues about how to present numerical results in a graphical form, what graphical canvas should be used to plot data points, histograms, functions and other objects.

The task of choosing the right graphical canvas may look daunting at first. It is further complicated by large choice of canvases and large number of objects which have to be visualized for numerical calculations. Below we summarize the canvases included to jHepWork:

HPlot   2D canvas and contour plots for `P1D`, `F1D`, `H1D` and other graphical primitives;

HPlotJa   2D canvas with interactive editor for drawing diagrams. Support for the `H1D`, `P1D`, `F1D`, graphical primitives and Feynman diagrams;

SPlot   a light-weight 2D canvas, supports `H1D`, `P1D` and arrays;

HPlot2D   contour (or density) plots in 2D for classes `P1D`, `H2D` and `F2D`;

HPlot3D   interactive 3D plots, `P2D`, `P3D`, `H2D` and `F2D`;

HPlot3DP   interactive 3D surfaces for parametric and non-parametric functions (`FPR`);

HChart   2D charts. Support for `P1D`, $X-Y$ charts, area, bar, histogram, pie charts;

HGraph   interactive interconnected graphs.

Usually, all such canvases originate from different base classes and created for different visualization tasks. Most of these canvas are also implemented as Java singleton classes (see our discussion later). In the following sections we will describe these canvases and help to identify the most appropriate canvas for representation of your results.

We should remind that jHepWork is a graphics-intensive program, and the burden of plotting graphs on various canvas on the CPU could be immense. The advise is to build graphical canvas after performing all CPU-consuming numerical calculations. In this book, we do not always use this rule since our examples are not too CPU consuming. But, in real situations, one should consider to reorganize analysis codes such that creation of graphical canvas goes after numerical calculations.

S.V. Chekanov, *Scientific Data Analysis using Jython Scripting and Java,*
Advanced Information and Knowledge Processing,
DOI 10.1007/978-1-84996-287-2_11, © Springer-Verlag London Limited 2010

10.1 HPlot Canvas

In Sect. 3.3.1 we have already discussed the HPlot class. We will remind that it can be used to show F1D, P1D and H1D objects on X–Y plane. In this section we will discuss this canvas in more detail.

First, we will remind how to build this canvas and make it visible:

```
>>> from jhplot import HPlot
>>> c1=HPlot('Canvas')
>>> c1.visible(1)
```

If the canvas should not be shown, use the method visible(0) for Jython—in this case it will be created in the computer memory. We remind that, Jython "1" means Java 'true' and "0" corresponds to Java 'false'. One can use also the method visible() instead of visible(1).

The size of the canvas on the screen can be customized as:

```
>>> c1=HPlot('Canvas',600,400)
```

This creates an initial canvas of the size 600 by 400 pixels (explicitly defined). One can resize the canvas frame later by dragging the edges of the canvas frame with the mouse. Most objects, such as titles, labels, symbols, lines etc. should be resized proportional to the canvas size.

The constructor:

```
>>> c1=HPlot('Canvas',600,400,1,2)
```

also creates a canvas of the size 600 by 400 pixels. In addition, the last two numbers tell that two plot regions (or pads) will be created inside the same canvas. If the last two numbers are, say 3×2, then 6 pads are created (3 pads are in X direction and 2 pads are in Y direction). One can navigate to the current pad using the cd(i1,i2) method, where i1 is the location of the pad in X, and i2 is the location in Y. For example, if one needs to plot a jHepWork object obj on the first pad, use:

```
>>> c1.cd(1,1)
>>> c1.draw(obj)
```

where obj could either be F1D, P1D or H1D. It should be also noted that one can plot a list of objects at once:

```
>>> a=[]
>>> a.append(f1) # add first  F1D
```

```
>>> a.append(f2) # add second F1D
>>> c1.draw(a)   # draw all functions in the list
```

One can navigate to the second pad using the method cd(). For example, if an object obj should be shown on the second pad, use

```
>>> c1.cd(1,2)
>>> c1.draw(obj)
```

By default, the HPlot canvas has the range between 0 and 1 for the X or Y axis. One should specify a necessary range using the method

```
>>> c1.setRange(Xmin,Xmax,Ymin,Ymax)
```

where Xmin (Xmax) and Ymin (Ymax) are the minimum (maximum) ranges for X and Y axes. Alternatively, one can set "auto"-range mode using the method setAutoRange(). After calling this method, the minimum and maximum values for the X-axis range will be determined automatically.

One can change color for many attributes of the canvas, as well as to annotate the canvas. First, one should import the classes Color or Font from the Java AWT library. How to use the class Color and Font has been discusses in Sect. 3.3.1. How to set user-defined annotations is shown in the example below:

```
>>> from java.awt import Color
>>> c1.setGTitle('Global canvas title',Color.red)
>>> c1.setNameX('X axis title')
>>> c1.setNameY('Y axis title')
>>> c1.setName('Current pad title')
>>> c1.visible()
>>> c.setAutoRange()
```

All the statements above are self-explanatory. One may add a background color to the canvas as:

```
>>> c1.setBackgroundColor(Color.yellow)
```

or one can specify custom fonts for the legends as:

```
>>> from java.awt import Font
>>> font=Font('Lucida Sans',Font.BOLD, 12)
>>> c1.setLegendFont(font)
```

Read the Java documentation of the HPlot class [1].

Finally, you can edit pad titles and titles of the axes using the mouse. Simply double click on the area with the text. We will discuss this in more detail in the following section.

10.2 Working with the HPlot Canvas

There are several important operations you should know while working with the HPlot canvas frame:

10.2.1 Find USER or NDC Coordinators

To determine the coordinate of the mouse in the USER coordinate system (i.e. determined by the range of X or Y axis) or the NDC (a user-independent, given by the actual size of the canvas on the screen) coordinate system, click on the middle mouse button. The mouse pointer should be located *inside* the area with the drawing pad. The located coordinates will be displayed at the bottom of the canvas frame.

10.2.2 Zoom in to a Certain Region

To zoom in, use the middle mouse button. The mouse location should be *below* the X-axis or on the *left* side of the Y-axis. Drag the mouse holding the middle button. A red line indicating a zoom region in X (or Y) will be shown. After releasing the middle mouse button, you will see an updated HPlot canvas with the new axis range. To set the axis range to the default, use the right mouse button and select a pop-up menu "Default axis range".

One can also click-and-drag the mouse pointer to create a zoom rectangle. To perform a zoom to rectangle, press the middle mouse button, hold, and drag the mouse to draw a rectangle. Release the middle mouse button to complete the rectangle.

10.2.3 How to Change Titles, Legends and Labels

If one needs to change titles, legends and labels, select the appropriate object and double click on the right mouse button. This brings up a window with all graph settings. One can change the location of a selected object by dragging it while holding the mouse button.

10.2.4 Edit Style of Data Presentation

Click on the right mouse button. You should see a pop-up menu with several sub-menu options. Click on "Edit" and this will bring up a new window which allows to make necessary changes. In particular, one can change:

- axis ranges, set auto-range, logarithmic or linear scales;
- ticks size, numbers of ticks, colors, axis colors etc.;
- labels, legends, title, the names of X and Y ranges;
- the presentation style of data points or a histograms. One can select the fill style, points, lines, colors for points. One can also remove (or add) statistical or systematical error bars. Refresh the canvas to update drawn graphics.

Of course, all these attributes can be changed using Jython macros. Read the API documentation of the HPlot class. We remind that one can access API documentation by calling the method doc() of the HPlot class (or any class).

10.2.5 How to Modify the Global Margins

To change the global title, which can be set using the setGTitle() method, navigate the mouse to the title location at the very top of the frame, and click on the right mouse button. A new pop-up menu will appear. Using this menu, one can increase or decrease the divider location, make the divider invisible, change fonts and colors.

One can edit left, right and bottom margins of the main frame using exactly the same approach: navigate to the frame border and use the mouse pop-up menu. One can access all attributes of the margins using this method:

```
>>> c1.panel()
```

This method returns the GHPanel class (an extension of the swing JPanel) which keeps attributes of all four margins. Here are several operations associated with this class (all sizes are given in pixels):

```
>>> # set margins as some location
>>> c1.setMarginSizeLeft(50)
>>> c1.setMarginSizeRight(20)
>>> c1.setMarginSizeBottom(50)
>>> c1.setMarginSizeTop(50)
>>>
>>> # get the margin size
>>> a=c1.getMarginSizeLeft()
>>> a=c1.getMarginSizeRight()
>>> a=c1.getMarginSizeBottom()
```

```
>>> a=c1.getMarginSizeTop()
>>>
>>> # set text of the global margins
>>> c1.setTextBottom('This is X')
>>> c1.setTextLeft('This is Y')
>>> c1.setTextTop('This is X')    # same as setGTitle()
>>> c1.setTextBottom('This is Y')
>>> # set global margin background
>>> c1.setMarginBackground(Color color)
>>> # returns JPanel of the global margins
>>> c1.getMarginPanelLeft()
>>> c1.getMarginPanelRight()
>>> c1.getMarginPanelTop()
>>> c1.getMarginPanelBottom()
```

Look at the jHPlot API documentation for more details.

10.2.6 Saving Plots in XML Files

One can save all drawings shown on the HPlot canvas into external files using the File menu. The plots can be restored later using the same menu. This works for most of the plotted objects, like HLabel, Primitives and other graphics attributes to be discussed below.

The output files have the extension *.jhp and contain a XML file which keeps all attributes of the HPlot canvas and the data files necessary to recreate figures without running Jython macros. Look inside of this file after unzipping it (below we show how to do this using a Linux/UNIX prompt):

```
>>> unzip [file].jhp
```

This creates the directory with the name [file] with [file].xml, where [file] is the specified file name and several data files. The general form of the file names is: plotXY-dataZ.d, where X are Y are the positions of the pads inside the HPlot canvas and Z indicates the data set number.

It should be noted that all data files are just the outputs from P1D objects (see Chap. 5). Therefore, one can easily read such files using the methods of the P1D class. This is useful in case if the automatic procedure from the File menu fails, and the user wants to re-plot the data using a different macro.

10.2.7 Reading Data

One can open a serialized (the extension ser) or PFile (the extension pbu) file for browsing Java objects stored inside this file using the menu [File] and [Read data] of the HPlot canvas. This file should be created using the class HFile as discussed in Sect. 11.3 (using the default compression). Select a file with the extension "ser" or "pbu" and open it. You will see a dialog with all objects inside the file. Select an object and click on the "Plot" button. If the object can be plotted, you will see it inside the canvas. Most objects can be visualized on the HPlot canvas, such as P0D, P1D, H1D histograms, functions and Java strings. The latter are converted into a HLabel object on the fly for drawing inside the HPlot canvas. In addition, one can store GUI dialogs based on the JFrame class of the Java swing library.

We remind that the browser is based on the HFileBrowser and PFileBrowser classes which can be called from the scrips. See Sects. 11.3.2 and 11.4.

10.2.8 Cleaning the HPlot Canvas from Graphics

All graphs on the same HPlot canvas can be removed using several methods. If one needs to clean canvas from plotted objects (histogram, function etc.), use the method cleanData(). Note: in this case, only the current plot defined by the cd(i1,i2) method will be updated. If one needs to remove all objects from all plots on the same HPlot canvas, use cleanAllData().

It is also useful in many cases to remove all user settings from a certain graph, as well as to remove input objects. In this case, use the method clear(). One can also use the method clear(i1,i2) to remove graphics on any arbitrary pad, since i1 and i2 specify the pad location. The method clearAll() removes drawings on all pads, but keeps the main canvas frame untouched. The method close() removes the canvas frame and disposes the frame object.

10.2.9 Axes

The axis range can be set automatically by calling:

```
>>> c1.setAutoRange()
```

This method tells that the canvas determines the X and Y ranges automatically from the objects passed to the method draw(obj).

A user can specify the X–Y range manually by calling the method:

```
>>> c1.setRange(xMin,xMax,yMin,yMax)
```

One can remove all drawn axes using:

```
>>> c1.removeAxes()
```

If only one axis should be drawn instead of all four, first remove all axes and then call the method:

```
>>> c1.setAxisY() # show only Y axis
>>> c1.setAxisX() # show only X axis
```

One can draw a small arrow at the end of axes as:

```
>>> c1.removeAxes()
>>> c1.setAxisX()
>>> c1.setAxisArrow(1)  #   arrow type 1
>>> c1.setAxisArrow(2)  #   another arrow type 2
```

If no mirror axes should be drawn, use these methods:

```
>>> c1.setAxisMirror(0,0) # no mirror axis on X
>>> c1.setAxisMirror(1,0) # no mirror axis on Y
```

If no ticks should be drawn, use:

```
>>> c1.setTickLabels(0,0) # no mirror axis on X
>>> c1.setTickLabels(1,0) # no mirror axis on Y
```

Finally, call the method update() to redraw the canvas.

10.2.10 Summary of the HPlot Methods

Table 10.1 shows the major methods of the HPlot class. This list is incomplete, therefore, use the code assist or the Java API documentation to find more methods.

10.2.11 Saving Drawings in an Image File

One can export a graph shown on the HPlot canvas (including all its pads) into an image using the method export('FileName'), where 'FileName' is a string representing the file name. With this method, the graphs can be saved to a variety of vector graphics formats as well as bitmap image formats.

The export() statement should always be at the end of your script, when all objects have been drawn with the draw(obj) method or after the update() statement.

In the example below:

Table 10.1 The main methods of the HPlot class. The following notations are used: 'i' denotes an integer value, 'd' means a double value, 'b' corresponds to a boolean ('b=1' means true and 'b=0' means false), 'c' is Java AWT color, 'f' is the Java AWT class. Finally, 'axis' defines axis: 'axis = 0' for X and 'axis = 1' for Y

Methods	Definitions
visible(b)	makes it visible (b=1) or not (b=0)
visible()	sets to visible
setAutoRange()	sets auto-range
cd(i1,i2)	navigates to a specific region (i1 × i2)
update()	updates a region defined by cd(i1,i2)
updateAll()	updates all regions
drawStatBox(H1D)	draws statistical box for a histogram
setMargineTitle(i)	defines the region size for the global title
showMargineTitle(b)	do not show the global title
setGTitle(string,f,c)	sets the global title with Font and Color
viewHisto(b)	shows Y starting from 0 (for histograms)
setLegendFont(f)	sets the legend font
setLegend(b)	draws the legend when b=1 (if b=0, do not draw it)
setLegendPosition(axis,d)	sets legend for axis
setLogScale(axis,b)	sets log scale (b=1) or not (b=0) for axis
setTicsMirror(axis,b)	sets ticks (b=1) or not (b=0) for axis
setGrid(axis,b)	shows grid (b=1) or not (b=0) for axis
setGridColor(c)	sets grid color
setGridToFront(b)	grid in front of drawing (b=1) or not (b=0)
setBox(b)	bounding box around the graph (b=1) or not (b=0)
setBoxOffset(d)	offset of the bounding box
setBoxFillColor(c)	fill color of the bounding box
setBoxColor(c)	color of the bounding box
setBackgroundColor(c)	background color of the graph
setRange(axis,min,max)	set the range for axis (axis = 0,1)
setRange(minX,maxX,minY,maxY)	ranges for X and Y
setAutoRange(axis, b)	sets auto-range for axis
setAutoRange(b)	sets auto-range for X and Y
setLabel(string,f,c)	sets a label at random position

```
>>> c1.export('file.ps')
```

we export drawings on the canvas HPlot (c1 represents its instance) into a PostScript image file. One can also export it into PNG, JPEG, EPS, PDF etc. formats using the appropriate extension for the output file name. Here are more examples:

```
>>> c1.export('file.eps') # create an EPS file
>>> c1.export('file.png') # create an PNG file
>>> c1.export('file.jpg') # create  a JPG file
>>> c1.export('file.pdf') # create  a PDF file
>>> c1.export('file.svg') # create  a SVG file
>>> ...
```

If you are not too sure which extension to use, look at the [File]-[Export] menu which can give you some ideas about the supported graphics formats. One can use this menu for exporting graphs into images without calling the method export() inside your scripts.

It is also useful to create an image file using the same name as that of your script, and in the same directory where the script itself is located. In this case, type:

```
>>> c1.export(Editor.DocMasterName()+'.ps')
```

where the method Editor.DocMasterName() accesses the file name of the currently opened script.

It is also possible to save the HPlot canvas to an image file with a pop-up dialog. One should use the method exportDialog(file) for this task.

10.3 Labels and Keys

10.3.1 Simple Text Labels

Labels can be shown on the HPlot canvas using the Text class. It is impossible to interact with such simple labels using the mouse, since this class is based on the standard Java 2D graphics. However, due to a low memory consumption, such labels can be rather useful. The Text class is located in the Java package jhplot.shapes. The example below shows how to access such labels:

```
———————————— Text label example ————————————
from jhplot.shapes import Text
from jhplot import *
c1=HPlot('Canvas with a text')
c1.visible()
lab=Text('Label in USER system', 0.5, 0.2)
c1.add(lab)
c1.update()
```

You may notice that, instead of the draw() method, we use the add() and update() methods. This could be rather handy since now we can add many objects to the same canvas and then trigger update of the canvas to display all added objects at once. The text label will be drawn in the USER coordinate system at $X = 0.5$ and $Y = 0.2$. For the NDC system, use the method setPosCoord('NDC').

```
>>> lab=Text('Text in the NDC system', 0.5, 0.2)
>>> lab.setPosCoord('NDC')
>>> lab=Text('Text in USER system', 0.5, 0.2)
>>> lab.setPosCoord('USER')
>>> c1.add(lab) # add to HPlot canvas
>>> c1.update() # update the canvas
```

As before, one can set the text fonts, color and transparency level using setFont(f) and setColor(c) methods, where 'f' and 'c' are Java AWT Font and Color classes, respectively. The transparency level can be set using the setTransparency(d) method ($0 < d < 1$, with $d = 1$ for completely transparent objects).

10.3.2 Interactive Labels

Once a HPlot is initialized, one can insert an interactive text label, which is significantly more memory consuming object than objects created using the Text class. Such labels are created using the HLabel class. One can drag this label using the mouse, adjust its position and edit the text using a GUI dialog after double clicking on the label text. As before, use the method add(obj), and then make it visible by calling the update() method. Here is a typical example:

```
>>> lab=HLabel('HLabel in NDC', 0.5, 0.2)
>>> c1.add(lab) # add it to HPlot object
>>> c1.update() # trigger update
```

In the code above, the HLabel object is inserted at the position 0.5 and 0.2 in the USER coordinate system of the HPlot canvas. One can observe this by clicking on the middle mouse button and by looking at text message at the bottom of the HPlot frame. The status panel at the bottom of the frame should indicate the mouse position in the USER system.

Alternatively, one can set the label location in the user-independent coordinate system ("NDC"). We remind that this coordinate system is independent of the window size and is defined with two numbers in the range from 0 to 1. Again, one can study this coordinate system by clicking on the middle mouse button.

The same label in the NDC system should be created as:

```
>>> lab=HLabel('HLabel in NDC',0.5,0.2,'NDC')
```

One can modify the label attributes using the usual `setFont(f)`, as well as `setColor(c)` method.

The position of the label can be adjusted using the mouse. A double click on the label brings up a label property window.

One can also show a multi-line interactive label on the `HPlot` canvas using the `HMLabel` class. It is very similar to `HLabel`, however, instead of a string, it takes a list of strings. Each element of such list will be shown on a new line. We will show a relevant example in Sect. 10.5.

It should be noted that to show a legend, global title or title for an axis, the `HLabel` class is not necessary; one should use the special methods of the `HPlot` canvas, such as `setGTitle()`, `setNameX()` and `setNameY()`.

10.3.3 Interactive Text Labels with Keys

Unlike the `HLabel` class, the `HKey` class creates an interactive label with a text and a key describing the shown data. It should be noted that it behaves differently than the legend which is automatically shown with the corresponding data set. The `HKey` object is not related to any data set and can be shown even if no data are plotted. This class is rather similar to `HLabel` and has all the methods which the `HLabel` class has. To make it visible, call the `update()` method. Here is a typical example using the Jython shell:

```
>>> ... create c1 canvas
>>> h1 =HKey('key type=32',55,62) # key at x=55 and y=62
>>> h1.setKey(32,7.0,Color.blue)  # key of type 32,size=7
>>> h1.setKeySpace(4.0)    # space between a key and text
>>> c1.add(h1)
>>> c1.update()
```

Various key types are shown in Fig. 10.1. This figure was generated by the script shown below:

```
─────────── Keys and their descriptions ───────────
from java.awt import Font,Color
from jhplot  import  *

c1=HPlot('Canvas',600,550)
c1.visible()
c1.setGridAll(0,0)
c1.setGridAll(1,0)
c1.setGTitle('HKey types')
c1.removeAxes()
c1.setRange(0,100,0,100)

for j in range(1,13):
```

```
        title='key type='+str(j)
        hh=HKey(title,15,97-7*j)
        c=Color(0,65 + j*10,0)
        hh.setKey(j,2.0,c)
        hh.setKeySpace(4.0)
        c1.add(hh)
h=HKey('key type=20',55,90)
h.setKey(20,7.0,Color.blue)
h.setKeySpace(4.0)
c1.add(h)

h1=HKey('key type=21',55,83)
h1.setKey(21,7.0,Color.blue)
h1.setKeySpace(4.0)
c1.add(h1)

h1 =HKey('key type=30',55,76)
h1.setKey(30,7.0,Color.blue)
h1.setKeySpace(4.0)
c1.add(h1)

h1 =HKey('key type=31',55,69)
h1.setKey(31,7.0,Color.green)
h1.setKeySpace(4.0)
c1.add(h1)

h1 =HKey('key type=32',55,62)
h1.setKey(32,7.0,Color.red)
h1.setKeySpace(4.0)
c1.add(h1)

c1.update()
```

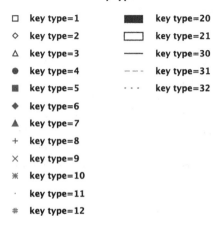

Fig. 10.1 Various types of
the keys used by the
setKey() method

10.4 Geometrical Primitives

The package `jhplot.shapes` can be used to display several (non-interactive) geometrical primitives, including the text label discussed before:

`Arrow(x1,y1,x2,y2)`	shows an arrow from `(x1,y1)` to `(x2,y2)`
`Circle(x1,y1,R)`	inserts a circle with the radius `R`
`Ellipse(x1,y1,rX,rY)`	inserts an ellipse with the radius `rX` (`rY` for Y axis)
`Text('text',x1,y1)`	inserts a text label
`Line(x1,y1,x2,y2)`	inserts a line from `(x1,y1)` to `(x2,y2)`
`Picture(x1,y2,file)`	inserts a PNG or JPG figure
`Rectan(x1,y1,w,h)`	rectangle with the width `'w'` and height `'h'`

In all cases, the objects will be drawn in the USER coordinate system. But one can also insert the primitives in the NDC system using the usual method `setPosCoord('NDC')` to be applied to the objects above. As before, the method `setPosCoord('USER')` sets the user coordinates.

To show all such graphical primitives on a canvas, use the `add()` method and execute `update()` when you want all objects to be shown. To add a different color or a line width, use additional arguments for the constructor. For example:

```
>>> from java.awt import Font,BasicStroke
>>> from jhplot.shapes import *
>>> stroke= BasicStroke(1.0)
>>> c=Circle(x1,y1,R,stroke,Color.red)
```

or, alternatively, one can use several "setter" methods:

```
>>> c=Circle(x1,y1,R)
>>> c.setFill(1)
>>> c.setColor(Color.red)
>>> c.setStroke( BasicStroke(1.0) )
>>> c.setDashed(3.0)
>>> c.setTransparency(0.5)
```

The last line in the above code makes the circle filled with red color and sets its transparency to 50%.

The arrow lines can be of two different types depending on the arrow style. The style can be set using the `setType(i)` method, where `i=1,2,3`. The length and the width of the arrows can be set as `setEndLength(d)` and `setEndWidth(d)`, where `'d'` is a double value.

To fill the primitives with a certain color, use the method `setFill(1)`. The color and the transparency level is set by the `setColor(c)` and `setFont(f)` methods, respectively.

10.5 Text Strings and Symbols

In the previous section, we have discussed several important classes to add annotations to the HPlot plots. In many occasions, you would need to shows special
symbols on HKey, HLabel, Text labels or HPlot methods designed to show
global titles and axis names.

The text attributes can be set manually or via the label property dialog. Subscripts
and superscripts in the text should be included as for the standard LaTeX text files,
i.e. use the "underscore" symbol (_) to show subscripts and the "hat" symbol to
show superscript. To indicate over-line, use the reserved word #bar{symbol}.

Below we give a small code example which makes a label with several special
symbols:

```
————————————————— Special symbols ————————————————
from jhplot import *
c1=HPlot()
c1.visible()
s1='&omega;,F^{2},F_{2},&gamma;&rarr;e^{+} e^{-}'
s2='g &rarr; q#bar{q}'
s=[s1,s2]
lab=HMLabel(s,0.3,0.7)
c1.add(lab)
c1.update()
```

This creates a multi-line label with the text:

$$\omega, F^2, F_2, \gamma \to e^+ e^-, \qquad g \to q\bar{q}$$

Symbols for the jHepWork labels must be encoded in HTML using the entity
reference notation of the ARENA project [2]. For instance, use ω symbol
to show the Greek letter ω. Figures 10.2 and 10.3 for lists of symbols supported by
jHepWork. These figures are generated by the example macros 'symbols1.py'
and 'symbols2.py' located in the directory 'macros/examples'.

10.6 SHPlot Class. HPlot Canvas as a Singleton

During the debugging stage, it is often necessary to execute a script (using the key
[F8]) and then manually close the HPlot canvas. If you do not close the HPlot
canvas, a new instance of the canvas will be created next time you execute the same
(or different) script. It will overlay on the existing canvas, but the worst thing is that
it will consume the computer memory.

jHPlot symbols I

Add '&' in front and terminate by ';'

d₃ subs	± plusmn	Å Aring	Ù Ugrave	í Iacute	A Alpha	Φ Phi
d³ super	² sup2	Æ AElig	Ú Uacute	î icirc	B Beta	X Chi
đ s̄ ā	³ sup3	Ç Ccedil	Û Ucirc	ï iuml	Γ Gamma	Ψ Psi
nbsp	´ acute	È Egrave	Ü Uuml	ð eth	Δ Delta	Ω Omega
¡ iexcl	µ micro	É Eacute	Ý Yacute	ñ ntilde	E Epsilon	α alpha
¢ cent	¶ para	Ê Ecirc	Þ THORN	ò ograve	Z Zeta	β beta
£ pound	· middot	Ë Euml	ß szlig	ó oacute	H Eta	γ gamma
¤ curren	¸ cedil	Ì Igrave	à agrave	ô ocirc	Θ Theta	δ delta
¥ yen	¹ sup1	Í Iacute	á aacute	õ otilde	I Iota	ε epsilon
¦ brvbar	º ordm	Î Icirc	â acirc	ö ouml	K Kappa	ζ zeta
§ sect	» raquo	Ï Iuml	ã atilde	÷ divide	Λ Lambda	η eta
¨ uml	¼ frac14	Ð ETH	ä auml	ø oslash	M Mu	θ theta
© copy	½ frac12	Ñ Ntilde	å aring	ù ugrave	N Nu	ι iota
ª ordf	¾ frac34	Ò Ograve	æ aelig	ú uacute	Ξ Xi	κ kappa
« laquo	¿ iquest	Ó Oacute	ç ccedil	û ucirc	O Omicron	λ lambda
¬ not	À Agrave	Ô Ocirc	è egrave	ü uuml	Π Pi	µ mu
shy	Á Aacute	Õ Otilde	é eacute	ý yacute	P Rho	ν nu
® reg	Â Acirc	Ö Ouml	ê ecirc	þ thorn	Σ Sigma	ξ xi
¯ macr	Ã Atilde	× times	ë euml	ÿ yuml	T Tau	o omicron
° deg	Ä Auml	Ø Oslash	ì igrave	ƒ fnof	Y Upsilon	π pi

Fig. 10.2 Symbols for jHepWork strings (Set I)

Several HPlot frames shown on the same desktop is useful in certain situations, for example, when one necessary to compare plots generated by different Jython scripts. But in many cases, the last thing you want is to create a new HPlot object and manually dispose it in order to prepare for the next script execution.

One can avoid the creation of new canvas frame by using the SHPlot class:

```
>>> from jhplot import *
>>> c=SHPlot.getCanvas()
>>> c.visible()
>>> c.setAutoRange()
```

Here 'c' represents a SHPlot object, which is a Java singleton extension of the HPlot class. You can work with this class exactly as with the HPlot canvas. For example, to set the overall size of the canvas to 600×400 pixels and to create 2×2 plotting pads, just type:

Fig. 10.3 Symbols for jHepWork strings (Set II). Some symbols could not be available for some platforms—this case is shown as open squares

jHPlot symbols II

Add '&' in front and terminate by ';'

τ tau	← larr	Σ sum	□ sub	" quot
υ upsilon	↑ uarr	− minus	□ sup	& amp
φ phi	→ rarr	□ lowast	□ nsub	< lt
χ chi	↓ darr	√ radic	□ sube	> gt
ψ psi	↔ harr	□ prop	□ supe	Œ OElig
ω omega	□ crarr	∞ infin	□ oplus	œ oelig
ϑ thetasym	□ lArr	□ ang	□ otimes	Š Scaron
ϒ upsih	□ uArr	□ and	□ perp	š scaron
ϖ piv	□ rArr	□ or	□ sdot	Ÿ Yuml
• bull	□ dArr	∩ cap	□ lceil	ˆ circ
… hellip	□ hArr	□ cup	□ rceil	˜ tilde
′ prime	□ forall	∫ int	□ lfloor	ensp
″ Prime	∂ part	□ there4	□ rfloor	emsp
‾ oline	□ exist	□ sim	□ lang	thinsp
/frasl	□ empty	□ cong	□ rang	zwnj
℘ weierp	□ nabla	≈ asymp	◇ loz	zwj
ℑ image	□ isin	≠ ne	♠ spades	lrm
ℜ real	□ notin	≡ equiv	♣ clubs	rlm
™ trade	□ ni	≤ le	♥ hearts	– ndash
ℵ alefsym	∏ prod	≥ ge	♦ diams	— mdash

```
>>> c=SHPlot.getCanvas('Canvas',600,400,2,2)
```

When you create the canvas using `SHPlot.getCanvas` method, the `HPlot` frame is instantiated only once, and only once. If the canvas frame already exists from the previous run, it will be cleared and then a new graph will be drawn on the same canvas. In this case, you do not close the canvas frame manually.

10.7 Visualizing Interconnected Objects

To visualize interconnected objects, one should use the `HGraph` class. It is very similar to `HPlot`, since it also extends on the `GHFrame` class used as a basis for the HPlot frame. This means that one can set titles, margins and plotting pads in exactly the same way as for the HPlot canvas. The example below shows how to build interconnected objects:

```
A graph with interactive objects

from java.awt import Color
from jhplot  import *

c1 = HGraph('Canvas')
c1.setGTitle('Connected objects')
c1.visible()

v1,v2,v3,v4 ='v1','v2','v3','v4'
c1.addVertex( v1 )
c1.addVertex( v2 )
c1.addVertex( v3 )
c1.addVertex( v4 )

# set positions
c1.setPos( v1, 130, 40 )
c1.setPos( v2, 60, 200 )
c1.setPos( v3, 310, 230 )
c1.setPos( v4, 380, 70 )

# set edges
c1.addEdge( v1, v2 )
c1.addEdge( v2, v3 )
c1.addEdge( v3, v1 )
c1.addEdge( v4, v3 )
```

The plot is fully interactive: one can drag the connected objects using the mouse to modify their locations and even edit their attributes. Figure 10.4 shows the result of the above script.

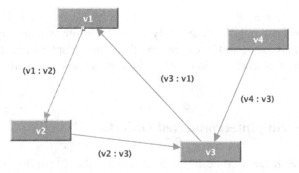

Fig. 10.4 Using the HGraph canvas to show interconnected objects

10.8 Showing Charts

The HChart class is also very similar to the HPlot class. The HChart canvas
allows to create various charts (pie chart, bar char, histogram chart, line chart, area
chart). All charts are based on the jFreeChart package library [3]. This is a simple
example showing how to create two pads with a pie-like and a bar-like charts:

```
————————————————————————— Chart examples —————————————

from jhplot import *
c1 = HChart('Canvas',600,400,2,1)
c1.setGTitle('Chart examples')
c1.visible()

c1.setChartPie()
c1.setName('Pie chart')
c1.valuePie('Hamburg',1.0)
c1.valuePie('London',2.0)
c1.valuePie('Paris',1.0)
c1.valuePie('Bern',1.0)

c1.cd(2,1)
c1.setChartBar()
c1.setName('Bar charts')
c1.valueBar(1.0, "First",  "category1");
c1.valueBar(4.0, "Second", "category2");
c1.valueBar(3.0, "Third",   "category3");

c1.update()
```

The result of the execution of this script is shown in Fig. 10.5.

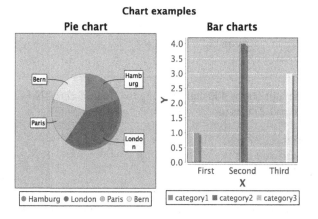

Fig. 10.5 A pie and a bar-like charts using the HChart canvas

When the HChart canvas is created, one can set the following types of charts:

```
>>> c1.setChartXY()       # create a XY chart
>>> c1.setChartPie()      # create a Pie chart
>>> c1.setChartPie3D()    # create a Pie chart in 2D
>>> c1.setChartLine()     # create a line chart
>>> c1.setChartAre()      # create an area  chart
>>> c1.setChartBar()      # create a bar  chart
>>> c1.setChartBar3D()    # create a 2D bar  chart
>>> c1.setChartHistogram() # create a histogram
```

Then one can add values using the value+ChartName() method. For example, to add a value to the bar chart use valueBar() method. Check the HChart API documentation for details. Finally, to display a chart, execute the usual update() method.

One can access many jFreeChart components via several "getter" methods. For example, c1.getChar() will return the JFreeChart class for further manipulations.

10.9 SPlot Class. A Simple Canvas

For simple tasks involving drawings of $X-Y$ plots, which do not require plotting many graphical objects, or interaction with a canvas using the mouse, one can use a light-weight canvas based on SPlot class ("S" originates from the word "simple"). The class SPlot is based on the PTPlot package [4] which was originally designed to make scatter plots on $(x-y)$ planes and simple histograms.

The methods of the SPlot canvas are similar to those of the HPlot class, but the number of such methods is not too large. However, this canvas requires much less computer memory and thus is better suited for applets or to show streams of data at runtime.

One can build the SPlot canvas as:

```
>>> from jhplot import SPlot
>>> c1=SPlot('Canvas')
>>> c1=visible()
```

This creates a default 600 × 400 canvas using the SPlot class. Now one can show the names for X and Y axes as:

```
>>> c1.setNameX('X')
>>> c1.setNameY('Y')
>>> c1.setAutoRange()
```

Both axes can be set to an auto-range mode, so you do not need to worry about setting proper axis ranges:

```
>>> c1.setAutoRange()
```

To set the ranges, use the method `setRange(xmin,xmax,ymin,ymax)`, where the arguments define the ranges for the *X* and *Y* axes.

To add a single point at a location (x, y), use the method:

```
c1.addPoint(0,x,y, b)
```

Usually, "b=0" (Java false) if points are not connected by lines. When "b=1" (Java value `true`), then the point should be connected to the next one plotted by calling the `addPoint()` method again. The data set is characterized with an integer number (0 in the above example). Finally, the method `setMarksStyle('various')` tells that each new dataset will be shown with different symbols.

To add an error bar for a point at the location (x, y), use the method:

```
>>> addPointErr(dataset,x,y,yLow,yHigh,con)
```

where `dataset` is an integer number, `yLow` and `yHigh` are the lower and upper errors for the *Y* axis, and `con` is a boolean value defining whether the points are connected ($= 1$) or not ($= 0$).

After filling the data points in a loop, one should call `update()` to make data visible on the canvas.

The method `addPoint()` adds a single point to the canvas and, as we already know, to call methods inside a loop in Jython is CPU and memory consuming. However, as for the HPlot canvas, one can also use the method `draw(obj)` to draw Java high level objects, such as histograms (H1D), data containers (P1D) or simply arrays of numbers with *x* and *y* values. Data can be shown with symbols connected by the lines if `setConnected(1,set)` is called with the first argument 1 (or boolean `true` for Java), where `set` specifies the data set identification number (integer). There is no need to call the update method after calling the `draw(obj)` method, since the plot will be updated automatically.

The plot can be zoomed into a specific rectangular area by clicking and dragging, to draw a rectangle where desired; this feature is also different from the HPlot canvas where you can rescale axis ranges one at the time. Finally, one can save the plot into an image using the usual `export(file)` method. Look at all other methods associated with this class using the code assist.

Below we give a small example showing how to plot two different sets of data points:

———————————— SPlot example ————————————

```
from java.util import Random
from jhplot  import *
```

```
c1 = SPlot()
c1.visible()
c1.setAutoRange()
c1.setNameX('Time')
c1.setNameY('Data')

p1=P1D('data1')
p2=P1D('data2')
r = Random()
for i in range(20):
    x=100*r.nextGaussian()
    y=200*r.nextDouble()
    p1.add(x,y)
    p2.add(x+x,y+y)

c1.draw(p1)
c1.draw(p2)
c1.addLegend(0,'Data1')
c1.addLegend(1,'Data2')
c1.update()
```

The execution of this script brings up a window with two sets of data points and the inserted legends indicating each data set.

In Sect. 10.15 we will discuss how to use the class SPlot to draw streams of data in real time, without using the draw(obj) method.

10.9.1 Henon Attractor Again

Let us give another example illustrating the SPlot canvas. In Sect. 5.5, we have shown how to build the Henon strange attractor using the P1D class. One feature of that script was that we could not see how (x, y) points are populating the canvas at runtime, i.e. we have to wait until the P1D container is filled and only then we could display the attractor. Being inconvenient, this also leads to a large memory usage.

Let us rewrite the same code using the addPoint() method of the SPlot class. We will update the graph axes after generating 100 events. Execution of the script shown below illustrates how points populate the attractor immediately after the execution of this script.

```
────────────────── Henon attractor ──────────────────
from jhplot import *

c1 = SPlot()
c1.setGTitle('Henon attractor')
c1.setMarksStyle('pixels')
```

```
c1.setConnected(0, 0)
c1.setNameX('x')
c1.setNameY('y')
c1.visible()
a = 1.4;  b = 0.3
x=0; y=0
for i in range(100000):
  x1=x
  x=1+y-a*x*x
  y=b*x1
  c1.addPoint(0,x,y,1)
  if i%1000==0: c1.update()
```

We will leave the reader here for more experimentation.

10.10 Canvas for Interactive Drawing

For complicated tasks which involve drawing diagrams together with the usual data-driven plots, one should use the HPlotJa canvas. From the point of view of displaying functions, arrays and histograms, this canvas is rather similar to the HPlot class, i.e. one can use the same method draw(obj) to display instances of lists, F1D, H1D and P1D objects. Yet, it has many advanced features such as:

- plots are more interactive. One can easily manipulate with different pads, make inserts, overlays etc.;
- one can open ROOT or AIDA files in an object editor described in Sect. 11.5.3;
- one can draw diagrams using Jython scripts or using the object editor.

One should emphasize that this canvas is very easy to use for making inset plots, i.e. showing one pad inside the other. This is technically impossible for the HPlot canvas, in which each pad is built using the JPanel Java class. In case of HPlotJa, the pads are located inside a single panel.

The HPlotJa canvas is based on the JaxoDraw package [5] designed to draw Feynman diagrams used in high-energy physics. The original package was significantly modified by adding the possibility to plot the standard jHepWork objects and adding vector graphics capabilities.

To build an instance of the HPlotJa canvas, use the following code snippet:

```
>>> from jhplot import *
>>> c1=HPlotJa('Canvas')
>>> c1=visible()
```

The methods associated with the HPlotJa canvas are similar to those of the HPlot class and we will not repeat them here. The largest difference with the

HPlot is that there are many methods related to drawing axes which cannot be called directly, but only via the method getPad() as shown in this example:

```
>>> from java.awt import Color
>>> from jhplot import *
>>> c1=HPlotJa('Canvas',1,2)
>>> c1=visible()
>>> c1.cd(1,2)
>>> pad=c1.getPad()          # get current pad
>>> pad.setRange(0,0,1)      # set axis range [0,1] on X
>>> pad.setFillColor(Color.red) # new  color
>>> pad. + [F4]                 # check other methods
```

As for the original JaxoDraw package, the HPlotJa canvas has a complete graphical user interface that allows to carry out all actions in a mouse click-and-drag fashion. To bring up the graphical editor, go to the [Option] menu and select [Show editor]. Now you can create and edit the graph objects using the mouse clicks. One can draw many graphical objects, lines, circles etc. One can remove, drag and resize all plotted objects, including the pad regions. Finally, one can edit properties of all plotted objects.

10.10.1 Drawing Diagrams

Once an object of the HPlotJa canvas is created, one can draw diagrams using JaxoDraw mouse click-and-drag fashion. To do this, you have to select the [Editor] option from the [Option] menu.

The power of Jython scripting allows to draw diagrams interactively or using Jython macro files. This will require importing the static methods of the Diagram class from the package jhplot.jadraw. Below we show a typical example of how to draw "gluon" and "fermion" lines typically used for representation of Feynman diagrams:

```
————————————— "gluon" and "fermion" lines ——————————
from jhplot import *
from jhplot.jadraw import *

c1=HPlotJa('Canvas',500,400,1,1,0)
c1.visible()
c1.showEditor(1)

# draw text box
g1=Diagram.TextBox("Gluons and fermions",0.25,0.15)
c1.add(g1)

# gluon line in NDC coordinates
```

Fig. 10.6 Drawing diagrams
on the HPlotJa canvas
using Jython

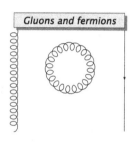

```
gl=Diagram.GlLine(0.3,0.2)
gl.setRelWH(0.0,0.5,'NDC')
cl.add(gl)

# gluon loop in NDC
gl=Diagram.GlLoop(0.5,0.4)
gl.setRelWH(0.0,0.1,'NDC')
cl.add(gl)

# fermion line
gl=Diagram.FLine(0.7,0.2)
gl.setRelWH(0.0,0.5,"NDC")
cl.add(gl)

cl.update()
```

It should be noted the way how the HPlotJa canvas is created: for this example,
we do not show axes, since the last argument in the constructor HPlotJa is zero
(Java false). Also, we use the add() and update() methods, as we usually do
when showing labels and graphical primitives.

Figure 10.6 shows the output of this example. One can further edit the diagram
using the editor panel. Please refer to the Java API documentation of the package
jhplot.jadraw to learn more about the classes and methods of this package.

10.10.2 SHPlotJa Class

Similar to the SHPlot class, one can create a singleton representing the HPlotJa
canvas object using the static class SHPlotJa. In this case, every execution of
a script does not create a new object of the canvas frame, but it just redraws the
existing one. The example below shows how to create such singleton:

```
>>> from jhplot import SHPlotJa
>>> cl=SHPlotJa.getCanvas()
>>> cl.setGTitle("Global title");
>>> cl.setNameX("Xaxis")
```

```
>>> c1.setNameY("Yaxis")
>>> c1.visible(1)
>>> ....
```

Of course, all methods of the `HPlotJa` canvas are also applicable to the `SHPlotJa` class.

10.11 HPlot2D Canvas

Although one can use the `HPlot` canvas to show 2D histograms, functions or arrays as contour or density plots, it is not the best way of doing this, because this canvas was not designed from the ground to support such types of plots. Instead, use the `HPlot2D` canvas for such tasks. This canvas is partially based on the SGT project [6].

The `HPlot2D` is very similar to the `HPlot` and shares many common methods. The initialization of this canvas looks also very similar:

```
>>> from jhplot import *
>>> c1=HPlot2D('Canvas')
>>> c1.visible()
```

This brings up a frame with the `HPlot2D` canvas.

Now let us walk through several examples which show how to use this canvas. First, let us draw a F2D function as a contour plot:

```
─────────── 2D functions on HPlot2D ───────────
from jhplot import *
from java.awt import *

f1=F2D('x^2+sin(x)*y^2',-2,2,-2,2)
c1=HPlot2D('Canvas',600,700)
c1.visible()
c1.setName('2D function')
c1.setNameX('X variable')
c1.setNameY('Y variable')

c1.setStyle(2)
c1.draw(f1)

lab1=HLabel('&omega; test',0.7,0.5, 'NDC')
lab1.setColor(Color.white)
c1.add(lab1,0.1)

c1.update()
```

Fig. 10.7 2D function shown
as a density plot

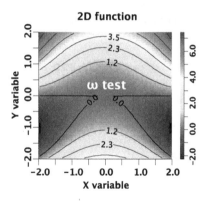

The resulting plot is shown in Fig. 10.7. The methods used for the canvas object `c1`
are rather similar to the methods of the `HPlot` canvas. The only new feature is that
when we add a text label, we use the method `add(obj,d)`, where `'d'` specifies
the label height.

We should mention that the `HPlot2D` canvas is fully interactive. One can move
and edit labels, click-and-drag the mouse pointer to create a zoom rectangle. One
can edit axis attributes by clicking on an axis (this pops-up a dialog where all modifications can be made).

There are several options for showing an object inside the canvas. They are con-
trolled with the method `seStyle(style)`, where

`style=0` draws data using a raster style;
`style=1` draws using a contour style;
`style=2` combined style (raster and contour);
`style=3` draws data using filled areas.

One can access several objects for modifications inside Jython scripts. First, one
can obtain the axis object as `getAxis(axis)`, where `axis` is either 0 (for *X*)
or 1 (for *Y*). Using the methods of this axis object, one can make necessary modifications, like setting new fonts, redefine colors etc. One can also access the color
bar-key as `getColorBar()`, which also has several useful methods.

One important method you should keep in mind is

```
>>> setRange(axis,min,max,bins)
```

which sets the range for *X* (axis = 0), *Y* (axis = 1) and *Z* (axis = 2). The variable
`bins` specifies how many divisions between `min` and `max` values should be used.
In case of the *Z*-axis, the variable `bins` specifies how many contour levels to draw.

Let us give another example. This time we will plot data using different styles.
We display a histogram and 2D array. In case of the histogram, we change the range
for *X* and *Y* axes using the `setRange()` method. We note that the last argument
for this method, which usually defines the number of bins between the minimum
and maximum values, does not do anything. This is because we plot the histogram

which has its own binning, and this cannot be changed when the histogram is shown.

```
──────────────── 2D data on HPlot2D ────────────────

# HPlot2D. Showing H2D,P1D and 2D function
# S.Chekanov

from jhplot import *
from java.util import *

h1=H2D('Data',30,-3.0, 3.0, 30,-3.0, 3.0)
p1=P1D('Data')

r=Random()
for i in range(1000):
  x=r.nextGaussian()
  y=r.nextGaussian()
  h1.fill(0.6*x-0.5, y)
  p1.add(0.4*x+1,    y+0.5)

c1=HPlot2D('Canvas',600,400,3,1)
c1.visible()

c1.cd(1,1)
c1.setName('H2D histogram');
c1.setStyle(0)
c1.draw(h1)

c1.cd(2,1)
c1.setName('H2D range')
c1.setStyle(1)
c1.setRange(0,-2.0,2.0,50)
c1.setRange(1,0.0,1.0,50)
c1.draw(h1)

c1.cd(3,1)
c1.setName('2D array')
c1.setStyle(1)
c1.draw(p1)
```

The output of this script is shown in Fig. 10.8.

10.12 3D Canvas

We have already discussed canvases for 3D graphics in Sects. 3.4.2, 6.1.1 and 8.2. Here we remind that one should use the classes HPlot3D and HPlot3DP for a 3D representation of mathematical functions and data. The first class is used for plotting

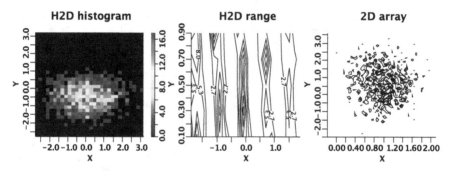

Fig. 10.8 2D data shown using different styles and ranges

H2D, P2D and F2D objects, while the second one for parametric equations based on the FPR class, see Sect. 3.8.

10.13 HPlot3D Canvas

This canvas is used for visualization of H2D, P2D and F2D objects, which can be drawn with the same draw(obj) method as for the 2D case. In case if two objects should be shown, say obj1 and obj2, on the same canvas, one should call the method draw(obj1,obj2). Table 10.2 lists the most important methods of the HPlot3D Java class:

We should remind that if you are debugging a script and do not want to create many pop-up windows with canvas frames after each script execution, you may want to instantiate a singleton using the class SHPlot3D. In this case, every new macro execution redraws the existing canvas, instead of creating a new canvas object. One can build a singleton as usual:

```
>>> c1=SHPlot3D.getCanvas('3D',600,400,2,2)
```

This creates four plotting pads inside a 3D canvas of the size 600 × 400.

10.13.1 HPlot3DP Canvas

The second 3D canvas, HPlot3DP, is used to draw surfaces defined by an analytic parametric or a non-parametric function of the type FPR, as shown in Sect. 3.8.

The HPlot3DP canvas can be constructed exactly in the same way as the HPlot3D class. Specifically, one can define the frame size during object instantiation and set any number of drawing pads. The major difference with HPlot3D is following: if several objects should be shown on the same pad, then one can use

Table 10.2 The main methods of the `HPlot3D` class. "b" indicates a boolean value (1 for true and 0 for false), while "i" is an integer value. The notation "d" denotes a float value. The attributes "c" and "f" correspond to the `Color` and `Font` AWT Java classes. "text" represents a string value. The character "a" is a shortcut to "axis" (a=0 for X, a=1 for Y, a=2 for Z)

Methods	Definitions
setRotationAngle(i)	set rotation angle to 'i' degrees
getRotationAngle()	get rotation angle
setScaling(i)	set scaling factor to 'i' (default is 12)
getScaling()	get scaling
setElevationAngle(i)	set elevation angle to 'i' degrees
getElevationAngle()	get elevation angle
setAxesFontColor(c)	set fonts for axes labels
setColorMode(i)	set color mode (from 1 to 4)
	0: wire-frame
	1: hidden
	2: color spectrum 3: gray scale
	4: dual shades
setDistance(d)	set distance to objects
getDistance()	get distance from object
setLabelFontColor(c)	set color for labels
setPenWidthAxes(i)	set line width for axes
getPenWidthAxes()	get pen with for axes
setNameX("text")	set name for axis X
setNameY("text")	set name for axis Y
setNameZ("text")	set name for axis Z
setLabelFont(f)	set fonts for labels
setLabelFont(f)	set fonts for axes numbers
setLabelColor(c)	set label color
setTicFont(f)	set fonts for ticks
setTicOffset(d)	set ticks offset
setRange(Xmin,Xmax,Ymin,Ymax)	set plot ranges
setRangeZ(Zmin,Zmax)	set range for Z
cd(iX,iY)	go to the pad (iX,iY)
clear()	clean the current region
clear(iX,iY)	clean the pad (iX,iY)
clearAll()	clean all pads
quite()	remove frame
update(iX,iY)	update the pad (iX,iY)
updateData(iX, iY)	update data shown in the pad (iX,iY)
updateData()	update data on the current pad
updateAll()	update all pads
getLabelOffset(a)	get label offset for axis
draw(obj)	draw an object (P2D, F2D..)

Table 10.3 The main methods of the `HPlot3DP` class. "b" denotes a boolean value (1 for true and 0 for false), while "i" is an integer value. The notation "d" indicates a float value. The attributes "c" and "f" correspond to the `Color` and `Font` AWT Java classes

Methods	Definitions
setFog(b)	sets the fog style for 3D
setAxes(b)	shows or not axes
setAxes(b1,b2,b3)	shows axes for X, Y, Z
setAxesColor(c)	axes color
setBackgColor(c)	background color
setNameX(text)	text for X axis
setNameY(text)	text for Y axis
setNameZ(text)	text for Z axis
setCameraPosition(d)	set camera position zoom in for positive "d" zoom out for negative "d"
setEyePosition(x,y,z)	set eye positions
clear(object)	clear the frame
cd(i1,i2)	navigates to a i1 $\times$ i2 pad
update()	updates the canvas
draw(obj)	draws an object (FPR)

the `draw(obj)` methods several times, one after other. The main methods of this canvas are given in Table 10.3.

One can zoom in the pad area using the right mouse button. In macro files, one can zoom in and zoom out using the following method:

```
>>> c1.setCameraPosition(d)
```

where d is a double value, which should be positive for zooming in, and negative for zooming out. One can change the location of axes and the object position using the method `setEyePosition(x,y,z)`. The graphs can be edited using the GUI dialog and the [Edit] menu by clicking on the mouse button.

Below we show the use of several methods of the `HPlot3DP` class:

```
———————————————— HPlot3DP canvas ————————————————
from java.awt import Color
from jhplot import *

c1 = HPlot3DP('Canvas',600,600)
c1.setGTitle('HPlot3DP examples')
c1.visible()

f1=FPR('u=2*Pi*u; x=cos(u); y=sin(u); z=v')
c=Color(0.5,0.2,0.5,0.5) # color+transparency
f1.setFillColor(c)
f1.setLineColor(Color.green)
```

Fig. 10.9 HPlot3DP canvas
and its graphical methods

HPlot3DP examples

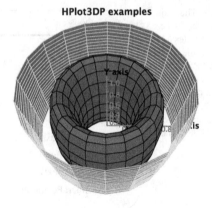

```
f2=FPR('u=2 Pi u; v=2 Pi v; r=0.6+.2cos(u); \
        z=.8 sin(u); x=r cos(v); y=r sin(v)')
f2.setFillColor(Color.blue)
f2.setFilled(1)

c1.setFog(0)
c1.setAxes(1)
c1.setNameX("X axis")
c1.setNameY("Y axis")
c1.setAxesColor(Color.gray)
c1.setAxesArrows(0)

print c1.getEyePosition()
print c1.getCameraPosition()
c1.setCameraPosition(-1.2) #zoom out
c1.draw(f2)
c1.draw(f1)
```

The above code brings up a frame with the image of a cylinder and a torus as shown
in Fig. 10.9. One can rotate objects and zoom into certain area using the mouse
button. In addition, one can further edit the figure using the [Edit] menu.

10.13.2 3D Geometry Package

The HView3D class also allows to draw 3D objects (cubes, shapes cylinders etc.)
which are already predefined Java classes, thus there is no need to define them using
parametric equations. Look at the Java API of the jhplot.v3d package. This is a
typical example of how to draw 3D shapes:

```
────────────── Creating 3D objects ──────────────
from java.awt import Color
from jhplot  import *
from jhplot.v3d import *

c1 = HView3D('Canvas',400,400)
c1.visible(1)
c1.setGTitle("3D objects in HView3D")

o=Cube(c1.getModel(),40) # a cube of the size 40
o.setRot(45,45,45) # rotate
c1.draw(o)

o=Sphere(c1.getModel(),30.,80,80) # a sphere eithr=30
o.setTrans(40,-20,10) # move it
c1.draw(o);

o=Cone(c1.getModel(), 30, 100, 50)
o.setTrans(-20, 30, 0)
o.setColor(Color.red)
c1.draw(o);

o=Cylinder(c1.getModel(), 40, 100, 2)
o.setTrans(-1, 0, 0)
o.setRot(0, 60, 0)
o.setColor(Color.yellow)
c1.draw(o)

c1.update() # draw all objects
```

The execution of this script brings up a window with 3D objects using their attributes
passed to the constructors.

10.14 Combining Graphs with Java Swing GUI Components

In Sect. 2.18 we have shown how easy to use the Java Swing components in Jython.
Jython scripting allows a creation of graphical user interfaces (GUI) using the Java
GUI widgets, such as windows, menus, buttons, etc. One can make them to respond
to physical events (keyboard, mouse etc.) with a few lines of Jython code. This is
an important when it is necessary to make "control panels" or "graphical user interfaces" for applications in order to manipulate with Java programs at run time. Such
applications can be deployed as self-contained programs in jar files after compiling
Jython codes into the Java bytecodes, the machine language of the Java virtual machine. But how one can use the jHepWork canvas together with such components?
We will briefly discuss such topic in this section.

Numerous methods of the canvases described in the previous sections are so rich that one can always find a necessary method suited for your GUI application. The most important method for the GUI development is that which returns the Swing JFrame object holding the entire canvas panel. In the example of Sect. 2.18, an instance of the Java class JFrame was created using the standard constructor. When you are working with the jHepWork canvases, there is no need to do this. What you will need is to use the getFrame() method which returns an appropriate JFrame instance:

```
>>> c1 = HPlot('GUI')
>>> jframe=c1.getFrame()
```

Once you have retrieved the frame object, you can add the necessary Swing GUI components (menu, buttons, sliders etc.) to control the displayed data.

Let us show an extensive example in which we illustrate how to attach a panel with two Swing JButton buttons and a TextArea component directly to the HPlot canvas object. One button will be used to generate a histogram populated with Gaussian numbers. The second will be used to erase graphs from the canvas. The text area displays a short message about what happens when you press these buttons.

```
————— Combining HPlot canvas with Swing components —————
from java.awt import *
from javax.swing import *
from java.util import *
from jhplot  import *

c1 = HPlot('GUI')
c1.setGTitle('Plot area with GUI')
c1.visible(1)
c1.setAutoRange()
fr=c1.getFrame()

h1 = H1D('Histogram',20, -2.0, 2.0)
h1.setFill(1)
h1.setFillColor(Color.blue)

pa0 = JPanel(); pa1 = JPanel()
pa2 = JTextArea('GUI test',6,20)
pa2.setBackground(Color.yellow)

def act1(event):
    r=Random()
    for i in range(100):
        h1.fill(r.nextGaussian())
    c1.draw(h1)
    pa2.setText('Generated 100 Gaussian numbers')
```

```
def act2(event):
   c1.clearData()
   pa2.setText('Clear plot')

pa0.setLayout(BorderLayout());
bu1=JButton('Gaussian', actionPerformed=act1)
pa0.add(bu1,BorderLayout.NORTH)
bu2=JButton('Clear', actionPerformed=act2)
pa0.add(bu2,BorderLayout.SOUTH)
pa0.add(pa2,BorderLayout.WEST)
fr.add(pa0,BorderLayout.EAST)
fr.pack()
```

After execution of this script, you will see the HPlot canvas with an attached panel with the Swing components, see Fig. 10.10. Press the buttons to see the output of this program.

By examining the script, you may wonder why do we need the method pack() at the very end of this script. We call this method to avoid changes in the size of the HPlot canvas, so the extra panel just extends the canvas without distorting its default size (which is 600 × 400 pixels). If you will want to append extra canvas by keeping the same size of the entire frame, remove the method pack() and, call any method which updates the frame (calling setVisible(1) can do this).

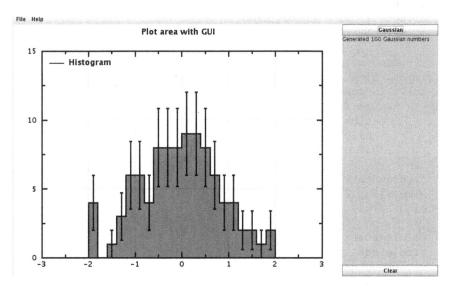

Fig. 10.10 HPlot canvas together with a Swing control panel

10.15 Showing Streams of Data in Real Time

Showing streams of data in real time is very often situation. Instead of collecting data in some container, such as Java or Jython array, and plotting data at once using the draw(obj) method, one can plot a fraction of data samples without waiting until all data set become available and populate a data holder to be used for drawing.

In this section, we will show how to fill histograms or data holders and update the canvas after each such operation. Let us illustrate this by using the Jython sleep() function: we will fill a histogram and update the plot after each random Gaussian number. Thus we will see how the histogram is filled at runtime. At the end of the loop, we will clean up the canvas and then re-plot the final histogram. How to clean the canvas from plotted data has been discussed in Sect. 10.2.8.

—————————— HPlot data stream ——————————

```
from java.awt import Color
from java.util import Random
from jhplot  import *

c1 = HPlot('Canvas')
c1.setGTitle('data stream')
c1.visible()
c1.setLegend(1)
c1.setAutoRange()

h1 = H1D('Updated histogram',20,-2.0,2.0)
h1.setFill(1)
h1.setErrX(0)
h1.setErrY(1)
h1.setFillColor(Color.blue)
h1.setColor(Color.blue)

r = Random()
import time
for i in range(1000):
        h1.fill(r.nextGaussian())
        time.sleep(1)
        c1.clearData()
        c1.draw(h1)

c1.drawStatBox(h1)
time.sleep(2)
c1.draw(h1)

c1.clearData()
time.sleep(2)
c1.setLegend(1)
c1.draw(h1)
```

One may notice that, before each draw(obj) statement, we clean the canvas from the data using the method clearData(). This is necessary in order to avoid generating many objects during plot drawings, since each call of the method draw(obj) creates a new object and this leads to an extensive memory usage.

For showing steams of data, it is good idea to use a light-weight canvas, such as SPlot discussed in Sect. 10.9. Here is an example in which we fill a canvas with data points and rescale the plot to fit all data:

─────────── A data stream using the SPlot canvas ───────────

```
from java.util import Random
from jhplot  import *

c1 = SPlot()
c1.visible()
c1.setAutoRange()

c1.setMarksStyle('various')
c1.setConnected(1, 0)
c1.setNameX('Time')
c1.setNameY('Data')

r = Random()
import time
for i in range(20):
    x=r.nextGaussian()
    y=r.nextGaussian()
    c1.addPoint(0,x,y,1)
    c1.update()
    time.sleep(1)
```

This example is rather fancy: you will see data points connected with various lines. When one uses the addPoint() method, there is no need in removing data from the canvas after each call to this method, since the method addPoint() does not create a new object after each call.

References

1. Chekanov, S.: The JHPLOT package. URL http://hepforge.cedar.ac.uk/jhepwork/api/jhplot/
2. The ARENA project: URL http://www.w3.org/Arena/tour/symbols.html
3. jFreeChart package: URL http://www.jfree.org/
4. Ptplot 5.6: Ptolomy project. URL http://ptolemy.berkeley.edu/java/ptplot/
5. Binosi, D., Theussl, L.: JaxoDraw: A graphical user interface for drawing Feynman diagrams. Comput. Phys. Commun. **161**, 76 (2004). URL http://jaxodraw.sourceforge.net/
6. Denbo, D.W.: Scientific Graphics Toolkit. URL http://www.epic.noaa.gov/java/sgt/

Chapter 11
Input and Output

jHepWork has a large choice for file input and output (I/O) for efficient data storage and processing. There are several mechanisms for I/O:

- streams to write or read data using the native JAVA platform. The Java libraries for I/O are located in the `java.io` package which can be imported using the usual `'import'` statement. We will not discuss the Java I/O streams in this book as they are described in detail in any Java book [1–5]. We have already discussed the class `DataOutputStream` in Sect. 2.16 in context of writing and reading data using Java called from Jython;
- native Python I/O discussed in Sect. 2.16;
- libraries included into the jHepWork Java package `jhplot.io`. In this chapter, we will discuss these libraries in more detail;
- third-party Python libraries. We will discuss the package `DIF` from these libraries in Sect. 11.10;
- third-party Java libraries included into the jHepWork. As example, we will discuss the Apache Derby relational database in Sect. 11.6.

File I/O, data streaming and format conversions are essential for any stage of data analysis. Unfortunately, in the standard manuals and tutorials, details of file I/O seem to be buried at the back. In this book, we will discuss this topic in great detail, focusing on I/O libraries included into the jHepWork program.

11.1 Non-persistent Data. Memory-based Data

In simple words, non-persistent objects contain data that are kept in the computer memory and cannot be restored after the end of the program execution. In contrast, persistent objects are stored in files and can be restored at any moment.

Most scripts discussed in this books hold objects in a non-persistent way. For example, when you create a histogram or an array with random numbers as:

S.V. Chekanov, *Scientific Data Analysis using Jython Scripting and Java,*
Advanced Information and Knowledge Processing,
DOI 10.1007/978-1-84996-287-2_12, © Springer-Verlag London Limited 2010

```
>>> from jhplot import *
>>> p=P0D('test')
>>> p.randomUniform(100000,0,1)
```

the object 'p' is created in the computer memory. Watch out the memory-status monitor at the bottom right corner of the jHepWork editor: you will see a notable increase in the used memory. By increasing the number of random numbers in the above example, one can eventually run out of memory, and an OutOfMemory-Error message will be thrown.

Thus, you are very limited by the available memory of your computer (and available memory assigned to the virtual machine). The memory assigned to store objects can be released manually: in Jython, to remove an object, say 'p', call the statement 'del p'. In addition, the Jython garbage collector takes care of reclamation of memory space in an automatic mode. One can also invoke the Jython garbage collector manually as:

```
>>> del p
>>> import gc
>>> gc.collect()
```

After calling the statement above, you will see a decrease in the used memory (again, look at the bottom right corner of the jHepWork IDE).

The computer memory is like the work table. It is fine to work on it using small portions of data because it is fast and effective. But in order to handle large pieces of data, you have to put data on "a shelf", i.e. copy them to a real file storage on a disk. Below we will discuss how to store jHepWork data in files and how to bring them back into the computer memory for fast operation.

11.2 Serialization of Objects

Before we have already discussed how to save separate objects into files. Most objects, like P1D or H1D, contain methods for writing them into files, which can be either text files or binary files.

These objects have also the methods to instantiate themselves from files. Look at the previous sections: Sect. 3.10 discusses this topic for functions, while Sects. 4.3, 5.4 and 6.4 for P0D, P1D and PND arrays, respectively. Section 8.6 describes the histogram I/O operations. In all these sections, we have illustrated how to save and restore all such objects using the standard Java serialization, including the serialization into a human-readable XML file format.

We also should remind that jHepWork objects can be saved into Jython lists, dictionaries or Java collection classes using the Serialized class. This is a generic example of how to save objects as lists:

```
>>> from jhplot.io import *
>>> list=[]
>>> list.append(object1)
>>> list.append(object2)
...
# write to a file
>>> Serialized.write(list,'file.ser')
...
# deserialize the list from the file
>>> list=Serialized.read('file.ser')
```

Similarly, one can use Jython dictionaries with the keys which are rather handy to extract objects:

```
>>> from jhplot.io import *
>>> map={}
>>> map['object1']=object1
>>> map['object2']=object2
...
>>> # write to a file
>>> Serialized.write(map,'file.ser')
...
>>># de-serialize the map from the file
map=Serialized.read('file.ser')
```

The class `Serialized` compresses data using the GZip format. One can also save and read objects without compression using the following methods:

```
>>> Serialized.write(obj,'file.ser',compression)
>>> obj=Serialized.read('file.ser',compression)
```

where a boolean `compression` should be set to zero ('false' for Java) if no compression is applied.

One can also use a XML format for all such output files. This format is implemented using the methods `writeXML(obj)` and `readXML()` from the same `Serialization` class. However, this is not the most economic approach, since string parsing is a CPU intensive and the XML files are significantly larger than files with binary data.

In all such approaches, there was a significant drawback: we write data at the end of numerical computations, so first we put the entire collection in a non-persistent state, and then move it into a persistent state. This can only work if the data are not too large and one can fit them in the computer memory.

Below we will discuss how to avoid such problems by writing objects with data directly into files sequentially, without building collections of objects first and then writing them using the `Serialized` class.

11.3 Storing Data Persistently

11.3.1 Sequential Input and Output

To load all data into a Jython or Java container and write it at once into a file using the `Serialized()` class is not the only approach. One can also write or read selected portions of data, i.e. one can sequence through data by writing or reading small chunks of data that are easier to handle in the computer memory.

This time we are interested in truly persistent way of saving objects: instead collecting them in the collections before dumping them into a file, we will write each object separately into a file. In this case, one can write and read back huge data sets to/from the disk.

The Java serialization mechanism for objects could be slower than the standard Java I/O for streaming primitive data types. As usual, there is a trade-off between convenience and performance and, in this book, we usually prefer convenience. In case of the Java serialization, restored objects carry their types, and such "self-description" mechanism is very powerful and handy feature as there is no need to worry about what exactly data have been retrieved from a file.

If the serialization is slow for your particular tasks, one can always use the I/O methods of jHepWork objects which are not based on the Java serialization. See, for example, Sects. 4.3 and 5.4 for such examples.

To write objects sequentially into a file and read them back, one can use the `HFile` class. A general call to create an instance of this class is:

```
>>> f=HFile(FileName,option,compression,buffersize)
```

where `FileName` is a string with the file name, `'option'` is a string which can be either `'w'` (write into file) or `'r'` (read from a file). If the boolean variable `'compression'` is set to zero ('false' in Java), objects inside the file will not be compressed. An integer value, `'buffersize'`, specifies the buffer size. The class `HFile` is constructed such that it writes data chunks to a memory buffer first, before recording them on the disk. This dramatically increases the speed of I/O operations, compare to the situations when writing one byte at a time. Selecting a correct buffer size is important; typically this number should be multiples of 1024 bytes. If the data chunks to be written are large, consider to increase the buffer size. This point will be illustrated in more detail later.

The following shortcuts for creation of a `HFile` object can be useful:

```
>>> f=HFile(FileName,option,compression)
>>> f=HFile(FileName,option)
>>> f=HFile(FileName)
```

In all these initializations, the buffer size is set to 12 kilobytes (KB). The second constructor assumes data compression, while the third assumes that the file is opened for reading objects only.

Let us give an example of how to store several Java objects in a file persistently. The example below shows how to create an output file with several Java objects. Some of them (like P1D) can contain data:

```
_____ Writing Java objects _____
from jhplot.io import *
from jhplot import *
from javax.swing import JFrame

# create a lsit with objects
list=[]
list.append(P0D('title'))
list.append(P1D('title'))
list.append(H1D('histogram',100,-5,5))
list.append(F1D('x*x',1,5))
list.append('String')
list.append(JFrame('Frame'))

f=HFile('output.ser','w')
for i in list:
    f.write(i)
f.close()
```

As we have mentioned before, data from the serialization contain a self-description. Let us read the file from the above example and check the types of the retrieved objects. In one case, we will use the Jython type(obj) method, in second we will use the Java-oriented type of check:

```
_____ Reading Java objects _____
from jhplot.io import *
from jhplot import *

f=HFile('output.ser')
while 1:
  obj=f.read()
  if obj == None:
      break
  print 'Type=',type(obj)
f.close()
```

The output of this script is:

```
Type= <type 'jhplot.P0D'>
Type= <type 'jhplot.P1D'>
Type= <type 'jhplot.H1D'>
Type= <type 'jhplot.F1D'>
Type= <type 'unicode'>
Type= <type 'javax.swing.JFrame'>
```

In case if an object belongs to a Java class, one can also access the information about this object using the getClass() method from the native Java API. This has already been discussed in Sect. 3.2.1, but here we will remind how to do this again:

```
>>> try:
>>>     cla=obj.getClass()
>>>     print 'Java=',cla.getCanonicalName()
>>> except:
>>>     print 'Non-Java class';
```

11.3.2 GUI Browser for Serialized Objects

If a serialized file has been created with the help of the HFile class using its default attribute (i.e. GZip compression), one can browser objects inside such file using a specially-designed GUI browser. Start the jHepWork IDE and select [Tools] and then [HPlot] menu, which brings up an empty HPlot canvas. Then select [Read data] using the menu [File] from the tool bar. This opens an object browser which will be discussed in more detail in Sect. 10.2.7.

In this section, we would rather show how to look at the serialized objects using Jython scripting. In our next example, we create a canvas and open a HFile file in the HFileBrowser dialog. You will see a list of objects stored in the file (together with their titles). One can select an object and plot it on the HPlot canvas. The selected object can be either a F1D function, a H1D histogram, P0D or P1D) array. This code snippet opens the file 'output.ser' created before in the object browser:

```
──────────────── Java object browser ────────────────
from jhplot.io import *
from jhplot import *

c=HPlot('canvas')
c.visible()
```

```
f=HFile('output.ser')
HFileBrowser(c,f,1)
f.close()
```

Select an entry, say "String" (which, obviously, means "Java String" object) and push the button "Plot". You will see the plotted text on the canvas. In exactly the same way, one can bring up a serialized JFrame (note that we did not set the size of this frame, so you will see only a small box). One can also plot other objects (histograms, jHepWork arrays) stored in the file 'output.ser'.

11.3.3 Saving Event Records Persistently

Now let us come to the question of how to store large data sets in a sequential form. The passed argument obj to the method write(obj) can be rather complex since it can keep a data record with various types of containers and their graphical attributes.

Below we will show a simple example of how to write 10000 events into a serialized file. Each event is a list with the event identification information (a string), POD array with random numbers and a histogram H1D. We print a debugging information after each 100 events, showing how the file is growing after inserting the data.

```
—————————————— Writing event records ——————————

from jhplot.io import *
from jhplot import *

def makeEvent(entry):
    'event with 3 objects'
    label='Event='+str(entry)
    p=POD(label)
    p.randomUniform(10,0,1)
    h=H1D(label,10,-1,1)
    h.fill(i)
    return [label,p,h]

f=HFile('output.ser','w')
for i in range(10000):
    ev=makeEvent(i)
    if (i%100 == 0):
        print ev[0]+' size=',os.path.getsize(file)
    f.write(ev)
f.close()
```

One should note that the data stream is flushed[1] after each 100 event by default. One can increase or decrease this number by using the method setFlush(i) of the HFile class, where 'i' is the number of entries after which data should be flashed to the disk. This is necessary since a serialization of multiple objects without resetting the stream causes an increase of the memory used by the Java virtual machine. We can call getEntries() to obtain the actual number of written entries and getBufferSize() to return the default size of the buffer used to write data.

Now, let us read the saved file back and restore all the objects from the list. We will read the events from the file in an infinite loop. When method read() returns None (no entries to read), than this indicates that the end of the file is reached and the loop should be terminated:

```
──────────────── Reading events ────────────────
from jhplot  import *
from jhplot.io import *

f=HFile('output.ser')
while 1:
    event=f.read()
    if event == None:
        print 'End of events'; break
    print event[0]
    p=event[1]
    h=event[2]
    # print p.toString()
    # print h.toString()
print 'No of processed events=',f.getEntries()
f.close()
```

We have commented out some debugging print statements intentionally: for a very large loop, the JythonShell console will be very quickly populated with large CPU consuming outputs.

We remind again that, when using the Java serialization, the retrieved objects from the HFile files are self-described. This can be checked in the above example by either uncommenting the debugging comments or checking the object types with the method type(obj).

11.3.4 Buffer Size for I/O Intensive Operations

We have mentioned before that, in order to optimize the execution speed of an application with I/O intensive operations, it is important to find a proper buffer size for

---

[1]For Java experts: flushing here means resetting ObjectOutputStream.

the `HFile` class. The buffer size depends on several factors, but one important factor is the size of the objects used for persistent storage on the disk. One simple way to optimize I/O is to benchmark the write/read operations. In the example below, we will create a relatively large array with random numbers and then write this array multiple number of times to a file. The objects will be written in an uncompressed way using the buffer size 1024 bytes.

```
───────────────── Benchmarking I/O operations ─────────────────
from jhplot.io import *
from jhplot import *
import time

p1=P0D('data')
p1.randomUniform(1000000,0,1)

buffer=1024
start = time.clock()
f=HFile('tmp.ser','w',0,buffer)
for i in range(1000):
   if (i%100 == 0):
      print 'pocessed=',i
   f.write(p1)
f.close()
print ' time (s)=',time.clock()-start
```

Note the required time to complete the execution of this code. Then change the buffer size. If you will set it to a small number (say, `buffer=2`) you will see a dramatic slow down in the execution of this script. Increasing the buffer size usually speeds up the I/O operations for large data sizes.

Of course, instead of the `P0D` array, one can use any jHepWork or Java object which implements the `Serializable` interface (almost all jHepWork objects do). After running the script above, do not forget to remove the temporary file `tmp.ser` (it is about 130 MB) as:

```
>>> import os
>>> os.remove('tmp.ser')
```

11.3.5 Input and Output to XML Files

Analogously, one can write and read data using the XML format. This can be achieved using the class `HFileXML`, replacing `HFile` by the `HFileXML` statement in the above example. You will see that the output files are significantly larger than in the case of using the `HFile` class; this time the data will be written to a human-readable XML format and no compression is used.

11.3.6 Non-sequential Input and Output

Data can also be written in a non-sequential order using a database-type approach. The database can be build using the HDataBase class.

A typical database associates a key of type string with each written record. This has some similarity with Jython dictionaries, but this time we are dealing with truly persistent approach. A data record can consist of only one "blob" of binary data. The file can grow and shrink as records are inserted and removed. The database operations do not depend on the number of records in the file, i.e. they will be constant in time with respect to file accesses. The index is small enough to be loaded into memory for efficiency.

Let us give a short example of how to create a small database. This will be a simple address book with two entries. First, we will create a binary database file, inserting two entries with the names of people as keys. Then we read the database again, check the number of entries and print all entries inside the file using the keys:

```
──────────── An object database: an address book ────────────
from jhplot.io import *
from java.util import *

# open a database for writing
f=HDataBase('addressbook.db','w')
f.insert('steve', 'Chicago')
f.insert('alexey','Minsk')
f.close()

# open database for reading
f=HDataBase('addressbook.db','r')
print 'Number of records=',f.getRecords()
print 'Is record exists?', f.isExists('alexey')
keys=f.getKeys()
while keys.hasMoreElements():
    next = keys.nextElement()
    print 'key=',next,' obj=',f.get(next)
```

The output of this script is shown below:

```
Number of records= 2
Is record exists? True
key= steve    obj= Chicago
key= alexey   obj= Minsk
```

The same approach can be used to store more complicated objects and associate them with the keys in form of strings. The example below shows how to build a file ('data.db') with object database which could be useful for writing arbitrary event records:

```
──────────────── An object database: Event records ─────────────
from jhplot  import *
from jhplot.io import *
import os.path

# build event from 3 objects
def makeEvent(entry):
    label='Event='+str(entry)
    p=P0D(label)
    p.randomUniform(10,0,1)
    h=H1D(label,10,-1,1)
    h.fill(i)
    return [label,p,h]

# write events into a database
f=HDataBase('data.db','w')
Events=1000
for i in range(Events):
    event=makeEvent(i)
    if (i%100 == 0):
        print event[0]+' size=',os.path.getsize('data.db')
    f.insert(str(i),event)
f.close()

f=HDataBase('data.db','r') # read the database
print 'extract event 26'; event=f.get('26')
print '=== '+event[0]+'===='
print event[1].toString()
print event[2].toString()
f.close()
```

Let us comment on this code. As before, for each insert(key,obj) method, we used the key defined by a string. Each string represents an event number but, of course, it can be any non-unique string. We access the event record No. 26 using the random access feature of this database and then print its objects.

One can always remove an object from the file with the remove(key) method or update an entry with the update(ob,key) method. Look at the Java API documentation of this class.

11.4 Compressed PFile Format

Despite the power of the Java serialization mechanism, it is important to note several its features when making decisions about input and output of numerical data:

- Java serialization can be slower in comparison with the usual I/O approach. It is good approach for storing complex objects, but it has unnecessary overhead for numerical data streams;
- It is not intended to be platform neutral. One cannot write or read data using other programming languages (such as C++);
- The size of the files created by the Java serialization can be larger than expected (even after compression). In fact, what we really need in many data-analysis applications is simply to write or read numerical values, rather than objects with all associated graphical attributes.

This poses a certain trade-off between the power of being able to write essentially any Java object into a Java serialized file and the shortcomings listened above. If Java serialization is too heavy approach for numerical calculations, consider an alternative I/O approach using the class called PFile. The methods of this class allow to write and read compressed data records constructed using the Google's Protocol Buffers package [6] (which gives the origin of the first letter "P" in the class name PFile). The Protocol Buffers program encodes structured data in a platform-neutral format, similar to XML. But, unlike XML, the Protocol Buffers approach is faster, simpler and the output file is smaller. This will be discussed in more detail in Sect. 11.8.

There are several advantages in using the class PFile: it is faster for writing compared to HFile and output data files generated by HFile are smaller. But, more importantly, one can use external programs to write and read files as long as a Protocol Buffers file which specifies data format is provided (again, see Sect. 11.8 for details). The class PFile is designed to keep data without their graphical attributes, which leads to smaller output files. However, only a limited set of objects is supported; at present, only objects from the package jhplot can be dealt by PFile.

In the previous section, we have considered constructors of the HFile class. An object of the class PFile can be created in a similar way. Below we list the major constructors to open a file 'data.pbu':

```
>>> f=PFile('data.pbu','w')        # open for writing
>>> f=PFile('data.pbu','r',map)    # open for reading and map
>>> f=PFile('data.pbu','r')        # as above with map=1
>>> f=PFile('data.pbu')            # as above with map=1
```

The file extension of the file is 'pbu', which originates from the underlying Protocol Buffers format. The only difference from the HFile class is additional boolean attribute 'map'. If this attribute is set to 1 ("true" in Java), we will create an association between a numerical record position and an object name which can usually be set using the setTitle() method. This is rather handy since one can retrieve an object by using its name (which can be obtained using the method getTitle()), rather than using its position in the PFile file. The price to pay is that it takes more resources in case of large files, since the constructor must pre-process records in order to make a map between object names and positions.

Let us give a detailed example. Below we generate two 1D arrays with 100 numbers and then create a two-dimensional array. We write all arrays in a compressed file:

```
────────── Writing jHepWork objects in PFile files ──────────
from jhplot.io import *
from jhplot import *

f=PFile('data.pbu','w')
x=P0D('X')
x.fill(100,0,1);  f.write(x)
y=P0D('Y')
y.fill(100,0,1);  f.write(y)

xy=P1D('XY',x,y); f.write(xy)
f.close()
```

Now let us read the objects from this file. As usual, we will open the file and check what is inside:

```
────────── Reading jHepWork objects ──────────
from jhplot.io import *
from jhplot import *

f=PFile('data.pbu')
print f.listEntries()
print 'No of entries=',f.getNEntries()

for i in range(f.getNEntries()):
    a=f.read(i+1)
    print 'entry=',i+1,type(a)
f.close()
```

We open the file in the read-only mode using the "name-mapping" option which will allow us to retrieve objects by calling their names. We also print all entries together with the record sizes using the method listEntries(). Finally, we read all records in a sequential order and verify their types:

```
1  ->  X  -->  809
2  ->  Y  -->  809
3  ->  XY --> 1613
No of entries= 3
entry= 1 <type 'jhplot.P0D'>
entry= 2 <type 'jhplot.P0D'>
entry= 3 <type 'jhplot.P1D'>
```

Since we have opened the file using the "mapped-name" option, one can take advantage of this by retrieving objects using their names, instead of the record positions:

```
――――――――― Reading objects using their names ―――――――
from jhplot.io import *
from jhplot import *

f=PFile('data.pbu')
x=f.read('X');    print type(x)
y=f.read('Y');    print type(y)
xy=f.read('XY');  print type(xy)
f.close()
```

Now we have convenient (and fast) access to records using their names. We should note that when the name-mapping option is enabled, objects inside PFile cannot contain duplicate names; check this by calling the method getTitle(). The name returned by this method will be used as a key for object mapping.

At present, all "named" objects can be processed by the PFile class, such as P0D-PND, F1D-F3D, FND, FPR, 1D and 2D histograms. One can also write strings.

One should point out again that the file format 'pbu' is platform-neutral. It has already been mentioned before that the record structure of the class PFile is constructed using a Protocol Buffers template file. This file, called HRecord.proto and located in the directory 'macros/system', can be used to build a C++ application which can read the files created by the PFile class (and vice versa). How to write more general data structures using the Google's Protocol Buffers will be discussed in Sect. 11.8.

If one needs to write jHepWork-specific objects from a C++ program, one can use the CBook C++ programming library [7] which is designed to write several major objects, such as H1D, H2D, P0D and P1D, into the 'pbu' files. Such files can later be opened by the Java PFile class, or can be analyzed and plotted using the PFileBrowser object browser to be discussed below.

11.4.1 Browser Dialog for PFile Files

The files created by the PFile class can be studied using the PFileBrowser dialog. This functionality is very similar to that of HFileBrowser class: Start the jHepWork IDE and select [Tools] and then [HPlot] menu. This brings up an empty HPlot canvas. Then press [Read data] using the option [File] on the tool bar menu. This pops us a file dialog for file loading. Select a file with the extension "pbu". This automatically opens an object browser which will be discussed in more detail in Sect. 10.2.7.

One can also call the browser from a code. In our next example we create a canvas and open a PFile file in the PFileBrowser dialog. This script shows a typical example:

```
──────────── Java PFile browser ────────────
from jhplot.io  import *
from jhplot  import *

f=PFile('test.pbu','w')
for i in range(10):
  p0= POD('Random='+str(i))
  p0.randomNormal(1000,0.0,1.0)
  f.write(p0)
f.close()

c1=HPlot('Browser')
c1.visible()
f=PFile('test.pbu')
PFileBrowser(c1,f,1)
```

Select one entry, and push the button "Plot". You will see a plotted histogram with all entries from the saved POD object. Similarly, one can plot H1D histograms, functions and multidimensional arrays.

11.5 Reading ROOT and AIDA Files

11.5.1 Reading ROOT Histograms

One can read and view files created by the C++ ROOT data analysis framework [8, 9]. At this moment, one cannot write ROOT files, but to read and extract histograms or graphs written in the ROOT format should not be a problem. For this, jHepWork uses the FreeHEP library.

First, open a ROOT histogram viewer as:

```
>>> from jhplot import *
>>> BRoot()                # open a viewer, or
>>> BRoot('Example.root') # open a ROOT file
```

The latter version of the constructor opens a ROOT file iwith the name Example.root' inside the histogram viewer. The BRoot class is a simple wrapper of the HistogramBrowser class from FreeHEP.

One can read ROOT files also without invoking a GUI browser. Reading 1D, 2D and 3D histograms can be done using the HRoot class which opens a ROOT file and extracts histograms:

```
──────────── Reading a ROOT file ────────────
from jhplot  import *
```

```
r=HRoot('Example.root')
print 'number of histograms=',r.getNKeys()
print 'ROOT version=',r.getVersion()
print 'List histograms=',r.toString()

h1=r.getH1D('mainHistogram') #  fetch a H1D
c1=HPlot('ROOT')
c1.setGTitle('ROOT histogram')
c1.visible()
c1.setAutoRange()
c1.draw(h1)
```

Here we first read the key of the ROOT histogram and then convert it into the H1D histogram. Then, the object h1 can be plotted in the usual way using the draw(h1) method of the HPlot class.

One can also use the class RootFileReader to do the same (but it takes more typing):

—————————————— Reading a ROOT file ——————————————

```
from jhplot  import *
from hep.io.root import *

r=RootFileReader('Example.root')
print 'Number of histograms=',r.nKeys()
print 'ROOT version=',r.getVersion()

key=r.getKey('mainHistogram') # read the key
his1=key.getObject()
h1=H1D(his1) # convert ROOT histogram to H1D

c1=HPlot('ROOT')
c1.setGTitle('ROOT histogram')
c1.visible(1)
c1.setAutoRange()
c1.draw(h1)
```

11.5.2 Reading ROOT Trees

The FreeHEP library contains a mechanism to read arbitrary ROOT objects, as well as the so-called ROOT trees. A ROOT tree can contain a list of independent branches, each can have its own definition. The RootIO package from the FreeHEP library allows to read ROOT files which can contain user defined objects [10]. The library is still under development and was tested with the ROOT 3.00 version. The RootIO package is included into the jHepWork library in a slightly redesigned style.

The topic of reading ROOT tree is rather common for high-energy physics, where data in the form of objects are stored sequentially in a machine independent way. Readers who are not familiar with this field can skip this section. For others, we will give a step-by-step instruction of how to read a file with ROOT trees containing user-defined objects (for example, particle tracks).

There are several steps in reading a ROOT file. First of all, one should look inside the file using the BRoot() method described above. This can help to understand the structure of the file.

Next, one should generate a Java interface library for the user-defined objects contained in a ROOT file. This interface library pre-processes a ROOT file and generates classes with the description of ROOT objects. We will illustrate this by using the ROOT file 'Event.root' which is located in

```
http://projects.hepforge.org/jhepwork/examples/data/
or
http://jwork.org/jhepwork/examples/data/
```

One can also find this file as well as the ROOT source file used to generate this file in [10].

Once the data file has been copied to a local directory, one should generate source files for our interface library by using this ROOT file. This can be done by executing the following code inside the jHepWork IDE:

```
―――――――――― Building an interface library ――――――――――
from hep.rootio.util import *

a=InterfaceBuilder('Event.root')
print a.message()
```

This script creates a directory 'hep' with all necessary Java files. These files need to be compiled into bytecode. There are many ways to do this. In case of jHepWork, one can compile all files in this directory as: (1) Navigate to the directory 'hep' using the project file menu, i.e. click on the button below the file menu (bottom left of the IDE) and select the directory 'hep'. Then, from the tool bar menu, select [Run] and [Javac all files]. You will see a message saying that all Java files have been compiled (check the existence of the files with the extension class in the directory 'hep').

Now you are ready to write a code which reads this ROOT file. Below we show how to access the ROOT tree and its branches inside the file 'Event.root'. Then we will plot squared transverse momenta of all particles (in this case, charged tracks).

```
――――――― Reading a ROOT tree with user-defined objects ―――――――
from hep.rootio.interfaces import *
from hep.rootio import *
from jhplot import *
```

```
reader = RootFileReader('Event.root')
tree =    reader.get('T')
branch = tree.getBranch('event')
n = branch.getNEntries()
print 'entries=',n

h1=H1D('Pt2',100,0,10)
for i in range(n):
    e = branch.getEntry(i)
    l = e.getTracks()
    print 'No=',i,' NTracks=',e.getNtrack(),' ',l.size()
    it = l.iterator()
    while(it.hasNext()):
        t = it.next()
        px=t.getPx()
        py=t.getPy()
        h1.fill(px*px+py*py)
c1=HPlot('Canvas')
c1.visible()
c1.setAutoRange()
c1.draw(h1)
```

The example reads all stored events, accesses tracks and extract their momenta. Then it plots squared transverse momenta of all stored tracks in the form of a 1D histogram.

We will not discuss this example further, since everyone who knows ROOT will not have any problems to understand it. We only note that one can also do all the above steps using a shell prompt. Look at the example in 'RootIO' stored in the 'example' directory.

11.5.3 Plotting ROOT or AIDA Objects Using jHepWork IDE

On can use the jHepWork IDE to read data files written in the ROOT or AIDA format in a GUI-driven fashion. One can do this as:

- Start the jHepWork IDE.
- Select the menu [Tools] and then [HPlotJa canvas]. You will see a new window which can be used for data visualization (based on the class HPlotJa). The HPlotJa class is described in Sect. 10.10.
- In the HPlotJa frame, go to the menu [Option] and select [Add pads]. You will see a sub-menu to build various pads (plotting regions), such as 1×1, $1 \times 2, 2 \times 1, 2 \times 2$. To create, for example, 4 plotting pads with $X-Y$ axes, select the sub-menu [pad 2x2].

- Open a data file using the [File] menu and sub-menu [Open data] of the HPlotJa window. The file extension should be '*.root' or '*.aida'. You will see an object browser from the right side of the HPlotJa frame. Using the mouse, locate some entry and then plot it using the right mouse button, which shows different choices of the pads where the object should be shown. One can plot data points either on one pad by overlaying the points from different objects, or one can plot data points on different pads.

One can do the same from a Jython script as:

```
>>> from jhplot import HPlotJa
>>> c1=HPlotJa('Canvas')
>>> c1.visible()
>>> c1.showBrowser('file.root')
>>> c1.showEditor(1)   # open an object editor
```

Once the script is executed, you should see a browser (on the left of the main canvas) listing all objects inside the file. If one clicks on an entry of this list, a histogram or data set will be plotted. One can also plot data on different pads using the mouse pop-up menu. One can learn more about how to read ROOT or AIDA files in Sect. 11.5.1.

11.6 Working with Relational SQL Databases

We cannot avoid the discussion of relational databases based on the SQL standard, since such databases are reality of our life.

So far we have discussed a "flat file" approach in which a binary or text file usually contains one record per line. A SQL-relational database has several advantages: one can implement a server-client mode, scalability (indexing of records), and concurrency when one needs to synchronize multiple updates at the same time.

However, I should warn you: in many cases you do not need such SQL databases at all. For data analysis, we often read data sequentially, i.e. we read data records from the top to the bottom all the way through. We are in less degree worry about a random access, concurrency or client-server mode as it is implemented in many relational databases. Straight file access is usually faster than executing a SQL query and less memory consuming.

Thus, in terms of performance, a data-analysis code may not benefit from switching to the SQL databases. Using a SQL database may or may not be worthwhile. Your decision should be based on complexity of the data access, where a database software needs to be installed, and many other factors.

Below will consider the Apache ("Derby") open-source relational database. More detail about Java implementation of this database can be found elsewhere [11].

11.6.1 *Creating a SQL Database*

First of all, let us prepare a module which keeps common information about our SQL database. This module will be necessary for creation and retrieval information in our further examples. We will consider an embedded Derby database i.e. a database for a simple single-user Java application. In this mode, only a single application can access the database at the time and no network access is required.

Let us prepare a file 'openDB.py' with the information necessary for creation of database, such as drivers, protocol, name of database table. The name of our database will be 'derbyDB', which will keep a single table with the name "location". We also set a user name and a password for this database. Below is a script which has all necessary information:

```
──────────── Common module: 'openDB.py' ────────────

from java.sql import *
from java.lang import Class
from java.util import Properties

framework = 'embedded'
driver = 'org.apache.derby.jdbc.EmbeddedDriver'
protocol = 'jdbc:derby:'
dbName='derbyDB'
table='location'

Class.forName(driver).newInstance()
props = Properties();
props.put('user', 'jhepwork')
props.put('password', 'secret')
```

This file will be imported for all Jython modules to be discussed below.

Now let us create a database called "derbyDB". Here we will use the Jython syntax to create a database and to insert several SQL statements into the table 'location', which will keep information about the street name (type string) and home number (integer).

The script below does the following: (1) loads the JDBC Java driver in order to start-up the Derby database. (2) Creates a table inside the database, first checking whether this table is already created. If it does exist, we remove it (see the statement 'drop table'+table. (3) Then we insert a few records with addresses.

```
──────────── Creating a SQL database ────────────

from openDB import *

scon=protocol+dbName+';create=true'
conn = DriverManager.getConnection(scon, props)
s = conn.createStatement()

try:
```

```
    s.execute('drop table '+table)
except SQLException:
  print 'no need to remove table'

s.execute('create table '+table+\
          '(num int, addr varchar(40))')
s.execute('insert into '+table+\
          ' values (1956,\'Webster St.\')')
s.execute('insert into '+table+\
          ' values (1910,\'Union St.\')')

s1='update '+table+' set num=?, addr=? where num=?'
ps = conn.prepareStatement(s1)
ps.setInt(1,180)
ps.setString(2, 'Grand Ave.')
ps.setInt(3, 1956)
ps.executeUpdate()
print 'Updated 1956 Webster to 180 Grand'
s.close()
conn.commit()

try:
  DriverManager.getConnection('jdbc:derby:;shutdown=true')
except SQLException:
  print 'all done'
```

This example shows how to use two alternative methods to insert and update the database: (1) one is based on the standard SQL statements, and the second using a prepared statement based on the method prepareStatement() for fast query.

It is important to close the database properly using the line:

```
'jdbc:derby:;shutdown=true'.
```

The DriverManager should raise only one exception : SQLException. The database should be shut down so it can perform a checkpoint and releases its resources.

After the execution of the script above, you will see a database directory derbyDB with the stored information.

11.6.2 Working with a Database

Now let us read the database entries. As before, we will use exactly the same common module 'openDB.py'. This time, however, we will set create=false for the argument of DriverManager.getConnection. Using two SQL queries,

we will print all database records and will search for a record with the string
`'Union St.'`. Finally, we close the database:

```
—————————————— Reading database ——————
from openDB import *

scon=protocol+dbName+';create=false'
conn = DriverManager.getConnection(scon,props)
s = conn.createStatement()

s1='SELECT num, addr FROM '+table+' ORDER BY num'
rs = s.executeQuery(s1)
while rs.next():
  print 'sorted='+rs.getString(1),rs.getString(2)

s2 = 'SELECT * FROM '+table+' WHERE addr=\'Union St.\''
rs = s.executeQuery(s2)
while rs.next():
  print 'Found=',rs.getString(1),rs.getString(2)
s.close()
conn.commit()

try:
  DriverManager.getConnection('jdbc:derby:;shutdown=true')
except SQLException:
  print 'all done'
```

11.6.3 Creating a Read-only Compact Database

Sometimes it is useful to create a read-only database and compress it into a jar file.
This allows a database to be distributed as a single file instead of multiple files
within a directory. In this case, we will not be able to modify it, since the database
will be represented by a single self-contained file.

The operation discussed above can easily be done with the Jython module `os`.
This module will be used to create the file `'derbyDB.jar'` with the database
directory by calling the external `jar` command (it comes with the Java installation):

```
————————— Compacting a database to a jar file ——————
import os
cmd='jar cMf derbyDB.jar derbyDB'
print os.system(cmd)
```

Now, how can we read the file `'derbyDB.jar'`? This can be done in same spirit
as shown in Sect. 11.6.2: use the same script, but add the extra line shown below:

```
protocol = 'jdbc:derby:jar:('+dbName+'.jar)'
```

after the `import` statement. Try to execute the corrected script. After this, you should be able to read the database stored in this jar file.

11.7 Reading and Writing CSV Files

The comma separated value files, CSV, are used to store information in a form of tables. This format is especially popular for import and export in spreadsheets. Data in such files are either separated by a comma or a tab, or by any other custom delimiter.

We have already considered how to read and write the CSV files using the Jython module `csv` (see Sect. 2.16.4). Below we will consider high-performance Java libraries to work with such files.

11.7.1 Reading CSV Files

jHepWork supports reading the CSV files as well as writing data into such files. Let us create a typical CSV file using any text editor:

```
───────────────────── File 'table.csv' ─────────
Sales,  Europe,  Russia,  USA
January,   10,   20,      20
February,  30,   20,      50
March,     10,   40,      100
April,     80,    7,      30
May,      300,  400,      90
June,      50,   10,      70
```

Dealing with CSV format couldn't be much easier when using the jHepWork Java classes—just open a CSV file using the class `CSVReader` and iterate over all its entries as shown in the example below:

```
───────────────────── Reading table.csv file ─────────
from jhplot.io.csv import *

r=CSVReader('table.csv',',')
while 1:
   line= r.next()
   if line == None: break
   print line
r.close()
```

As one can see, the method `next()` returns a list of strings for each line of the CSV file. File reading stops when the end of the file is detected. If you want to read arrays of lists, use the method `readNext()` instead. One can easily convert strings inside the list representing each row to floats and integers, as `float(str)` (returns a float value) or `int(str)` (returns an integer).

One can read all entries at once, instead of looping over all rows. This can be done using the methods:

`l=r.readList()` read all elements into a 2D list;
`l=r.readAll()`   read all elements into list, each row is represented as an array.

Let us take a closer look into the `CSVReader` class. It has several constructors:

```
>>> CSVReader(file, separator, quotechar, line)
>>> CSVReader(file, separator, quotechar)
>>> CSVReader(file, separator)
>>> CSVReader(file)
```

where `file` is the string with a file name, `separator` is the character used for separating entries in each row, `quotechar` is the character to use for quoted elements. Finally, `line` is an integer number representing the line number to skip before start reading. When no `separator` is given, then the default separation is done using a comma. If no `quotechar` is given, the default escaping quotes is ["], and if no `line` is given, the default end of line is the new line symbol.

Finally, one can display a CSV file in a spreadsheet using the `SPsheet` class discussed in Sect. 12.4.2. Once `CSVReader` object has been created, just pass it to the `SPsheet` object as:

```
─────────── Showing entries from 'table.csv' file ───────────
from jhplot import    *
from jhplot.io.csv import *
r=CSVReader('table.csv',',')
SPsheet(r)
```

Now let us move to the next subject where we will discuss how to write data into the CSV files.

11.7.2 Writing CSV File

To write data to CSV files is as easy as reading them. Instead of the `CSVReader` class, one should use the `CSVWriter` class. It has the following constructors:

```
>>> CSVWriter(file, separator, quotechar, line)
>>> CSVWriter(file, separator, quotechar)
```

```
>>> CSVWriter(file, separator)
>>> CSVWriter(file)
```

Here, `file` is an output file name, and other arguments are exactly as for the `CSVReader` class.

Let us show an example illustrating how to write lists of objects line by line:

```
———————————— Writing 'out.csv' file ——————————————
from jhplot.io.csv import *

w=CSVWriter('out.csv',',')

w.writeNext( ['Test1','20','30'] )
w.writeNext( ['Test2','100','50'] )
w.writeNext( ['Test3','200','100'] )

w.close()
```

Check the output file. It has the following structure:

```
"Test1","20", "30"
"Test2","100","50"
"Test3","200","100"
```

One can write the entire file at once using the `writeAll(list)` method.

11.8 Google's Protocol Buffer Format

Many situations require an analyzer to write structural data into a format which can be understood by variety of programming languages. Imagine an experimental apparatus producing a data stream. It is very likely that its code for data I/O is written in C/C++, since most hardware drivers are implemented in this system language. But, at the end of the day, this is a user who should analyze the data and who may prefer a human-friendly language (like Java or Jython) for final analysis code. Therefore, we should find a way to read structural data produced by experiments using these higher-level languages. Or, imagine an opposite situation: an analyzer produces data using Java or Jython, and an application implemented using a lower-level language should read this data during communication with a hardware.

One way to deal with such kind of problems is to use a self-described file format, such as XML. But there are several problems with this format when dealing with large data volumes: (1) programs are slow for loading; (2) there is a significant penalty on program's performance; (3) data files are large due to tags overhead.

Taking into account that the XML format is too cumbersome to use as an encoding method for large data files, one can use the ROOT format as an alternative. This may seem to be rather heavy approach (installed ROOT takes several hundreds of megabytes!), and to read most recent ROOT files in Java is not too easy: in fact, the ROOT framework was not designed from the ground to be friendly for other programming languages.

The Google team has recently released the Protocol Buffers package [6] which deals with serializing structured data in a platform-neutral format. This data-interchange format is used by Google for persistent storage of data in a variety of storage systems. It is also well tested on many platforms. The Protocol Buffers is self-describing format, which is equally well supported by C++, Java, Python and by other languages using third-party packages. In comparison with the XML format, the Protocol Buffers files are up to a factor 100 times faster to read, and file sizes are significantly smaller due to a built-in compression. This format is also very promising due to the offered backwards compatibility: new fields created by new protocol versions are simply ignored during data parsing.

Even more. The Protocol Buffers helps to abstract from a language-specific description of data structures. More specifically, this means that an analyzer only needs to produce a file describing his/her data records, and then the Protocol Buffers program generates a C++ or Java code for automatic encoding and parsing data structures.

Below we will discuss the Protocol Buffers in more detail. The program is included into jHepWork; look at the library protobuf.jar in the directory lib/system. To develop applications in C++, the reader is assumed to install the Protocol Buffers (at least version 2.2) from the official web site [6].

11.8.1 Prototyping Data Records

For our next example, we will assume the following experiment. We perform N measurements, each measurement is characterized by an identification number, a string (name) and arbitrary array with some other data. In each measurement, it is assumed that we observe n number of particles. This number is not fixed, and can vary from measurement to measurement. Finally, for each particle, we measure particle's energy and electric charge. We also assign a string with particle name.

Let us prototype an event record using the Protocol Buffers syntax, which is then can be used to generate a C++ or Java code. The code is shown below:

```
────────── Data prototype. File 'experiment.proto' ──────────
package proto;
option java_package = "proto";
option java_outer_classname = "Observations";

message Event {
   required int64   id = 1;
   required string  name=2;
   repeated double  data = 3 [packed=true];
```

```
message Particle {
  required string   name=1;
  required sint32   charge=2;
  required double   energy=3;
}
repeated Particle particle = 4;
}
message Experiment {
repeated Event event = 1;
}
```

This file contains the description of all objects in self-described platform-independent way, so later one can read or write data from a variety of languages. As one can see, the class Event envelops the entire data. Instead of using the class name, the Protocol Buffers uses the word "message". Each event has 4 records, id (integer), name (string), data (list of doubles) and the class Particle. Each such record (or "message") has numbered fields (in this example, running from 1 to 4). Such integer values are unique "tags" used in the binary encoding.

The value types can be numbers, booleans, strings or other message types, thus different messages can hierarchically be nested. The required field tells that such field should always be present. One can also set the default value using the line [default=value], where value is a given default value. The message field can have other two types: optional—a message can have zero or one of this field (but not more than one), and repeated—a message can be repeated any number of times, preserving the order of the repeated values. The fields can include enumerators for tight association of a specific value to a variable name.

In the above example, the message data can be repeated (similar to a list), so we can append any number of double values in each record. The field [packed=true] is used for a more efficient encoding of multiple data entries. Finally (and this is very important), the message of the Particle type can also be repeated, since we can have multiple number of particles in events.

As you may already have guessed, the message Particle keeps information about a single particle, such as its name, charge and energy. This message is constructed in exactly the same way as the outer messages. The only difference is that now we should specify the fields which are the most appropriate for defining particle properties.

A scalar message field can have one of the following types shown in Table 11.1. More types are given in [6].

11.8.2 Dealing with Data Using Java

Now let us generate a Java code using the above prototype file. Assuming that the Prototype Buffer is properly installed, and assuming that we are working in the Unix/Mac environment, we generate the Java code as:

Table 11.1 Scalar value types used in `.proto` files and their Java and C++ equivalents	Scalar message field types	
	int	int (Java and C++)
	double	double (Java and C++)
	int	int (Java and C++)
	bool	boolean (Java) and bool (C++)
	string	String (Java) and string (C++)
	fixed32	int (Java) and int32 (C++)
	fixed64	long (Java) and int64 (C++)

```
———————— Generating Java code ————————

protoc --java_out=.  experiment.proto
```

This creates a directory `'proto'` with the `Observations.java` file. The `-java_out` option tells to generate Java classes to be used for data encoding. Similar options are provided for other supported languages. The dot after this option tells to generate the output in the current directory. The output of this command is the Java file `Observations.java` located in the directory `proto`, as specified in the original `experiment.proto` file (see the Java-package statement).

Let us write a test code which generates 100 events with 10 particles in each event. We will create the file `WriteData.java` and copy it into the directory `'proto'` together with recently generated file `Observations.java`:

```
——— Writing data into a file. 'WriteData.java' file ———

package proto;
import proto.Observations.Experiment;
import proto.Observations.Event;
import java.io.FileOutputStream;

class WriteData {
public static void main(String[] args) throws Exception {
   Experiment.Builder exp = Experiment.newBuilder();
   FileOutputStream output =
                new FileOutputStream("data.prod");
   for (int e=0; e<100; e++){
     Event.Builder ev = Event.newBuilder();
       ev.setId(e); ev.setName("collision");
       ev.addData(1); ev.addData(2);
         for (int i=0; i<10; i++){
           Event.Particle.Builder p =
                 Event.Particle.newBuilder();
           p.setName("proton");
           p.setCharge(1); p.setEnergy(1);
           ev.addParticle(p);
         };
```

```
      exp.addEvent(ev); } ;

   exp.build().writeTo(output);
   output.close();
   }
}
```

Next we need to compile all Java files and build a jar library. For this, we define
an environmental variable JHEP as a directory with the installed jHepWork and then
compile and build a jar library.

─────────────── Compiling Java code ───────────────

```
javac -1.6 -cp $JHEP/lib/system/protobuf.jar proto/*.java
jar -cf proto.jar proto/*
```

This produces the file proto.jar with the compiled classes.

Let us test the above code. We will run the WriteData class in the usual Java
fashion.

─────────────── Running WriteData program ───────────────

```
java -cp $JHEP/lib/system/protobuf.jar:proto.jar \
        proto.WriteData
```

The result of this command is the output file 'data.prod' with our structural
data written in a binary form. Let us verify what is written by adding the code
ReadData.java into the directory 'proto'.

─────────────── Reading data. 'ReadData.java' file ───────────────

```
package proto;
import proto.Observations.Experiment;
import proto.Observations.Event;
import java.io.FileInputStream;

class ReadData {

public static void main(String[] args) throws Exception {
  Experiment   exp=
    Experiment.parseFrom(new FileInputStream("data.prod"));
    for (Event ev: exp.getEventList()) {
      System.out.println("Event id:" +ev.getId());
      System.out.println("Event name:" +ev.getName());
        for (Event.Particle p: ev.getParticleList()) {
          System.out.println(p.getName());
          System.out.println(p.getEnergy());
    } }
} }
```

It opens the file `data.prod` and fetches all objects in a loop, until all events are processed. For each event, the code extracts particles and their attributes. Let us compile this code by invoking the compilation command `javac` as was done before, assuming that it is located in the same directory `'proto'`. This will produce a new jar file. Then, run the code using the command:

```
──────────── Running ReadData program ────────────
java -cp $JHEP/lib/system/protobuf.jar:proto.jar \
        proto.ReadData
```

This code reads the data and prints all events with particle attributes.

11.8.3 Switching to Jython

In Sect. 12.12 we will discuss how to unwrap a Jython code into Java. In this example we will show the opposite case: We will rewrite our example code `'WriteData.java'` into a compact Jython script:

```
──────── Writing data into a file. 'WriteData.py' file ────────
from proto.Observations import Experiment
from proto.Observations import Event
from java.io import FileOutputStream

exp=Experiment.newBuilder()
for e in range(100):
    ev = Event.newBuilder()
    ev.setId(e); ev.setName('collision')
    ev.addData(1); ev.addData(2)
    for i in range(10):
        p=Event.Particle.newBuilder()
        p.setName('proton')
        p.setCharge(1); p.setEnergy(1)
        ev.addParticle(p)
    exp.addEvent(ev)
exp.build().writeTo(FileOutputStream('data.prod'))
```

As you can see, it is much smaller than the equivalent Java program. To run this code, we must copy `proto.jar` file to a place where Java can see it (for example, in the directory `lib/user`) or to include its location into Java CLASSPATH. Then one can execute this script using jHepWork IDE.

Similarly, one can rewrite the Java code used to read the data as:

```
──────── Reading data into a file. 'ReadData.py' file ────────
from proto.Observations import Experiment
from proto.Observations import Event
```

```
from java.io import FileInputStream

exp=Experiment.parseFrom(FileInputStream('data.prod'));
for e in exp.getEventList():
   print e.getId()
   print e.getName()
   for p in e.getParticleList():
       print p.getName(),p.getEnergy();
```

Again, run this code and you will see exactly the same output as in the case of Java.

11.8.4 Adding New Data Records

Now let us illustrate how to add a new event record to the existing file. To make our codding shorter, we will use Jython and add an event (but without any particle) as:

```
                    —— Adding new records. 'AddData.py' file ——
from proto.Observations import Experiment
from proto.Observations import Event
from java.io import FileOutputStream
from java.io import FileInputStream

exp=Experiment.newBuilder()
try:
   input=FileInputStream('data.prod')
   exp.mergeFrom(input); input.close()
except FileNotFound, e:
    print 'The file was not found, going to backup file'

ev = Event.newBuilder() # add new event
ev.setId(99); ev.setName('new entry')
exp.addEvent(ev)

output=FileOutputStream('data.prod')
exp.build().writeTo(output)
output.close()
```

In this example we simply read the file and then write a new file with the additional event.

11.8.5 Using C++ for I/O in the Protocol Buffers Format

Now we come to the main issue of how to generate the same file using a C++ program, so we can read it using Java, Jython or any other language. Of course, we also need to test the opposite situation when reading the Java-generated file 'data.prod' by a C++ program.

Below we will show the necessary steps to produce a C++ code which generates exactly the same data file as that shown in the previous section, and then how to read this file (or the file generated by the Java code above). We will use the same prototype file 'experiment.proto' as before. Let us generate a C++ code to be used for data encoding:

```
─────────────────── Generating C++ code ───────────────────
protoc --cpp_out=. experiment.proto
```

Note the option cpp which tells to generated the C++ encoding code. After execution of this line, we should find two files, experiment.pb.cc and experiment.pb.h in the same directory. As in case of Java, they contain the necessary information to be used for data serialization.

Let us write a small test program which creates the output file.

```
─────── Writing data into a file. 'write_test.cc' file ───────
#include <iostream>
#include <fstream>
#include <string>
#include "experiment.pb.h"
using namespace std;

int main(int argc, char **argv)
{
  GOOGLE_PROTOBUF_VERIFY_VERSION;
  proto::Experiment exp;

  for (int e=0; e<100; e++){
   proto::Event* ev = exp.add_event();
   ev->set_id(e); ev->set_name("collision");
   ev->add_data(1); ev->add_data(2);
     for (int i=0; i<10; i++) {
     proto::Event::Particle* p=ev->add_particle();
     p->set_name("proton");
     p->set_energy(1); p->set_charge(1);
     } };

    cout << "Write data.prod" << endl;
    fstream output("data.prod",
            ios::out | ios::trunc | ios::binary);
    if (!exp.SerializeToOstream(&output)) {
      cerr << "Failed to write address book." << endl;
      return -1; }

    output.close();
    google::protobuf::ShutdownProtobufLibrary();
```

```
        return 0;
}
```

For those who know C++ this code should look rather simple. Moreover, the code logic is rather similar to that of the Java code. Since we declared the package `proto` in the prototype file, one should use the corresponding C++ namespace for the class declarations. The first statement in the `main()` method verifies that we have not accidentally linked against the library which is incompatible with the version of the headers.

Next, we will need to compile all these files. For this, we use GNU gcc compiler installed by default on Unix/Linux systems

────────────── Compiling the source codes ──────────────

```
gcc write_test.cc experiment.pb.cc \
    -o write_test `pkg-config --cflags --libs protobuf`
```

Finally, try to execute the program `write_test`. We will see the created file `'data.proto'`. Check its size. It should have the same size as in the case of Java example. Surely, this is a good sign, since it likely indicates that the new file is correctly produced and, in fact, is totally identical to that generated previously using Java.

Let us move on and write a C++ code which will read the data from this file.

────────── Reading data from a file. 'read_test.cc' file ──────────

```
#include <iostream>
#include <fstream>
#include <string>
#include "experiment.pb.h"
using namespace std;

int main(int argc, char **argv)
{
 GOOGLE_PROTOBUF_VERIFY_VERSION;
 proto::Experiment exp;

 fstream input("data.prod", ios::in | ios::binary);
 if (!exp.ParseFromIstream(&input)) {
   cerr << "Failed to parse data file" << endl;
   return -1; }

 for (int i = 0; i < exp.event_size(); i++) {
   const proto::Event& ev = exp.event(i);
   cout << " ID: " << ev.id() << endl;
   cout << " Name: " << ev.name() << endl;

  for (int j = 0; j < ev.particle_size(); j++) {
    const proto::Event::Particle& p = ev.particle(j);
```

```
   cout << " - name: " << p.name() << endl;
   cout << " - energy: "   << p.energy() << endl;
} }
google::protobuf::ShutdownProtobufLibrary();
return 0;
}
```

The file reads the data and loops over all event entries. We can compile all source files as usual:

```
─────────────── Generating C++ code ───────────────
gcc read_test.cc experiment.pb.cc \
    -o read_test `pkg-config --cflags --libs protobuf`
```

and run the executable program as read_test. We will see the print messages included for debugging.

As a final exercise, try to use the Java or Jython code developed in the previous subsection for reading file created by our C++ program.

11.8.6 Some Remarks

The Protocol Buffers is easy-to-use format to organize sequential data in a platform-independent way. It is a good format for handling individual messages within a large data set which consists of large number of small pieces. Each piece can also be a structural data in the form of "messages".

We have already discussed that the class PFile (see Sect. 11.4) is completely based on the Protocol Buffers format. Each record is compressed using the ZIP file format. The records are implemented using the Protocol Buffers file HRecord.proto located in the directory 'macros/system'. The file HRecord.proto can be used to construct applications in C++ which can read and write files generated by the PFile Java class.

11.9 EFile Data Output

The usage of the Protocol Buffers encoding is simplified by the class EFile designed to deal with stacking Protocol Buffers messages into compressed files. Each message is written independently, which allows to write and read large data volumes organized in separate data records.

The data structure of each data record is based on the file HEvent.proto located in the directory macros/system. Since this file is rather short, we will show it here:

```
──────────────────── File 'HEvent.proto' ────────

message HEvent {
  optional int64    id = 1;
  optional string   name=2;
  repeated double   da = 3 [packed=true];
  repeated int64    ida= 4 [packed=true];
  message Entry {
    optional string   name=1;
    repeated double   da=3  [packed=true];
    repeated int64    ida=2 [packed=true];
  }
  message Group {
    optional string   name=1;
    repeated Entry   entry=2;
  }
  repeated Group   group = 5;
}
```

This file specifies the structure of a single "event" record which can be written in a sequential order using the class EFile. As usual, the strength of this approach is that one can write and read structured data using any programming language, as long as we can generate the source file with the language-specific implementation of the template shown above.

The organization of each event record given by the file HEvent.proto is following: The outer message represents an "event", which is characterized by an integer value id, name (string) and arrays of double (da) and integer (ida) values. In addition, each event HEvent holds a group of objects in a form of arrays. Each group is characterized by its name and, in turn, contains an array of other objects, called "entries". Each entry holds arrays of double (da) and integer (ida) values. In addition, each entry can have a name (string).

Surely, the structure given by the file HEvent.proto is not guaranteed to be universal and may not fit to possible situations. But the file can easily be modified and C++ or Java code source can be regenerated using this Protocol Buffers file depending on circumstances.

Yet, the data structure discussed above can be appropriate for a large range of situations. We will give here two examples where such event structure is adequate. In case of astrophysics, each HEvent can characterize a certain cluster of galaxies. Each cluster consists of galaxies which can be represented by an array of "Group" messages. In turn, the "Group" message can contain a description of stars in form of "Entry" messages. Finally, each "Entry" record can store information about a star which can be characterized by arrays of integer or double values.

Let us give another example. For particle physics, each HEvent record can represent a result of collision of two particles. Then, the "Group" message will characterize either "tracks" (charged particles) or all particles measured by a calorimeter (we will call them "cells"). Next, each entry can keep details of either a track or a cell characterized by its momenta, energy, location, etc.

Let us come to a specific example which illustrates a concrete program implementation of this approach. In case of jHepWork, the generated class from the above prototype is called `PEventFile`. Using this class and the class `EFile` which implements an I/O layer for each `HEvent`, one can create a file holding multiple records as shown in this example:

```
————————————— Writing events into a file —————————————

from jhplot.io import *
from jhplot.io.PEventFile import *

f=EFile('tmp.nbu','w')
for i in range(1000):
    ev = HEvent.newBuilder()
    ev.setName('collision')
    ev.setId(i); ev.addDa(1)

    tracks = HEvent.Group.newBuilder()
    tracks.setName('tracks')
    for j in range(20):
       track=HEvent.Entry.newBuilder()
       track.setName('pion'+str(i))
       track.addDa(1); track.addDa(2)
       tracks.addEntry(track)
    ev.addGroup(tracks)

    cells = HEvent.Group.newBuilder()
    cells.setName('cells')
    for j in range(20):
       cell=HEvent.Entry.newBuilder()
       cell.setName('cell'+str(i))
       cell.addDa(1);cell.addIda(1)
       cells.addEntry(cell)
    ev.addGroup(cells)

    f.write(ev) # write event
f.close()
```

As the reader may notice, we use the second example from particle physics. For each collision, we write two groups of objects, "tracks" and "cells". Then we filled "entries" characterized by one double and one integer number. Of course, nothing prevents us from adding more groups, more entries or more primitive values for each object. By convention, we will create an output file with the extension 'nbu'.

Now let us read this file. This part is rather simple:

```
————————————— Reading events —————————————

from jhplot.io import *
from jhplot.io.PEventFile import *
```

```
f=EFile('tmp.nbu')
print f.getVersion()

for i in range(1,f.size()):
    ev = f.read()
    print 'Read =',ev.getId(),ev.getName()
      # get groups
    for j in range(ev.getGroupCount()):
        g=ev.getGroup(j)
        print g.getName()
        # get entries
        for k in range(g.getEntryCount()):
            e=g.getEntry(k)
            print e.getDa(0)
f.close()
```

The above example shows how to loop over all HEvent entries, retrieve groups and then entries from each group. The method getVersion() checks the version of EFile implementation used to generate the file. The total number of HEvent messages is given by the method size().

As was already said, Java and Jython allow to write arbitrary structural data volumes in a simple and transparent way using their own classes and methods. But the strength of the above approach is that data files can also be generated by other programming languages using the Protocol Buffers template. More specifically, in case of C++, the CBook [7] library can be used to write data structures using the HEvent.proto prototype file implemented in C++. Therefore, one can generate data using a C++ code and read data using the EFile Java class exactly as shown in the previous example.

11.10 Reading DIF Files

The DIF ("Data Interchange Format") format has a limited support in jHepWork. Files in this format have the extension '.dif' and can be used to import or export spreadsheets between various applications.

The current implementation of this package [12] is only for read-only mode. There is no possibility for saving data into this file. To read such files, one should import the module dif located in the 'macros/system' directory. In the example below, we read the file 'nature.dif', extract all information into Jython tuples and print the entries:

_____ Reading the file nature.dif _____

```
import dif

f=open('nature.dif','r')
d = dif.DIF(f)
```

```
print d.header
print d.vectors
print d.data
```

Below we give some explanation:

d.header   dictionary of header entities, with names shifted to lowercase for ease
            of handling
d.data   list of tuples with data
d.vectors   list of vector names (or, better, column names)

Read more information about Python implementation of this module in [12].

11.11 Summary

It may well be that you get confused by the significant number of I/O formats supported by jHepWork. Below we will summarize the jHepWork I/O and give some guiding tips about what exactly to use in certain situations.

11.11.1 Dealing with Single Objects

If we need to save one object at the time (like a histogram, function or a P-type array), use their native methods. For example, to write an object into a file, use the toFile() method, see Sects. 5.4 and 8.6. For reading data, one can use the corresponding constructors which accept the name of the file with input data for object initialization. This also includes a compression mechanism, see the method readGZip() in Sect. 5.4. The good thing about this approach is that: (1) it is fast; (2) data can be produced using other programming languages, as long as our codding strictly follows the native jHepWork format.

A slower approach would be to use Java serialization, i.e. serializing separate objects with the method writeSerialized(). By default, files will be compressed. One can also write and read data stored in XML files using the method writeXML() and readXML(). This will be the slowest approach, but the advantage is that the XML files are human-readable and human-editable. Finally, it is rather easy to read such files using applications implemented in other programming languages.

11.11.2 Dealing with Collections of Objects

To read and save collections of jHepWork objects, use the serialization method writeSerialized() or readSerialized(). This can be done "in one go", if all Java objects are put either in a list or map. This approach also allows to deal with the XML-type formats. As we have already mentioned, this is human-readable, self-describing and platform-neutral protocol.

For large and complicated data structures, it makes sense to use the HFile class, see Sect. 11.3. In most cases, objects can be viewed using a GUI browser. Data in such files are compressed by default. As before, one can serialize data in XML using the class HFileXML.

An arbitrary collection of structural data can also be written using the Google's Protocol Buffer format discussed in Sect. 11.8. It is a compact way of encoding data in a binary format. The main advantage of this format is that data can be processed by programs implemented in other programming languages.

If we are interested in the jHepWork objects only, use the class PFile. This approach has several advantages compared to the Java serialization as discussed in Sect. 11.4. The class PFile writes data in a form of compressed records based the Google's Protocol Buffer format. This approach encodes high-level jHepWork objects (like histograms) in a platform-independent way. This means that data can be generated by C++ programs using the package CBook [7] and processed using the PFile class.

Similarly, one can use the class EFile (see Sect. 11.9) which deals with a rather broad range of data structured in "events" and "records". Both PFile and EFile can be used to read a particular event or record, without uncompressing the whole file. This feature is particularly useful for reading data in parallel using multi-core processors (this will be discussed in Sect. 16.4).

Similarly, one can use the class HBook and the CFBook package, see Sect. 8.6.1, which are mainly designed for XML-type outputs (but more optimized to keep large data sets, removing unnecessary tags for arrays). No compression is used.

As discussed in this chapter, if one needs to find an association between a key and a data object, one can use databases. In the simplest, non-persistent case, use Jython dictionaries or Java maps. For storing information in a persistent state, one can use a database based on the class HDataBase, see Sect. 11.3.6. Note that one can use a similar database approach using the PFile class. As discussed earlier, it can also be used for a direct association between a stored object inside a file and its textual key (which is just object title). For complicated input queries, one can use the Apache Derby SQL-type databases, see Sect. 11.6.

The format CSV is only useful for keeping data in simple tables. The only advantage of this approach is that it is simple and data can easily be converted by a spreadsheet-like application.

Finally, one can use a pure-Python approach or calling Java classes, see Sect. 2.16. The pure-Python method is generally not recommended for large data files, but it is very useful when it comes to the simplicity.

References

1. Richardson, C., Avondolio, D., Vitale, J., Schrager, S., Mitchell, M., Scanlon, J.: Professional Java, JDK 5th edn. Wrox, Birmingham (2005)

2. Arnold, K., Gosling, J., Holmes, D.: Java(TM) Programming Language, 4th edn. Java Series. Addison-Wesley, Reading (2005)
3. Flanagan, D.: Java in a Nutshell, 5th edn. O'Reilly Media, Sebastopol (2005)
4. Eckel, B.: Thinking in Java, 4th edn. Prentice Hall, Englewood Cliffs (2006)
5. Bloch, J.: Effective Java, 2nd edn. The Java Series. Prentice Hall, Englewood Cliffs (2008)
6. Google Protocol Buffers project. URL http://code.google.com/p/protobuf/
7. Chekanov, S.: CBOOK histogram library. URL http://jwork.org/jhepwork/cbook
8. Brun, R., Rademakers, F., Canal, P., Goto, M.: Root status and future developments. ECONF C0303241 (2003) MOJT001
9. Brun, R., Rademakers, F.: ROOT: An object oriented data analysis framework. Nucl. Instrum. Methods A **389**, 81 (1997). URL http://root.cern.ch/
10. Johnson, T.: FreeHEP RootIO. URL http://java.freehep.org/freehep-rootio/
11. Apache Derby: URL http://db.apache.org/derby/
12. Gonnerman, C.: Navy DIF file handler. URL http://newcenturycomputers.net/projects/dif.html

Chapter 12
Miscellaneous Analysis Issues Using jHepWork

In this chapter, we will consider several third-party libraries and classes which, besides being educational, are shown to be very helpful for working with data and their use directly translates to an increased productivity.

This chapter encompasses miscellaneous topics, such as how to retrieve files from the web, extract data points from images, display tables and spreadsheets, mixing Java and Jython codes in one program and many other.

12.1 Accessing Third-party Libraries

12.1.1 Extracting Data Points from a Figure

jHepWork includes a remastered `Dexter` package. This is a tool for extraction of data points from figures in raster formats (GIF, JPEG or PNG). To start this program, run the commands:

```
>>> from debuxter import *
>>> a=JDebux(FileName)
```

As before, the `FileName` indicates the file name of a GIF, PNG or JPEG figure. If you are lucky to have a PDF, EPS or PS figure, enlarge it as much as possible before converting it to the raster formats. There are plenty of tools around which can be used for such transformation. For example, for Linux/Unix, use the command `convert`.

After execution of the above lines, a frame with the `Dexter` program pops up with the inserted figure. After calibration of the X and Y axes, one can locate data points with the mouse and print their (x, y) values. Read the documentation of this tool which can accessed by clicking on the help menu.

S.V. Chekanov, *Scientific Data Analysis using Jython Scripting and Java,*
Advanced Information and Knowledge Processing,
DOI 10.1007/978-1-84996-287-2_13, © Springer-Verlag London Limited 2010

12.1.2 Cellular Automaton

A cellular automaton was introduced by von Neumann and Ulam as a simple model for self-reproduction [1]. On an abstract level, the automaton is the model of a spatially distributed process. It consists of an array of cells, each of which is allowed to be in one of a few states. During simulation, each cell looks to its neighbors and determines what state it should change to using a predefined simple rule. Such steps are repeated over the whole array of cells, again and again. The evolution of such system with time can be very complex. The most famous two-dimensional cellular automaton is the "Game Of Life" invented by mathematician J. Conway in the 1960's.

We will consider the cellular automaton in two-dimensions when each cell is characterized by x and y values. To start the cellular automaton, one should first create an instance of the HCellular class.[1] Once the instance is created, it should be easy to check the available methods of this instance as:

```
>>>from jhplot  import *
>>>c=HCellular()
>>>print c.getRules()
  [Aggregation, Aqua, AquaP2,
   BlockVN, Check24, Check29,
   Check35, Check25ByGA, CyclicCA8,
   CyclicCA14, VN, Life, Life2,
   Generation, GMBrain, Hodge,
   Ising, Stripe]
```

Each rule is defined by a string value which specifies the Java class used for the initialization. Learn about the rules using the documentation help by calling c.doc().

The rules can be applied to the instance HCellular using the method setRule(str). The code shown below illustrates how to initiate the well-known "Game Of Life":

```
>>> c.setRule('Aggregation')
>>> c.visible()
```

The first method sets the rule 'Aggregation' from the list of available rules. As for any canvas, the method visible() brings up a window with the 2D cellular automaton.

The rule class and the initial configuration can be accessed using the methods below:

---

[1]This instance is based on the Cambria [2] package.

```
>>> print c.getRule()
>>> print c.getInitString()
```

One can change the initial configuration for a cellular automation with the
setInitString(str). The most convenient way is to write a configuration
to an external file and then read it using Jython as discussed in Sect. 2.16.

12.2 Downloading Files from the Web

In this section, we will show how to download files from the Web and display them
in the jHepWork IDE (of course, in case of simple textual files). This can be done using the Jython module 'web.py' located in the directory 'macros/system'.
Essentially, it is rather similar to the well-known "wget" program available on the
Linux and Unix platforms, but it is much simplified and implemented using Jython
(thus it does not have platform dependencies). In case of the HTTP protocol, a file
can be downloaded using the Jython prompt as:

```
>>> from web import *
>>> wget('http://www.jython.org/Project/news.html')
```

This command retrieves the file 'news.html' from the server to the current directory and shows the progress status during the download.

12.3 Macro Files for jHepWork Editor

To show a text file in the jHepWork editor, one can use JythonShell of the IDE.
Simply type the line:

```
>>> view.open('news.html',0)
```

which opens the file "news.html" (downloaded using the previous example) in a
new tab of the editor. The object view represents an "editor" text component of the
jHepWork IDE. This instance is imported by default during the IDE startup. One
can find the methods of this editor instance using the dir(view) method.

   In order to find what can be done with the object view, run the program "test"
using JythonShell. It illustrates the major operations with the text editor and how to
manipulate with the example text loaded to the editor using various macros.

   Let us show an example of how to work with text files using the object view.
Assuming that a file is loaded into the editor, use these commands from JythonShell:

```
>>> view.selectAll()      # select all text in the editor
>>> mydoc=view.getText()  # get text to mydoc string
>>> view.setText(' ')     # clear the editor
>>> view.setText('text')  # set new text
```

In this example, mydoc is a string with the text from the editor. To find and replace a string are rather easy as shown in this code snippet:

```
>>> mydoc=mydoc.replace('old','new');
>>> view.setText(mydoc)
```

One can also find the current caret position and insert a text at a specified position as:

```
>>> cc=view.getCaretPosition()
>>> view.insertString('inserted text', 4)
```

where "4" is the position where the text will be inserted. One can move the caret to a specific line using the method view.goToLine(line), where line is the line number.

As usual, learn about the view class using the usual approach:

```
>>> type(view)
>>> dir(view)
```

12.4 Data Output to Tables and Spreadsheets

12.4.1 Showing Data in a Sortable Table

In some cases, instead of plotting on canvases, it is convenient to show data in a form of tables. jHepWork provides an excellent support for this. Use the HTable class as in the example below:

```
>>> from jhplot import *
>>> HTable(obj)
```

where obj is an object of either H1D, P1D or F1D classes. After executing this statement, one should see a frame with a table populated with the numbers representing one of such objects (by default, the numbers are formatted using the Java scientific format). Using this table, one can sort data by clicking at the top of each column or search for a particular value or string. Note: it is impossible to modify the data as the class `HTable` was designed only for examining data containers.

One can pass a boolean value telling how you expect the data to be formatted. If the data have to be shown as formatted strings using the Java scientific format, use this constructor:

```
>>> HTable(obj,1)
```

If data values should not be formatted, set "0" (Java false) instead of "1".

One can call the `HTable` object differently: if a data container is created, one can use the `toTable()` method. For example, to display a `H1D` histogram, one can use:

```
>>> HTable(h1)   # where h1 is H1D histogram
```

or, alternatively,

```
>>> h1.toTable()
```

As before, if `toTable()` is used with the argument "0", the numbers will not be formatted using the scientific format.

The `HTable` viewer can be filled by the standard Java `Vector` class. In this case, it accepts three arguments:

```
>>> HTable('name', nameVector, fillVector)
```

where `nameVector` is a vector with column names (string type) and `fillVector` with input data. Let us give an example illustrating this:

```
─────────────── Table with particles ───────────────
from java.util import Vector
from jhplot import *

v1=Vector()
v1.add('Name')
v1.add('Mass')
v1.add('Error')

vv=Vector()
vv.add('Electron')
vv.add(0.5109)
vv.add(4.0E-8)
```

```
v2=Vector() # fill with data
v2.add(vv)

vv.add('Proton')
vv.add(938.27)
vv.add(8E-5)
vv=Vector()
v2.add(vv)

HTable('Particles',v1,v2);
```

One can see that the second vector contains rows of vectors with the input data. The number of columns will be determined automatically by the HTable object.

12.4.2 Spreadsheet Support

For more flexibility, one can export data to a spreadsheet using the SPsheet class which is designed to build a spreadsheet-like table, so the data can be modified and edited in the usual way as for any Excel-like application.

To create an empty frame with the spreadsheet object, use

```
>>> from jhplot import *
>>> SPsheet()
```

One can open a CSV file by passing CSVReader object. This class was discussed in Sect. 11.7.1 where we also have shown a small script used to open a CSV file in the spreadsheet.

The functionality of the SPsheet class goes beyond simple manipulations with the CSV files: if one needs to build a spreadsheet using data stored in other formats, use the constructor:

```
>>> SPsheet(obj)
```

where obj represents an object of the type H1D, P1D or F1D.

12.5 Accessing External Java and Jython Libraries

If one needs to include Java external libraries deployed as jar files, one should first create a new directory inside the 'lib' directory and then put the jar files inside this directory. Next time when you run the script 'jhepwork.sh' (for Linux, Unix, Mac) or the script 'jhepwork.bat' (for Windows), your external classes should be available in the JythonShell and for the jHepWork IDE editor. The start-up script scans all directories inside the 'lib' directory recursively.

Unlike Java jar files, external Jython macros should be put into the directory 'macros/user'. This directory is imported automatically when a custom script is executed using the jHepWork IDE.

For more complicated Python packages, one should use the directory 'packages' inside the directory 'python'. This directory is always scanned by the jHepWork IDE and is included into the Jython 'os.path' list.

One can add other directories visible for the Jython, but you should always import the directory path in the initialization file 'jehepsys.py'. This file is located in the directory 'macros/system' and called every time when you run a custom script using the jHepWork editor.

12.6 Working with a jHepWork Project

All examples of this chapter are located in the directory 'proj' inside the directory 'macros/examples'. You may first look inside before you proceed further. One can access the examples via the menu [Tool]-[jHPlot examples].

12.6.1 Pure Jython Project

To start a project, click on the button indicating your current directory (it is located at the bottom of the jHepWork IDE editor, on the left part of the status bar). Select some directory and populate it with some Jython files. Each file should contain the line:

```
>>> import os
>>> os.path.append(ProjDir)
```

This adds the project directory to the system path of Jython. One can check this by calling the line print sys.path which prints the system path. Execute the scripts by clicking on the [run] button.

We have already mentioned that if a Jython package is copied into the directory 'python/packages', then you do not need to worry about appending the location of this directory into the os.path, since this directory is included automatically by jHepWork.

12.6.2 Pure Java Project

In case of a pure Java project, use the example located inside the directory proj.
Click on the project button again (look at the bottom of the jHepWork IDE), and
select the directory proj using the file browser. Then open the main class file
example_java.java by double clicking using the mouse button. This is a
simple Java program which calls the external library, Calc.java, located in the
'example2' directory.

First, build Java library files located inside the proj/example* directories. To
do this, select [Compile and jar] project files from the [Run] menu. After
executing it, a jar file 'classes.jar' will appear in the main project directory.
It contains all the classes built from the source files located in the project library
directories.

If example_java.java is already opened in the jHepWork IDE, press
[Run] (or [Run java]) using the tool bar. You will see the output from the
class Calc in the BeanShell prompt.

One can further adjust the class path of external libraries by modifying the Bean-
Shell scrips in run_java* files in the director 'macros/user'.

12.6.3 Mixing Jython with Java

One can mix Jython and Java codes for CPU intensive tasks, such as large loops over
primitive data types, reading large files and so on. Such tasks must be implemented
in Java to achieve the best possible performance. Jython scripts can be used as a
glue language for various Java libraries with the source files located inside the main
project directory.

First, build a jar file which contains the byte-codes of Java classes. Again, for
this example, set proj as your project directory and select the menu [Compile
and jar] project files from the [Run] tool-bar menu. After executing it, a jar file
classes.jar appears in the main project directory, which contains the classes of
the source files located in the project directory.

Then, open the Jython file 'example_java.py'. This script calls the class
Calc and prints its output. Execute this Jython script by clicking on the [Run]
button on the tool-bar menu. The output result will be the Calc class located in the
directory example2/Calc.java.

One can modify the file example2/Calc.java and recompile the project
again using the sub-menu [javac all files] from the [Run] menu. One can
also rebuild the jar library. Note that, to trigger changes, you will need to reload the
BeanShell console using the small reload button on the BeanShell tab. Then run the
example_java.py again. You should see a different output from the class Calc.

12.7 Working with Images

12.7.1 Saving Plots in External Image File

We should remind that in order to export a canvas into an image file, one can use the
export(FileName) method after calling draw(obj) or update() methods.
This was discussed in Sect. 10.2.11. Below we will continue with this example.

12.7.2 View an Image. IView Class

If an image is saved in JPG, GIF or PNG format, one can view it using the IView
class. Below we will show how to use it.

Assume we have created a script which plots a histogram or any other object on
the HPlot canvas (or any other canvas). We assume that the instance of this canvas
is c1. If you want to create an image file and open it immediately after creation in
the IView frame, put these lines at the end of your Jython script:

```
>>> file=Editor.DocMasterName()+'.png'
>>> c1.export(file)    # export to a PNG file
>>> IView(file)        # view the image
```

where the method Editor.DocMasterName() accesses the file name of the
currently opened script. Execution of this file brings up a frame with the PNG image
inside.

12.7.3 Analyzing and Editing Images

If an image is saved in one of the raster formats, such as JPG, GIF, PNG, one can
edit it (crop, resize, add a text, etc.) using the IEditor class. This class is built
on the top of the ImageJ program [3], which is of the most advanced image editor
implemented in Java.

Let us show a typical example when a canvas, like HPlot or HPlot3D, is first
saved in an image file and then the IEditor class is used to edit and analyze the
created image:

```
>>> file=Editor.DocMasterName()+'.png'  # file name based
>>> c1.export(file)            # export to a PNG file
>>> IEditor(file)              # edit the created image
```

where c1 is a canvas object and Editor.DocMasterName() is the name of the
currently opened document in the IDE. The example shows how the image created
by the c1 object is redirected to the IEditor program for further editing.

12.8 Complex Numbers

We already know that Jython has support for complex numbers which comes from the Python language specification, see Sect. 2.2.4. We remind that one can define a complex number in Jython as x+yj, where j indicates an imaginary part.

In addition, jHepWork libraries provide a comprehensive support for usual complex-number arithmetic operators and necessary standard mathematical functions. Unlike Jython, the complex numbers and operations are implemented in Java, thus one can use these libraries in pure Java projects.

Complex numbers in jHepWork are based on the class Complex from the Java package jhplot.math. Unlike Jython, the imaginary part is denoted by the conventional character 'i'. A complex number is defined by the constructor Complex(real,img), where real is a real part and img is an imaginary part of a complex number. Either part can be zero.

Below we show a simple example of how to deal with the complex numbers using the Java jhplot.math package:

```
>>> from jhplot.math import *
>>>
>>> a=Complex(10,2)
>>> print a.toString()
10.0 + 2.0i
>>> b=Complex(20,1)
>>> cc=a.minus(b)
print cc.toString()
-10.0 + 1.0i
>>> cc=a.times(b)
>>> print cc.toString()
198.0 + 50.0i
```

There is a large number of methods associated with the class Complex. Look at the API description of the class Complex or use the jHepWork code assists while working with this class.

12.9 Building List of Files

Let us come back to the example given in Sect. 2.17 where we wrote a small Jython script to transverse all subdirectories in order to collect files with a given file extension. This time, the jHepWork Java library will be used which significantly simplifies the code, essentially reducing it to a single-line statement.

Instead of creating the function wolk() shown in Sect. 2.17, one can use a static class from jHepWork called FileList as in this example:

```
>>> from utils import *
>>> files=FileList.get('dir',redex)
```

This creates a list `files` with the file names in the input directory, `'dir'`, and a Java regular expression string `redex` using the syntax from the standard `java.util.regex` package. For the example discussed in Sect. 2.17, we should use `redex='.dat$'`.

There are a few advantages in using this approach: (1) it is significantly faster than that based of the Jython loop implemented in the `walk()` function; (2) one can use a very powerful Java regular expression engine; (3) Finally, it requires only one line of the code to scan all directories to build the file list!

12.10   Reading Configuration Files

To facilitate efficient data processing, especially if one needs to run the same program multiple number of times with different initial conditions, it is often necessary to pass some initial values to this program at run time. Similarly, this is necessary for computer simulation programs, when a program reads input parameters from a file. Changing the simulation conditions will only require editing text in the input file rather than editing and recompiling the source code.

It is very convenient to use the so-called Java configuration or property files. Unlike binary files, the configuration files should be readable by a human, so the Java serialization mechanism is not too convenient (unless you use a XML serialization, and know how to go around in editing such XML files).

Below we will show several approaches how to read the configuration files. One approach will be based on a pure Python class, while the second one will use Java.

12.10.1   Configuration Files Using Jython

Let us create a small file with the name `'jythonapp.conf'` with several entries as in this example:

```
————————————— Configuration file example —————————————
# this is a comment
# Nr of events to process
events = 1000
# release version
release = 1.1
# input file
input = data.txt
```

This configuration file has the format: 'name = value'. White spaces between these elements are ignored during reading such files. A configuration file can contain comments: by default, the '#' character at the start of each line with a comment text.

How would you read such files in Jython? We will use a small Jython module based on the package ConfReader [4] for this purpose. This code reads the configuration file, parses all text lines and prints the input values and their types:

```
────────────────── Reading configuration file ──────────────────
from Conf import *
try:
  config = ConfReader('jythonapp.conf')
except IOError:
  print 'Cannot read configuration file'

config.set('release', post=float)
config.set('events', post=int)
config.set('input', default='/tmp')

try:
    config.parse()
except ConfigMissingError, why:
    print 'Missing config', why
except ConfigPostError, why:
    print 'Postprocessing Error', why

print config.release, type(config.release)
print config.events, type(config.events)
print config.input,  type(config.input)
```

The parameters of the configuration file are retrieved with the help of the set(str,option) function. It takes two arguments. The first argument is a string which contains a parameter name. The second, option, should be in the form post=value (to be discussed below). The example contains several exception statements which have briefly been discussed in Sect. 2.15.

Running the script shown above prints:

```
1.1 <type 'float'>
1000 <type 'int'>
data.txt <type 'str'>
```

As you can see, the program correctly identifies the input values and their type.

Let us come back to the set(str,option) function. As mentioned, the second argument of this function should be in the form post=value, where the attribute value can have the following values:

default   the value returned if the config is not found
post   post-processing function to use; can be a lambda form or any function

required set to 1 if config is required
list   set to 1 if config should always be returned as list. Multiple entries will
          be appended to the list, if this is not a list but a set; otherwise, the last entry
          will be used

The above example is based on the module 'Conf.py', which is located in the
directory 'system' inside the 'macro' directory. The module is rather flexible:
you can change the default syntax of your configuration file rather easily.

Optionally, the configuration file can be arranged in sections of data, which can
be used to organize references to multiple resources. Please read the documentation
given in the file 'Conf.py'

12.10.2 Reading Configuration Files Using Java

One can also read the configuration files using the Properties class from the
Java package java.util. The Properties can be used to save a property file
in a stream. Each key and its corresponding value in the property list is a string.

Below we show an example of how to read the configuration file created in the
previous section. We will use the strings 'release', 'events' and 'input'
as the keys. Then we print the loaded strings in the ISO 8859-1 character encoding.
Since we expect 'events' to be an integer number, we convert it to the integer
type and modify its value. Finally, we set a new property using 'events' as the
key and save it to a file with the optional comment 'New settings' (one can use also
'None' in case of no comments).

```
————————————— Reading a configuration file —————————————
from java.util import Properties
from java.io  import *

p=Properties()

# read property file
p.load(FileInputStream("jythonapp.conf"))
print  p.getProperty("release")
print  p.getProperty("input")
print  p.getProperty("events")

# save new file with x10 more events
ev=int(p.getProperty("events"))
ev=10*ev
p.setProperty("events",str(ev))
p.store(FileOutputStream("new.conf"),"New settings");
```

One should note that, in case of Java, the file extension of the property files is
'properties' by convention, rather than that given in this example.

12.11 Jython Scripting with jHepWork

Throughout the book, we have considered a programming style with the direct Jython calls to the jHPlot Java classes. Java, however, does not support operator overloading, which is one of important features of object-oriented programming. We will remind that "operator overloading" means that one can define custom types and operations. For example, one can redefine operators like "+" or "−" depending on the types for which these operators are used.

The operator overloading is useful since this feature can significantly simplify a program and makes it more concise and readable. One can easily see a problem if one wants to do a lot of elementary mathematical operations with objects, like to add them, subtract etc. In this case, one should call Java methods which, usually, have long names. For the simplest mathematical operations, such as addition and subtraction, it is more convenient to use + and −.

jHepWork allows to use the operator overloading by calling the corresponding Jython classes, which are directly mapped to the Java jHPlot classes. Such Jython classes inherit all methods of the Java jHPlot classes. At the same time, the most common arithmetical operators are overloaded. The package which allows to do this is called shplot (where the first letter "s" means "scripted" or "simple" hplot), in contrast to name jhplot ("Java j-hplot")

Below is an example of how to build Jython histograms and to perform some common operations:

─────── Jython scripting with an operator overloading ───────

```
from java.util import Random
from  shplot import *

c1=hplot('scripting',1,2)
c1.visible()
c1.setAutoRange()

h1=h1d('histogram1',200,-5,5)
h2=h1d('histogram2',200,-5,5)

r=Random()
for i in range(500):
    h1.fill(r.nextGaussian())
    h2.fill(0.2*r.nextGaussian())
h1=h1+h2                     # add 2 histograms
c1.draw(h1)                  # draw
h1=h1*2                      # scale by a factor 2
c1.draw(h2)

c1.cd(1,2)                   # go to the next plot
c1.setAutoRange()
h1=h1-h2                     # subtract 2 histograms
h1=h1/h2                     # divide  2 histograms
```

```
h1=h1*10                    # scale by factor 10
h1=h1/100                   # divide by 100
c1.draw(h1)
```

One should note that all Jython classes which inherit Java jHPlot classes have exactly the same names as the corresponding Java classes from the `jhplot` package. The only difference is that Jython classes have to be typed in lower case. This means that the statement such as:

```
>>> from  jhplot import *
>>> c1=HPlot('scripting')
```

creates an instance of the Java class `HPlot`. However, when one uses the lower case for the same class name, the corresponding Jython instance of this object is created:

```
>>> from  shplot import *
>>> c1=hplot('scripting')
```

In this case, the major arithmetical operators are overloaded since `'c1'` is a pure Jython object.

Analogously, one can look at the Jython equivalent of the Java H1D class:

```
>>> from  shplot import *
>>> h1=h1d('test',200,-5,5)
>>> h1.   # press [F4] to see all methods
```

One may notice that all `jHPlot` Java methods are inherited by the Jython class `h1d`. Thus, even if you write a code using Jython `shplot` classes, one can easily access the Java classes:

```
>>> from  shplot import *
>>> h1=h1d('test',200,-5,5) # build a Jython histogram
>>> h1.setFill(1)           # accesses a Java method
```

In the above example, `setFill(1)` was applied directly to the Java class H1D. If one needs to subtract, add, divide, scale the data, use -, +, /, * or *factor, instead of the long `oper(obj)` or `operScale(obj)` statements of the Java H1D class.

One can go even further in redesigning the Jython coding. When plotting an object on a canvas, like a histogram, one can say that a histogram was added to the canvas. So, why not, instead of the usual `draw(obj)` method, use the operation +? Yes, this is also possible: Here is a code which shows how to draw three histograms on the same canvas using a single line:

```
>>> c1+h1+h2+h3
```

Of course, the same operation can be done with any other class, like H1D or P1D, which can also be drawn on the canvas. Note: subtraction of objects (which may correspond to the removal of objects from the canvas) is not implemented for the hplot Jython class.

In the same spirit, data can be shown in a table. Below we display histogram values in a pop-up table:

```
>>> htable()+h1
```

where the Jython htable class is a direct mapping of the HTable Java class.

As before, one should use the jHepWork code assist to learn more about Jython/Java methods. This time, the code assist will be applied to Jython objects, rather than to Java ones.

Note that not all Java classes have been mapped to the corresponding Jython classes. Please look at the directory 'macros/shplot' to find out which Jython modules are available. In the same directory, one can find several useful examples (their names contain the string _test).

12.11.1 Jython Operations with Data Holders

In case of Jython scripting, all major operations for P0D, P1D and PND are overloaded as well:

```
>>> from  shplot import *
>>> p1=p1d('test1')   # a Jython p1d based on P1D
>>> p2=p1d('test2')
>>> p1=p1+p2           # all values are added
```

Note that, for the latter operation, all y-values are added (and errors are propagated respectively), while x values remain to be the same. This operation cannot append arrays, instead, use the merge() method.

Here are some other examples:

```
>>> p1=p1+p2     # all values are added
>>> p1=p1-p2     # subtraction
>>> p1=p1*p2     # multiplication
>>> p1=p1/p2     # division
>>> p1=2*p1      # scale by 2
>>> p1=p1/10     # scale by 0.1
```

To draw objects on a canvas, use "+" as before:

```
>>> c1+p1    # add p1d object to a  canvas
```

Similarly, one can work with one-dimensional (P0D) or multidimensional (PND) arrays.

12.12 Unwrapping Jython Code. Back to Java

In most cases, jHepWork libraries are just Java libraries deployed as jar files, so you may choose to write a Java code instead of Jython scrips.

One can develop Java codes using the jHepWork IDE exactly as when working with Jython scripts. Any Java IDE, such as Eclipse or NetBeans, is also good for such task. Applications based on the jHepWork libraries can be converted to a Java jar file and deployed together with other jar libraries which come with jHepWork.

For a Java application, you should remember to create the main() method, which must be encapsulated into a class with the same name as the file name. Unlike Jython, you should explicitly declare all variables and put the Java new operator when instantiating a class. Do not forget changes to be made for loops. Lastly, put semicolons at the end of each statement. Our next example illustrates the difference between the Jython and Java syntax. In the case of Jython scripting, let us write our example code as:

```
>>> p0=P0D('test')
>>> for i in range(5):
    ...   p0.add(i)
    ...   print i
```

The same code snippet in Java looks like:

```
P0D p0 = new P0D('test');
for ( i=1; i<5; i++ ) {
        p0.add(i);
        System.out.println(i);
        }
```

One should put these lines into a file (with the same name as the class name), compile it using javac compiler and run it as any Java application. One can also put this Java bytecode into a jar file. Comment lines should start from the usual Java comment symbols instead of the symbol "#". If you are using an external Java IDE, do not forget to add the directory 'lib' from the jHepWork installation into CLASS-PATH, so all jHepWork libraries should be visible for Java.

Let us show a more detailed example. In Sect. 8.1 we have considered how to plot three histograms using Jython. Now we will rewrite this code in Java:

```
──────────────── File 'Example.java' ────────────────
import java.awt.Color;
import java.util.Random;
import jhplot.*;

class Example
{
public static void main(String[] args)
{
  HPlot c1 = new HPlot("Canvas",600,400,2,1);
  c1.visible(true);
  c1.setAutoRange();
  H1D h1 = new H1D("First",20, -2.0, 2.0);
  h1.setFill(true);
  h1.setFillColor(Color.green);
  H1D h2 = new H1D("Second",100, -2.5, 2.5);
  Random r = new Random();

  for (int i=0; i<500; i++)  {
        h1.fill(r.nextGaussian());
        h1.fill(r.nextGaussian()); }

  H1D h3 = new H1D("Third",20, 0.0, 10.0);
  h3.setFill(true);
  for (int i=0; i<50000; i++)
        h3.fill(2+r.nextGaussian());

  c1.cd(1,1);
  c1.setAutoRange();
  c1.draw(h1);
  c1.draw(h2);

  c1.cd(2,1);
  c1.setAutoRange();
  c1.draw(h3);
  }
}
```

Save these lines into the file 'Example.java' and open it inside the jHepWork IDE. Now you can: (1) Compile and run the program by pressing the [Run] button. (2) Or one can just compile the file loaded to the editor into bytecode using the menu [Run] → [Javac current file]. Then you can run the compiled code as: [Run] → [Run Java]. As you will see, the output will be identical to that shown in Fig. 8.1 of Sect. 8.1.

We should note that the code above can be compiled manually, without any IDE. One would only need to import all Java jar libraries from the jHepWork installation directory. For an Unix-type environment (like Linux or Mac OS), you can compile the code using the prompt:

```
bash> a_compile.sh Example.java
```

where 'a_compile.sh' is a bash script which can look like:

```
——————————— Compilation script 'a_compile.sh' ———————————
#!/bin/bash
args=$#
if [ $args == 0  ]
then
    echo "did not specify input file!"
    exit 1;
fi

for i in lib/*/*.jar
do
    CLASSPATH=$CLASSPATH:"$i"
done

CP=$CP:$CLASSPATH
javac -classpath "$CP" $1
echo "File $1 compiled!"
```

The script scans all subdirectories inside the directory "lib" of the jHepWork installation directory, and appends all jar files into the CLASSPATH variable.

Analogously, one can run the compiled class file 'Example.class' using the same script, but replacing the statement javac by java, and passing the name 'Example' (not Example.java!).

One can easily convert this program into a Java applet, build your own Java library or to deploy it as an application. We will discuss how to build a Java applet in the next section.

12.13   Embedding jHepWork in Applets

jHepWork numerical libraries can easily be deployed over the Web in form of applets. This is certainly one of the most significant advantage of Java compare to other programming languages.

In this section, we will show how to write a small applet which brings up a HPlot canvas filled with a histogram at run-time. If you have already some experience with Java, the code below will be rather trivial:

—————————————— file 'Histogram.java' ——————————————

```java
import java.applet.*;
import java.awt.*;
import java.util.Random;
import jhplot.*;

public class Histogram
    extends Applet implements Runnable {

 private static final long serialVersionUID = 1L;
 private HPlot c1;
 private Thread thread;
 private Random rand;
 private H1D h1;
 private int i;

 public void init() {
 c1 = new HPlot("Canvas");
 c1.setGTitle("Gaussian random numbers");
 c1.visible(true);
 c1.setAutoRange();
 h1 = new H1D("Random numbers",20, -2.0, 2.0);
 h1.setColor(Color.blue);
 h1.setPenWidthErr(2);
 h1.setFill(true);
 h1.setFillColor(Color.green);
 c1.setNameX("X axis");
 c1.setNameY("Y axis");
 c1.setName("100 numbers and statistics");
 rand = new Random();
 i=0; }

 public void start() {
     (thread = new Thread(this)).start(); }

 public void stop() {
     thread = null;
     c1.drawStatBox(h1); }

public void run() {
 try {
  while (thread == Thread.currentThread()) {
    Thread.sleep(50);
    h1.fill(rand.nextGaussian());
    c1.clearData();
    c1.draw(h1);
    i++;
    if (i >100) stop();
```

```
        }
    } catch (Exception e) {} }
}
```

Although it is a bit lengthy (it is not a Jython script!), the code itself is rather simple: We create the usual `HPlot` and `H1D` objects and set some attributes. Then we fill a histogram with random numbers in a thread. We animate the canvas, i.e., we plot random numbers in a loop with a time delay. After the number of entries reaches 100, we stop filling the canvas and display the statistics.

Compile this code as discussed in the previous section, either inside the jHep-Work IDE or with the help of the compilation script discussed in the previous section:

```
bash> a_compile.sh Histogram.java
```

After compilation, you should see the file 'Example.class'. The next step is to embed it into a HTML file or a PHP script which can be used for a Web browser.

Our preference is to use a PHP script which is more flexible as it allows automatic scanning of all jar files inside a directory. First, copy the library directory 'lib' with the jHepWork jar libraries to a location which can be accessible by the web server. Then, write a script:

```
────────────────── PHP script 'applet.php' ──────────────────
<?php
require("list_files.php");
$list=list_files("lib/system/") . ", " .
      list_files("lib/freehep/");
$html_body  = "";
$html_body .= <<<EOT
<html>
<body>
<h1>Histogram example</h1>
<APPLET
  CODE="Histogram.class"
  WIDTH=0 HEIGHT=0 ARCHIVE="$list">
  Please use a Java-enabled browser.
</APPLET>
</body>
</html>
EOT;

print $html_body
?>
```

This PHP script creates the necessary HTML file with our applet "on-the fly". The only unusual variable is $list which is a comma-separated list of jar files from the directory 'lib'. This list is created by the function 'list_files.php', which returns a list of files in a certain directory. This function is shown here:

────────── PHP script 'list_files.php' ──────────

```php
<?php
function list_files($dir)
{
$list="";
if ($handle = opendir($dir)) {
    while (false !== ($file = readdir($handle))) {
        if ($file != "." && $file != "..") {
            $ff= $dir . $file;
            $list .= $ff . ", ";
        }
    }
    closedir($handle);
}
return $list;
}
?>
```

Now you are ready. Copy the three files above:

```
applet.php, list_files.php,  Example.class
```

into a Web-server accessible directory. Make sure that this directory contains the directory 'lib' from the jHepWork installation directory (one can also make a link to this directory). Consider to remove unnecessary jar files, if they are not used to run the applet (this will make the execution faster). Then, point a Web browser to the file 'applet.php'. The PHP module 'list_files.php' will scan all jar files inside the directory 'lib' and will pass it to the 'applet.php' file. This file will build a proper HTML file to be displayed inside a Web browser. You should see a HPlot frame with filled histogram entries. After 100 entries, the animation will stop and you will see the statistics for the filled histogram.

References

1. von Neumann, J.: The General and Logical Theory of Automata (1963), p. 288
2. Suzudo, T.: Cambria package
3. ImageJ Java library. URL http://rsb.info.nih.gov/ij/
4. Nottingham, M.: Confreader—Configuration File Reading Class (1998)

Chapter 13
Data Clustering

13.1 Data Clustering. Real-life Example

Clustering algorithms are important tools for an unsupervised classification of data (compared to a "supervised" classification which will be discussed in the following chapter). The main idea is to classify a given data set through a certain number of clusters by minimizing distances between objects inside each cluster. A detailed discussion of a clustering analysis can be found, for example, in several books [1–3], but many details could also be found from many other (less costly) sources.

The data clustering in jHepWork is based the jMinHEP Java library [4] and includes:

- The K-means algorithm (single and multi-pass), when classification is done by minimizing the sum-of-square distances between data points and the so-called geometric centroids assuming that the number of clusters is fixed a priory. The clustering is done in a single pass starting from seed centroids positioned at random location. The algorithm assigns each object to a group of objects that have the closest centroid, and then it recalculates the positions of new centroids. Then this step is repeated until the centroids no longer move.

  This algorithm has the disadvantage that it depends on the initial conditions, thus jMinHEP has also an option to re-run the algorithm multiple number of times until a stable solution is found. Typically, you will need to define a number of expected clusters, a maximum number of iterations and/or a precision with which the clustering is done. To get started, you may rely on the default values for the precision and the maximum number of iterations.

- The C-means (fuzzy) algorithm. Unlike the K-means approach, this method of clustering allows data points to belong to two or more clusters. The algorithm usually uses a "fuzziness" coefficient (typically, it is set to 2) and an accuracy of the calculations.

  It is always a hard task to plot such clusters, since each point has a membership probability to belong to a particular cluster. However, one could supply some probability (say, 0.7) to illustrate data points which have a membership probability above the specified value.

S.V. Chekanov, *Scientific Data Analysis using Jython Scripting and Java*, 335
Advanced Information and Knowledge Processing,
DOI 10.1007/978-1-84996-287-2_14, © Springer-Verlag London Limited 2010

- The agglomerative hierarchical clustering algorithm. In this case, no a predetermined number of clusters is set. The algorithm tries to determine the so-called "natural grouping". The clustering continues until all points are clustered into a single cluster.

The clustering can be done with the package jminhep.algorithms. One should follow these steps to perform the clustering:

- Create a data container using the class DataHolder. This data holder keeps data in a multidimensional phase space, i.e. each data point can be characterized by an arbitrary number of values.
- Each point is represented by the DataPoint class from the package jminhep.cluster. Data points can be defined in a multidimensional phase space. One should fill this data holder from an external source, or fill it on-the-fly by simulating events. We will show how to do this in the example below.
- Create and initialize a clustering algorithm class and pass it the DataHolder() constructor.
- Run the algorithm in a loop and retrieve the output.

Then one can print or plot the retrieved centroids of clusters by projecting multi-dimensional space into $X-Y$ plane, and optionally, one can plot the original points used for the clustering.

13.1.1 Preparing a Data Sample

First, let us learn how to fill data into a DataHolder container. Once this data holder is ready, we will save it into a file, and then we will read this file in order to illustrate the clustering analysis.

One single point should be represented by a DataPoint object. For example, a point with five components can be filled as:

```
>>> from jminhep.cluster import *
>>> p=DataPoint([1,2,3,4,5])
>>> print p.getDimension()
>>> print p.toString()
```

Then one should add each point represented by the DataPoint class to the DataHolder container. Below we add two data points:

```
>>> from jminhep.cluster import *
>>> d=DataHolder('data')
>>> p=DataPoint([1,2,3,4,5])
>>> d.add(p)
>>> p=DataPoint([5,6,7,8,9])
>>> d.add(p)
```

One can apply the method add(obj) multiple number of times to fill in the DataHolder container. One can get the data back to the usual arrays with the getArrayList() or getArray() methods.

Here are a few useful methods of how to retrieve information about the container:

```
>>> from jminhep.cluster import *
>>> size=d.getSize()     # get size
>>> p=d.getRow(n)        # get DataPoint at index "n"
>>> min=d.getMin()       # get min
>>> max=d.getMax()       # get max
>>> print d.toString()   # print
>>> d.clear()            # clear data
```

Now let us generate data with three clusters. The data points will be generated in 3D and clustered around three cluster centers, at the positions $(-8, 5, 10)$, $(10, 20, 5)$ and $(10, 1, 20)$. This can be done using Gaussian numbers as:

```
──────────────── Creating a data set ────────────────
from java.util import Random
from jhplot.io import *
from jminhep.cluster import *

data = DataHolder('Clusters in 3D')
r = Random()

for i in range(20):    # 1st cluster
  a =[]
  a.append(   r.nextGaussian()- 8)
  a.append( 2*r.nextGaussian()+ 5)
  a.append( 3*r.nextGaussian()+10)
  data.add( DataPoint(a) )

for i in range(30): # 2nd cluster
  a =[]
  a.append( 10*r.nextGaussian()+10)
  a.append(  2*r.nextGaussian()+20)
  a.append(  5*r.nextGaussian()+ 5)
  data.add( DataPoint(a) )

for i in range(10): # 3rd cluster
  a =[]
  a.append( 3*r.nextGaussian()+10)
  a.append( 2*r.nextGaussian() +1)
  a.append( 4*r.nextGaussian()+20)
  data.add( DataPoint(a) )
```

```
Serialized.write(data,'data.ser')
```

We should note that the sizes of the clusters are different—this is given by the scaled factors used to multiply the random values generated by the method r.nextGaussian(). The second cluster is rather broad—its radius spans 10 (arbitrary) units.

We have written the data in a compressed serialized file. One can also try to use the writeXML() method to write data into a XML file for easy viewing. We should remind: if you need to write a lot of data, the best way to do this is to use the HFile or HFileXML classes, see Sect. 11.3. We can write each DataPoint object persistently in a loop. This can be done rather straightforwardly, since DataPoints objects can be serialized. As explained before, this approach allows to create rather large files since we do not store all data in the computer memory before writing them to a disk.

So, the data are filled with the help of the DataPoint class in three dimensions. But how one can display such data, especially in cases when there are more than three dimensions? One can project the DataHolder into a P1D array and plot it using the standard HPlot class. For example, if our data are represented by a filled DataHolder object, one can view 1st and 2nd components as:

```
>>> from jhplot  import *
>>> c1=HPlot()
>>> c1.setAutoRange()
>>> c1.visible()
>>> p1=P1D(data,0,1) # fill 1st and 2nd component
>>> HTable(p1)        # show data in a table
>>> c1.draw(p1)       # draw 1st and 2nd component
```

13.1.2 Clustering Analysis

Assume we have prepared an object data of the type DataHolder. Now one can perform a realistic clustering analysis. Generally, to run any clustering algorithm, one needs to follow these steps:

```
>>> from jminhep.algorithms import *
>>> alg = [Algorithm]Alg(data)
>>> alg.setClusters(NumberOfClusters)
>>> alg.setOptions(some options)
>>> .. more options ..
>>> alg.run() # run over the data
>>> print 'algorithm: ' + pat.getName()
>>> .. get results
```

In the above example, '[Algorithm]' is the name of clustering algorithm which could be:

KMeansAlg(data)   the standard K-Means cluster algorithm;
KMeansExchangeAlg(data) K-Means cluster algorithm using the exchange mode;
FuzzyCMeansAlg(data)   Fuzzy (C-means) cluster algorithm;
HierarchicalAlg(data)   an agglomerative hierarchical clustering algorithm.

Then one should set the expected number of clusters and other options. For example, for the C-means algorithm, one should set the number of expected clusters, the numerical precision of clustering and a fuzziness parameter. It is also possible to set the maximum number of iterations to stop clustering if no appropriate solution can be found. It is also useful to set the probability association if you need to know which points belong to which cluster (this is not necessary for the K-means algorithm). The description of each mode can be found with the method getName().

To run the algorithm over the data, call the method run(). This method should be invoked for a fixed number of clusters. One may also consider to use the *runBest()* method. In this case, the program will attempt to determine the number of clusters by re-running the algorithm many times over the same data set. Then it calculates the most optimal solution by minimizing the so-called "compactnesses" of the cluster configuration. The smaller compactness is, the higher chance that a particular cluster solution is the most optimal.

After the clustering (i.e. after running the method run()), one can find the "compactness" of the cluster configuration by calling getCompactness(). The centroid positions can be retrieved with the method getCenters() and the data point association can be obtained using several built-in methods.

We now have all the machinery to perform a clustering analysis and below we show a detailed example. Let us read the prepared data and run the K-means algorithm over this data. First, we prepare a function printAnswer() designed to print all relevant information about the cluster configuration. Then we will run the K-means algorithm for: (1) one-pass using three clusters; (2) 10 passes with different random seed locations. The algorithm returns the cluster configuration with the smallest compactness; (3) Running the K-means algorithm using multiple number of passes, and determining the best possible configuration with the smallest compactness.

```
──────────── Running the K-means algorithm ────────────
from jhplot.io import *
from jminhep.algorithms import *

data=Serialized.read('data.ser')

def printAnswer(alg):
  print 'Name='+alg.getName()
  print 'No of final clusters:' +str(alg.getClusters())
```

```
   print 'No of points: ' +str(alg.getNumberPoints())
   print 'Compactness: ' +str(alg.getCompactness())
   centers = alg.getCenters()
   print centers.toString()

alg=KMeansAlg(data) # run K-means algorithm
alg.setClusters(3)
alg.setOptions(1000,0.001)
alg.run()
printAnswer(alg)

alg.run(10) # run 10 times with different seeds
printAnswer(alg)

alg=KMeansExchangeAlg(data)
alg.setEpochMax(200)
alg.runBest()    # find smallest compactness
printAnswer(alg)
```

The clustering results will be printed by the `printAnswer()` function. The short-ened output from execution of the above example is given below:

```
Name=kmeans algorithm fixed cluster mode
No of final clusters:3
No of points: array('i', [30, 13, 17])
Compactness: 1.487
0   (-2.55, 3.93, 12.75)[0]
1   ( 2.40, 19.69, 6.22)[0]
2   (16.76, 19.80, 4.99)[0]

Name=kmeans algorithm for multiple iterations
No of final clusters:3
No of points: array('i', [21, 10, 29])
Compactness: 0.891
0   (-8.24,  6.0,   8.23)[0]
1   ( 8.70,  1.09, 20.40)[0]
2   (11.23, 19.78, 5.91)[0]

Name=K-means clustering for best estimate
No of final clusters:5
No of points: array('i', [15, 10, 5, 20, 10])
Compactness: 0.843
0   (17.52,19.77,4.36)[0]
1   (6.57, 19.27,9.48)[0]
2   (-2.46,20.64,1.11)[0]
3   (-8.18,5.36, 8.92)[0]
4   (8.700,1.09, 20.40)[0]
```

One may immediately see that the answers from all three algorithms are different. This is not too surprising—the clustering analysis is an inherently ambiguous task, especially when dealing with overlapping clusters.

Try to make some modifications to the input data to reduce the cluster overlaps. One can simply set all multiplicative factors for the method $r.nextGaussian()$ to a smaller value (say, 1), see the code snipped given in Sect. 13.1.1. This reduces the cluster size, and thus overlaps between the clusters. After re-running the above code with the K-means clustering, you will see that all three algorithms give the same answer.

Now let us visualize the clustering results. In the code below, we will read the data prepared in the previous subsection, and plot them (only 1st and 2nd columns). This time, however, we will run the C-means algorithm with a fixed number of expected clusters.

```
———————————— C-means algorithm and visualization ————————————

from java.awt import Color
from jhplot  import *
from jhplot.io import *
from jminhep.algorithms import *

data=Serialized.read('data.ser')

c1=HPlot()
c1.visible()
c1.setAutoRange()
p1=P1D(data,0,1) # 1st and 2nd component
c1.draw(p1)

alg = FuzzyCMeansAlg(data)
alg.setClusters(3)
alg.setOptions(1000, 0.001, 1.7)
alg.setProb(0.7)
alg.run()
print 'algorithm: ' + alg.getName()
print 'Compactness: ' + str(alg.getCompactness())
print 'No of final clusters:' + str(alg.getClusters())
centers=alg.getCenters()
print 'No of points in cluster=',alg.getNumberPoints()
p2=P1D(centers,'Centroids',0,1)
p2.setColor(Color.red)
p2.setErrAll(0)
p2.setSymbol(9)
p2.setSymbolSize(15)
c1.draw(p2)
```

The resulting plot is shown in Fig. 13.1. The figure displays the input data projected in 2D (black dotes) and the crosses show the centers of three clusters. You may wonder: why the cluster centers are shifted from the visually expected positions?

Fig. 13.1 The result of the
C-means clustering algorithm

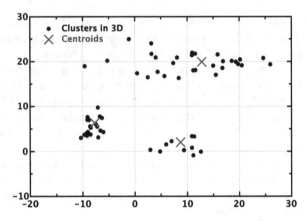

The answer is simple: the clustering was done in 3D, while Fig. 13.1 shows only a 2D slice of the data.

13.1.3 Interactive Clustering with JMinHEP

One can always perform the clustering analysis using the JMinHEP program [4] which allows to select the cluster algorithm and initial conditions using an interactive menu.

One can start the jMinHEP cluster program using the `HCluster` class that executes the main frame of the program. One can append data (in a form of `DataHolder` object) to this canvas during the initialization.

```
───────────── Building a data set ─────────────

from jhplot  import *
from jhplot.io import *
data=Serialized.read('data.ser')
c1=HCluster(data)
```

Data can also be loaded using the menu of this frame.

References

1. Spath, H.: Cluster Analysis Algorithms. Wiley, New York (1980)
2. Kaufman, L., Rousseeuw, P.: Finding Groups in Data: An Introduction to Cluster Analysis. Wiley, New York (2005)
3. Jain, A., Dubes, R.: Algorithms for Clustering Data. Prentice Hall, New York (1988)
4. Chekanov, S.: The JMINHEP package. URL http://hepforge.cedar.ac.uk/jminhep/

Chapter 14
Linear Regression and Curve Fitting

Very often, an empirical data set has to be explained using a model based on a mathematical equation. To find a proper function and adjust free parameters of this function that most closely match the data is the primary goal of curve fitting.

We start this chapter with the simplest linear case and then consider curve fitting using arbitrary functions. The section below discusses the elements of the linear regression which models a relationship between two variables by fitting a linear equation to observed data. Then, in the following sections, we will discuss non-linear regression.

14.1 Linear Regression

Linear regression is a method of finding a linear correspondence between two data sets. The method is based on fitting a collection of data points by a simple linear form, $Y = A + BX$. The variable X is considered to be an explanatory variable, and the other, Y, is considered to be a dependent variable. Sometime, the variable A is called the intercept, and B is the slope of the line.

Before going into the heart of the matter, it is useful to remind that it is always a good practice to visualize a data set before attempting to fit it using the linear function. For example, in case of two-dimensional data, one can make a contour plot (see Sect. 5.2.3) in order to determine the strength of the relationship between two variables.

14.1.1 Data Set

First, let us create a data set in order to illustrate a linear regression analysis. We assume that an explanatory variable X is distributed in accordance with a Gaussian distribution. We also assume that the "dependent" variable is a function of the explanatory variable and an additional "noise" simulated using random Gaussian numbers. Look at this code:

S.V. Chekanov, *Scientific Data Analysis using Jython Scripting and Java,*
Advanced Information and Knowledge Processing,
DOI 10.1007/978-1-84996-287-2_15, © Springer-Verlag London Limited 2010

```
───────────── Creating a data set ─────────────
from jhplot  import *
from jhplot.io import *
from java.util import Random

p1=P1D('data')
r = Random()
for i in range(100):
   x=10*r.nextGaussian()
   y=20*x +200*r.nextGaussian()+50.0
   p1.add(x,y)
Serialized.write(p1,'data.ser')
```

In the example above, we fill a P1D using random values x and y (the latter depends on x) and store it in a serialized file for the use in the next section.

14.1.2 Analyzing the Data Set

For the linear regression analysis of the created data, we will use the class LinReg from the package jhplot.stat which provides all necessary tools for the linear regression fits. To initialize the linear regression calculation, we should pass our data in the form of P1D to the LinReg() constructor:

```
>>> from jhplot.stat import *
>>> r=LinReg(p1d)
```

We should note that one can also use the usual Java arrays or Jython lists for X and Y values for the input, but here we will concentrate only on the P1D case.

After calling the constructor LinReg with the input P1D object, the calculation is already done and the next step is to extract the linear regression results. First of all, we are interested in the intersect and the slope values of the fit. Both values, including their statistical uncertainties, can be extracted using the following methods:

```
>> print r.getIntercept(), '+/-',r.getInterceptError()
>> print r.getSlope(),'+/-',r.getSlopeError()
```

Next, we could be interested in how to display the fitted values. This can be done using the method getResult(), which returns a F1D function with the results of the linear regression.

```
>>> f=r.getResult()
```

Next, the prediction and the confidence levels of the linear regression can be extracted as:

```
>>> d=r.getCorrelation()
>>> p1d=r.getResiduals()
```

The first method returns the usual correlation coefficient,

$$\rho = \frac{\text{cov}(X, Y)}{\sigma_X \sigma_Y}$$

where $\text{cov}(X, Y)$ is the covariance and σ is the standard deviation. We remind that the correlation coefficient varies from -1 to $+1$, with -1 being for a perfect negative correlation between X and Y, and $+1$ for a perfect positive correlation.

The second method returns a P1D with the residuals, i.e. the vertical distances of each point from the regression line.

In addition, one can also calculate the 95% confidence interval for the regression line given by the method getResult(). This band encloses the true best-fit linear regression line, leaving a 5% chance that the true line is outside the confidence interval boundaries. The confidence interval is the area that has a 95% chance of containing the true regression line. This typically means that many points could be far away from the confidence-interval band.

Finally, we can calculate the 95% prediction interval. The prediction interval is the area in which you should expect 95% of all data points to fall. As we would expect, the prediction interval is further away from the best-fit line than the confidence bands.

Both the prediction and the confidence intervals can be obtained using the following two methods:

```
>>> a1=r.getPrediction()
>>> a2=r.getConfidence()
```

These two methods return two P1D objects representing the lowest and the highest level for the prediction and the confidence level. We remind than one can easily plot them with the method c1.draw(obj), where obj is either a1 or a2, since this method can be used for displaying lists of P1D objects. There are several other convenient methods which can be found by reading the API documentation of this package.

Now let us illustrate how to analyze the data set generated in the previous section. First, we read the data from the serialized file. Then we plot the data, the linear regression fit result, the confidence interval and, finally, the predictions:

```
———————————————— Linear regression analysis ————————
from jhplot.io import *
from jhplot     import *
from jhplot.stat import *
```

```
from java.awt import Color

p1=Serialized.read('data.ser')

c1 = HPlot('Canvas',700,500,2,1)
c1.visible()
c1.setGTitle('Linear regression')
c1.setAutoRange()

r = LinReg(p1)
print r.getIntercept(), '+/-',r.getInterceptError()
print r.getSlope(),'+/-',r.getSlopeError()
print 'Correlation=',r.getCorrelation()

c1.cd(1,1)
c1.draw(p1)
c1.draw(  r.getResult()  )
c1.draw( r.getConfidence() )

c1.cd(2,1)
c1.setAutoRange()
c1.draw(p1)
c1.draw(  r.getResult()  )
c1.draw( r.getPrediction() )
```

The output of the script is shown below:

```
Intercept= 48.041 +/- 17.578
Slope=      20.248 +/-  1.728
Correlation= 0.76383
```

(as before, the output numbers are shown here with a reduced precision). At the same time, the script generates two plots shown in Fig. 14.1.

14.2 Curve Fitting of Data

Now let us come to a non-linear case, when the relationship between X and Y values cannot be explained by the linear behavior. In this case, we need to fit data points using an arbitrary curve.

Moreover, each data point can represent a group of events or measurements, thus points can have statistical (and systematical) errors. Obviously, the latter situation must be considered in a different way, since points should not have the same statistical weight during the fit: points with larger errors should be taken into account with a smaller weight.

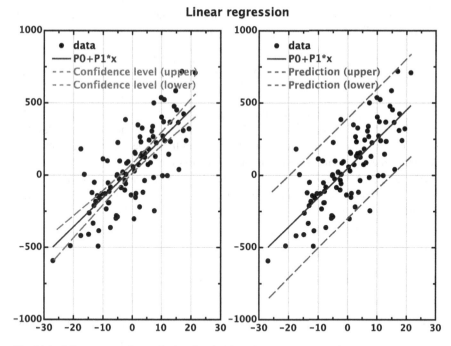

Fig. 14.1 A linear regression analysis using the class jhplot.stat.LinReg

A typical case is a histogram. Look at the example shown in Fig. 8.1. Each histogram height in a certain bin contains a statistical error. As we already know, in case of a simple event-counting measurement, such error is equal to a variance σ_i in a bin i, with the variance value proportional to the square root of the number of entries in each bin. Thus, one can introduce a weight factor w_i, which reflects the degree of influence of each data point on the final parameter estimate. Typically, the weight w_i for the bin i is expressed in terms of the variance σ_i as $w_i = 1/\sigma_i^2$.

Another example is when data points are represented by a P1D with arbitrary statistical or systematical errors. If errors are known, one can perform a fit by attributing a weight factor for each point. Of course, errors can be arbitrary, i.e. they do not need to be exactly the same as in the case of the counting experiment.

To start the curve fitting, we should initially define:

- A fit method which defines a distance measure between data points;
- A mathematical function describing our model.

To find a proper function which is expected to describe data after adjusting free parameters is non-trivial task. First of all, we should start with an initial guess for the functional form and initial free parameters. The number of free parameters should be as small as possible, and the quality of the fit should be sufficiently satisfactory, so one can claim that the analytical model describes the data. The degree of the fit quality is usually characterized by the χ^2 value:

$$\chi^2 = \sum_{i=0}^{N}\left[\frac{y_i - f(x_i)}{\sigma_i}\right]^2 \tag{14.1}$$

where $f_i(x)$ is a given function with free parameters that should be found after the fitting procedure, y_i is a data value in Y, σ_i is the data point variance (i.e. error in a bin) in the bin i. The fitting procedure tries to find the values of free parameters of the input function, so the χ^2 has the minimum possible value.

As mentioned before, it is usually a good idea to plot data first to see how the fit function may look like and what initial parameter values may have sense to produce the desired result.

14.2.1 Preparing a Fit

First of all, let us initialize the curve fitting package by creating an instance of the HFitter class.

```
>>> from jhplot   import *
>>> f=HFitter()
>>> print f.getFitMethod()
['chi2','cleverchi2','uml','bml','leastsquares']
```

The example shows how to print the available methods of the HFitter class. The most popular method is the so-called chi2. This method, which is based on the minimization of (14.1), is the default method used by the HFitter() class if it instantiated without passing any arguments. The χ^2 method works the best if data are represented by discrete points with Gaussian uncertainties. The main assumption is that the points y_i are normally distributed around a function $f(x_i)$, uncorrelated and have a variances σ_i. If the number of events in each bin is larger than a certain minimum number (say, several hundred entries), then the distribution of the expected events per bin is approximately Gaussian.

Next, the cleverchi2 method does not include the variance of experimental data points. It is based on the minimization of the following function:

$$\chi^2 = \sum_{i=0}^{N}\left[\frac{(y_i - f(x_i))^2}{|f(x_i)|}\right] \tag{14.2}$$

As before, y_i is the data-point value (or the height of the bin i), x_i is the center of the ith bin, $f(x_i)$ is the value of the function calculated in the ith bin.

Other two options, uml and bml, are unbinned and binned maximum likelihood fits, respectively. If the dataset is binned (i.e. when we deal with a histogram), a binned maximum likelihood should be used.[1]

The maximum likelihood method should be used if data are represented by integer numbers of events which follow a Poisson statistics, i.e. for rare events. Experimentally, if you see a distribution with a small number of events in each bin, the best bet would be to try to use the maximum likelihood method. If the amount of data is very small, a larger bin size should be tried. In this case, it is preferable to avoid data binning at all: the fit can be done using unbinned maximum likelihood method.

The binned maximum likelihood method minimizes the quantity:

$$bml = \sum_{i=0}^{N} \left[f(x_i) - y_i \ln(f(x_i)) \right]$$

while the unbinned method is based on:

$$bml = -\sum_{i=0}^{N} \left[\ln(f(x_i)) \right]$$

Lastly, the leastsquares method assumes that the best-fit curve is the curve that has the minimal sum of the deviations squared (least square error) from a given set of data. Unlike the methods discussed above, this method does not use errors for data points, i.e.

$$ls = \sum_{i=0}^{N} \left[y_i - f(x_i) \right]^2 \tag{14.3}$$

This method was considered for the linear regression and it will not be discussed it anymore.

14.2.2 Creating a Fit Function

Let us assume we know which method should be used for the curve fitting. The next question is how to find a function which expectedly describes our data. The package contains several pre-built functions in the HFitter catalog, which can be printed as:

---

[1]Note that for a small number of events, the binning can result in a loss of information and larger statistical errors for the parameter estimates [1]. On the other hand, a benefit of the binning is that it allows for a goodness-of-fit test.

Table 14.1 The list of implemented fit functions of the HFitter class

Name	Definitions
e	Exponential
g	Gaussian
g2	Double Gaussian
lorenzian	Lorenzian
moyal	Moyal
landau	Landau
pow	power-law, $a*(b-x)^c$
P0	$y = a$
P1	$y = a + b*x$
P2	$y = a + b*x + c*x^2$
Pn	polynomial of nth order

```
>>> from jhplot  import *
>>> f=HFitter()
>>> print f.getFuncCatalog()
['e', 'g', 'g2', 'lorentzian', 'moyal',
 'p0', 'p1', 'p2', 'p3', 'p4', 'p5',
 'p6', 'p7', 'p8', 'p9', 'landau', 'pow']
```

This command prints the names of the implemented fit functions as described in Table 14.1.

There are several ways to create a function object to be used for the actual fit. The simplest approach would be to pick up a predefined function from the catalog. Alternatively, one can use a simple script to define any custom function (see later).

Generally, there is no need to include a custom function to the catalog using the method addFunc(name, func), where func is a function of the type IFunction which was briefly discussed in Sect. 3.7. To use this method is rather straightforward when working with Java: One needs to implement a function class using the examples located in jhplot.fit package. We will discuss this later.

14.2.2.1 Using Built-in Fit Functions

Let us create a Gaussian function using the predefined key 'g' and print all attributes of this function:

```
>>> f.setFunc('g')
>>> func=f.getFunc()
>>> print func
BaseModelFunction:  Title=g, name=g
```

```
Dimension: 1, number of parameters: 3
Codelet String: codelet:g:catalog
Variable Names: x0,
Parameters: amplitude=1.0, mean=0.0, sigma=1.0,
Provides Gradient:true,Provides Parameter Gradient:true
Provides Normalization: false
```

One may find this information a bit cryptic, but the important information can still be found: This function is implemented in one dimension, has three free parameters (amplitude, mean, sigma). One can also see the initial values assigned to this function during the initialization.

It should be noted that the object `func`, which is returned by the method `getFunc()`, belongs to the class `IFunction`. Let us illustrate some of them:

```
print 'Metatype and implementation:',func.codeletString()
print 'Parameter names:',func.parameterNames()
print 'List parameters', func.parameters()
print 'No of free parameters',func.numberOfParameters()
print 'Function title:', func.title()
print 'Function value:', func.value([1])
```

The last method is rather important: It accepts a list of values at which the function is evaluated. In case of an one-dimensional function, the list contains only one value. Usually, the printed information is sufficient to understand the function implementation. If you are not sure how the function is implemented, evaluate it at a fixed value.

One can also set free parameters to certain values. For example, one can set the mean of a Gaussian to ten using this method:

```
>>> func.setParameter('mean',10)
```

One can also set all three parameters by using a list:

```
>>> func.setParameters([100,10,1])
```

which sets the amplitude, the mean and the sigma to 100, 10, 1, respectively.

The real power of the pre-built fit functions comes when one needs to go beyond the simple functions already existing in the catalog. One can easily construct new functions using simple operations. For example, one can build a Gaussian function plus a second-order polynomial using the string 'g+p2', i.e.:

```
>>> f.setFunc('g+p2')
>>> func=f.getFunc()
```

```
>>> print func.title()+' has: ',func.parameterNames()
'amplitude', 'mean', 'sigma', 'p0', 'p1', 'p2'
```

Now you can add this function to the catalog as discussed above:

```
>>> from jhplot import *
>>> f=HFitter()
>>> f.setFunc('g+p2')
>>> func=f.getFunc()
>>> f.addFunc('g+p2',func)
>>> print f.getFuncCatalog()
array(java.lang.String, [ ... 'g+p2' ..])
```

14.2.2.2 Building Functions from a String

One can also generate a custom IFunction object using an analytical expression, function dimension and the list of parameters. Let us illustrate how to build a parabolic function from a string:

```
>>> f.setFunc('parabola',1,'a*x[0]*x[0]+b*x[0]+c','a,b,c')
>>> func=f.getFunc()
>>> print func.title()+' has: ', func.parameterNames()
parabola has:   ['a', 'b', 'c']
>>> print func.codeletString()
parabola:verbatim:jel:1:a*x[0]*x[0]+b*x[0]+c:a,b,c:
```

The setFunc() method takes several arguments: the function title, the number of dimensions, a string representing the function, and the names of the free parameters. The only independent variable of this function is 'x[0]'. In case of 2D functions, one can add a second variable 'x[1]'. The method codeletString() prints the implementation details of this function.

14.2.2.3 Building Functions from a Script

Finally, one can build a IFunction using Jython scripts as discussed in Sect. 3.7. In this case, one can include any logic you want into its definition (like if-else statements), or even call external Java libraries to access special functions. We will remind that one should create a custom class which is based on ifunc. This can look as:

```
>>> from shplot import *
>>> class [name](ifunc):
```

```
>>>    def value(self, v):
>>>        [equation]
>>>        return [value]
```

where [name] is the function name, [equation] its functional form. The purpose of the method value() is to return the calculated function value. Then one can instantiate an object of this class which now will have all properties of the IFunction class. This object can be used for the fits as usual.

14.2.2.4 Preparing a Fit Function

During the fit procedure, the fit optimization program tries to find the best possible parameters of the function which is expected to describe input data. It starts fitting with some initial values. It is always a good practice to set initial values to some sensible numbers. This can be done using two methods discussed before, either the method setParameters(list) or the method setParameter(name,v), where 'v' is an input value. One can also use shorter names, like setPar(name,v) of the class HFitter (there should be no any confusion here, since only one function is allowed for the fit).

```
>>>    f.setPar('mean',10)
```

This sets the parameter 'mean' to the value 10, assuming that setFunc('g') has been called first as shown previously (i.e. 'f' represents an instance of a Gaussian function).

During the minimization, one can restrict parameter variations to a certain range:

```
>>>    f.setParRange('mean', -10, 10)
```

This sets the range $[-10, 10]$ for the Gaussian mean.

Finally, in some situations, one may constrain certain fit parameters. Let us give an example by creating a double-Gaussian function. During the fit, we want to keep the mean of the first Gaussian exactly to be the same as for the second one:

```
>>>    f.setFunc('gauss2',1,N*( a*exp(-0.5*(mean0-x[0])\
>>>    *(mean0-x[0])/(s0*s0))+(1-a)*exp(-0.5*(mean1-x[0])\
>>>    *(mean1-x[0])/(s1*s1) ))',\
>>>    'N,a,mean0,s0,mean1,s1')
>>>    f.setParConstraint('mean0=mean1')
```

The method setParConstraint() does the trick. Make sure that you have correctly assigned the names of the parameters during function creation.

14.3 Displaying a Fit Function

Once the function is defined, one can display it in the usual way. This can be done using either F1D or F2D functions, after passing an IFunction object to the function constructor:

```
>>> ff=F1D(f.getFunc(),min,max)
>>> c1.draw(ff)
```

where c1 is a HPlot canvas. How to display a F1D function was discussed in Sect. 3.3. Analogously, one can build and display functions in 2D.

14.3.1 Making a Fit

Next, we will fit our data with the prepared function. This part is easy: if you have P1D, H1D, H2D or PND containers filled with data, simply execute the method fit(obj), where the obj is one of the objects mentioned above:

```
>>> f.fit(obj)
```

In case of 1D data, you may also restrict the fitting range with the method setRange(min,max) applied to the object f. After the fit, one can get the resulting function after the minimization as:

```
>>> func=f.getFittedFunc()
```

The method returns a IFunction function with the parameters determined by the fit minimization.

Another way to retrieve the results is to use this method:

```
>>> result=f.getResult()
```

The object result is a rather sophisticated as it keeps essentially everything you need to retrieve the fit results. Below we list the most important methods of this class and comment some of them:

```
>>> result.fittedParameters()
>>> result.errors()
>>> result.constraints()
>>> result.covMatrixElement(0,1)
>>> result.engineName()
>>> result.errors()            # parabolic errors
```

```
>>> result.errorsMinus()    # error -
>>> result.errorsPlus()     # error +
>>> result.fitMethodName()
>>> result.fitStatus()
>>> result.fittedFunction()
>>> result.fittedParameter("mean")
>>> result.fittedParameterNames()
>>> result.fittedParameters()
>>> result.ndf()              # number of degrees of freedom
>>> result.quality()          # chi2/ndf for Chi2 method
```

We should point out that the quality of the fit can be obtained via the method quality(), which represents the χ^2 per degree of freedom. The smaller its value is, the better the fit minimization.

It is time now to give a small example before going any further. Below we generate a Gaussian distribution and then we fit it with a Gaussian function.

──────────────── Gaussian fit ────────────────

```
from jhplot  import *
from java.util import Random

f=HFitter()
f.setFunc('g')
f.setPar('amplitude',50)

h1 = H1D('Data',50, -4, 4)
h1.setPenWidthErr(2)
h1.setStyle('p')
h1.setSymbol(4)
h1.setDrawLine(0)

r = Random()
for i in range(1000):
    h1.fill(r.nextGaussian())

c1 = HPlot('Canvas')
c1.visible()
c1.setAutoRange()
c1.draw(h1)

f.setRange(-4,4)
f.fit(h1)
ff=f.getFittedFunc()

r=f.getResult() # get fitted results
Pars    = r.fittedParameters()
Errors = r.errors()
Names   = r.fittedParameterNames()
```

```
print 'Fit results:'
for i in range(ff.numberOfParameters()):
  print Names[i]+' : '+str(Pars[i])+' +- '+str(Errors[i])

mess='&chi;^{2}/ndf='+str(round(r.quality()*r.ndf()))
mess=mess+' / '+str(r.ndf())
lab=HLabel(mess, 0.12, 0.69, 'NDC')
c1.add(lab)

f2 = F1D('Gaussian',ff,-4,4)
f2.setPenWidth(3)
c1.draw(f2)
print 'Quality=',r.quality(), ' NDF=',r.ndf()
```

The output of this program is shown below:

```
Fit results:
amplitude : 62.578 +- 2.465
mean : -9.484E-4 +- 0.032
sigma : 0.986 +- 0.023
Quality= 0.990  NDF= 34
```

In addition, the script generates Fig. 14.2 and shows the χ^2/ndf value inside the interactive label.

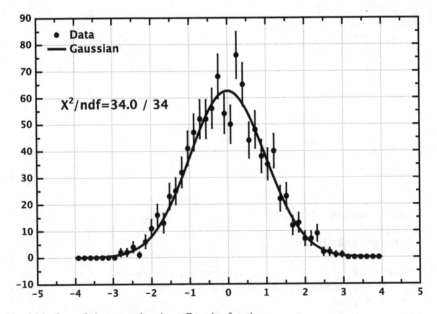

Fig. 14.2 Curve fitting example using a Gaussian function

14.4 Real-life Example. Signal Plus Background

With the curve fitting class in hand, now we will give a rather realistic example of how to perform a fit of data using a Gaussian plus a background function. As usual, we will divide the example into two parts: In one part, we will generate a histogram with data, and in the second part, we will read this histogram and perform a fit.

14.4.1 Preparing a Data Sample

Let us prepare our input for the curve fitting, which will consists of two histograms, H1D and H2D. The 1D histogram is filled with two Gaussian random numbers, one is for a signal and the second is for background. The signal distribution is modeled also by a Gaussian distribution by setting its width to some large value. Also, we fill a 2D histogram with Gaussian numbers, but this time, we shifted their means from zero and increased the widths. We will store these histograms in a dictionary, in which a string will be used as a key. Then, we write the dictionary into a XML file:

```
———————————— Histograms for curve fitting ————————————
from java.util import Random
from jhplot  import *
from jhplot.io import *

h1 = H1D('Data',50, -7, 7)
h2 = H2D('Data 3D',40,-10,10,40,-10,10)

r= Random()
for i in range(10000):
   if (i<5000): h1.fill(r.nextGaussian())
   h1.fill(5*r.nextGaussian()+5)
   h2.fill(2*r.nextGaussian(),2*r.nextGaussian()+3)

d={'h1':h1,'h2':h2,'description':'Gaussian+background'}
Serialized.writeXML(d,'data.xml')
```

Later we will read the 'data.xml' file and will use the keys 'h1' and 'h2' to retrieve the histograms from the dictionary.

14.4.2 Performing Curve Fitting

Let us now to fit the data prepared by the above script. After reading the serialized file, we will retrieve the histograms and fit them using a Gaussian function with a first-order polynomial describing the background under the signal peak:

```
                    ───── Signal+background fit ─────
from jhplot  import *
from java.util import Random
from java.awt import Color
from jhplot.io  import *

d=Serialized.readXML('data.xml')
h1=d['h1']; name=d['description']

f=HFitter()
print f.getFuncCatalog()
f.setFunc('g+p1')
f.setPar('p0',10); f.setPar('amplitude',100)

c1 = HPlot('Canvas')
c1.visible()
c1.setAutoRange()

h1.setPenWidthErr(2)
h1.setStyle('p')
h1.setSymbol(4)
h1.setDrawLine(0)

c1.draw(h1)
f.setRange(-7,7)
f.fit(h1)

ff=f.getFittedFunc()
r=f.getResult()
fPars  = r.fittedParameters()
fErrs  = r.errors()
fNames = r.fittedParameterNames()
print 'Fit results:'
for i in range(ff.numberOfParameters()):
  print(fNames[i]+':'+str(fPars[i])+' +- '+str(fErrs[i]))
print 'Chi2/Ndf=',r.quality()*r.ndf(),'/',r.ndf()

f2 = F1D('Gaussian',ff,-7,7)
f2.setColor(Color.blue)
f2.setPenWidth(3)
c1.draw(f2)
```

This code snippet prints the fit results:

```
Fit results:
  amplitude : 570.3 +- 11.3
  mean : 0.08 +- 0.02
```

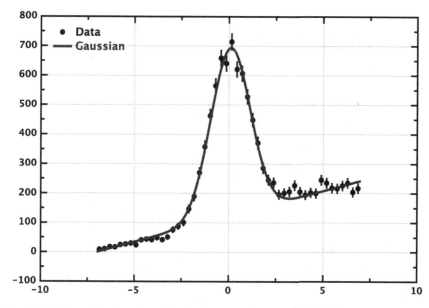

Fig. 14.3 Fitting a signal together with a background using the χ^2 minimization

```
sigma : 1.04 +- 0.02
p0 : 121.1 +- 2.2
p1 : 17.2 +- 0.3
Chi2/Ndf= 9/45
```

The output plot is shown in Fig. 14.3.

14.4.3 Fitting Multiple Peaks

This time we will learn how to fit multiple peaks. For this, we will prepare a histogram with three Gaussian peaks plus a background. Unlike the example discussed before, we will create the background distribution using random numbers generated in accordance with the analytical function $10 + 10 * x$ (see Sect. 9.5 for details). Below we show how to do this:

```
————————— A histogram with multiple peaks —————————
from java.util import Random
from jhplot  import *
from jhplot.io import *
from jhplot.math.StatisticSample  import *

xmin,xmax=0,20
h1 = H1D('Data',200,xmin,xmax)
```

```
f=F1D('10+10*x',xmin,xmax)
p=f.getParse()
max=f.eval(xmax)

r= Random()
for i in range(10000):
  a=randomRejection(10,p,max,xmin,xmax)
  h1.fill(a)
  h1.fill(0.3*r.nextGaussian()+4)
  h1.fill(0.6*r.nextGaussian()+10)
  h1.fill(0.8*r.nextGaussian()+15)

Serialized.write({'h1':h1},'data.ser')
```

Note that, for each iteration, the object a is an array with 10 numbers between xmin and xmax. The Gaussian peaks are located at 4, 10, 15 units (note the additive factors) and have different widths, given by the scaling factors, 0.3, 0.6, 0.8. The histogram is saved in a serialized file using a dictionary with the key h1.

Now let us fit this histogram. First, we will read the histogram using its key and plot it. Then we fit a region around the first Gaussian, making sure that we fit only the specified range before the second peak starts. Then we add an additional Gaussian to a new fit function and use the results of the previous fit for initialization of our new function: Our code which fits all Gaussian peaks and the background is given below:

─────────── Fitting multiple peaks ───────────

```
from jhplot  import *
from jhplot.io import *
from java.awt import Color

d=Serialized.read('data.ser')
h1=d['h1']

c1 = HPlot('Canvas'); c1.setRange(0,20,0,2000)
c1.visible(); c1.draw(h1)

f=HFitter()
f.setFunc('p1+g') # first fit
func=f.getFunc()
f.setPar('mean',4); f.setPar('amplitude',100)
f.setRange(0,7)
f.fit(h1)

ff=f.getFittedFunc(); r=f.getResult()
fPars  = r.fittedParameters()

f.setFunc('p1+g+g') # Next Gaussian
```

```
func=f.getFunc()
func.setParameters(fPars.tolist()+[500,10,0.5])
f.setRange(0,13)
f.fit(h1)

ff=f.getFittedFunc(); r=f.getResult()
fPars   = r.fittedParameters()

f.setFunc('p1+g+g+g')  # Next Gaussian
func=f.getFunc()
func.setParameters(fPars.tolist()+[500,15,0.5])
f.setRange(0,20)
f.fit(h1)

f2 = F1D('Fit function',f.getFittedFunc(),0,20)
f2.setPenWidth(1)
f2.setColor(Color.blue)
c1.draw(f2)
```

In this example, the list [500,10,0.5] simply specifies the initial parameters for the new Gaussian function, and it is added to the list created from the array fPars holding the parameter values from the previous fit. Similarly, we fit the third peak. Finally, we extract the resulting function and plot it. If you are not sure which parameter names are necessary to use at each step, print the outputs of the methods func.parameterNames() and func.parameters().

The result of the fit is shown in Fig. 14.4.

One can access the fit results using the same approach as in the previous examples: Insert the line r=f.getResult() at the end of your code and use the

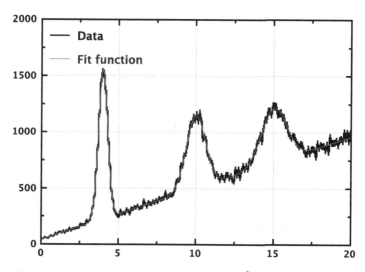

Fig. 14.4 Fitting multiple peaks and a background using the χ^2 minimization

methods of the object r to print the fit quality (χ^2/ndf), fitted parameter values and their statistical uncertainties.

14.4.4 Fitting Histograms in 3D

Let us turn to the H2D histogram which was filled with data but not yet analyzed. This time, we have to prepare a fit function with two independent variables. As before, we will fit our 2D function using the HFitter class and extract the resulting function using the same methods as for the 1D case. The output function should be shown on the HPlot3D canvas. Look at the example below:

```
──────────── Fitting a 2D histogram ────────────
from jhplot  import *
from jhplot.io  import *

d=Serialized.readXML('data.xml')
h1=d['h2']
c1 = HPlot3D('Canvas',800,400,2,1)
c1.visible()

f=HFitter()
f.setFunc('g2D',2, 'N*(exp( -0.5*((mu0-x[0])*\
          (mu0-x[0])+0.5*(mu1-x[1])*(mu1-x[1]))\
          /(s0*s0) ))','N,s0,mu0,mu1')
f.setPar('N',100);    f.setPar('s0',1.0)
f.setPar('mu0',0.0); f.setPar('mu1',1.0)
f.fit(h1)

ff=f.getFittedFunc()
r=f.getResult()
Pars   = r.fittedParameters()
Errors = r.errors()
Names  = r.fittedParameterNames()
print 'Fit results:'
for i in range(ff.numberOfParameters()):
   print Names[i]+': '+str(Pars[i])+' +- '+str(Errors[i])

print 'chi2/ndf='+str(round(r.quality()*r.ndf()))\
                  +'/',r.ndf()

f1 = F2D(ff,-10,10,-10,10)
c1.setGTitle('Fitting 2D data')
c1.cd(1,1)
c1.draw(h1)
```

2D data fitted with a 2D Gaussian

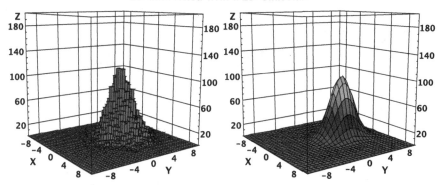

Fig. 14.5 Fitting a 2D histogram (*left*) and showing the fit result (*right*)

```
c1.cd(2,1)
c1.draw(f1)
```

The output figure is shown in Fig. 14.5.

14.5 Interactive Fit

One can fit 1D data (H1D histograms or P1D arrays) using an interactive dialog based on the HFit class. Assuming that c1 represents the HPlot canvas and h1 is a H1D histogram, one can start this fitter dialog as:

```
>>>a=HFit(c1,'c1',h1,'h1')
```

The execution of this line brings up a fit dialog where one can select the necessary analytical function (it must exist in the catalog). Note that we pass not only the objects, but also the variable names for the HPlot and H1D class. This is necessary if one needs to generate the JAIDA source code automatically. One should mention that one can also pass a P1D object instead of the histogram.

The fit panel allows a user to add custom functions, fit methods and perform the fit. One can also set up the initial parameters for the selected fit function using the [Settings] button. Once the fit is acceptable, one can generate the output fit parameters, which will be inserted directly to the editor in a form of Jython code. One can also automatically generate the JAIDA source code which corresponds to the fit. Once inserted, it can further be corrected using the editor. Then, HFit() statement can be removed and the generated JAIDA code can be run manually.

One can also specify a user-defined 1D function and add it to the HFit dialog. This is an example of how to insert a first order polynomial function to be used for the curve fitting:

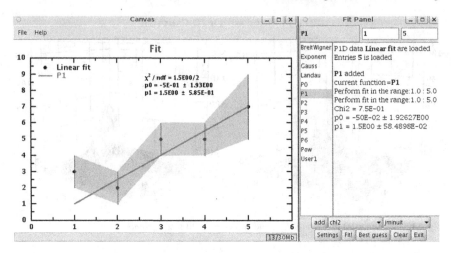

Fig. 14.6 Fitting data with statistical errors using an interactive fitter

```
>>>a=HFit(c1,'c1',h1,'h1')
>>>a.addFunc('User1','Tool tip','a*x[0]+b','a,b')
```

After script execution, a new function with the name User1 will be added to the
list of known functions. Select it with the mouse and click on the button [add].
One can add several functions for the fit. Then, by pressing the button [fit], you
will see the fit result.

Let us give a complete example of how to fit data points with statistical errors
using a custom function:

```
─────────── Fitting data interactively ───────────
from java.awt import Color
from jhplot  import *

c1 = HPlot('Canvas')
c1.visible()
c1.setAutoRange()

c1.setGTitle('Fit',Color.blue)
c1.setGrid(0,0)
c1.setGrid(1,0)

p1 = P1D('Linear fit')
p1.add(1.0,3, 1.0)
p1.add(2.0,2, 1.0)
p1.add(3.0,5, 1.0)
p1.add(4.0,5, 1.0)
p1.add(5.0,7, 2.0)
```

```
p1.setErr(1)
p1.setErrFillColor(Color.yellow,0.3)
c1.draw(p1)

a=HFit(c1,'c1',p1,'p1')
a.addFunc('User1','My fit function', 'a*x[0]+b','a,b')
```

The result of this script, after pressing the button [add] followed by [fit] on the HFit dialog, is shown in Fig. 14.6.

References

1. Eidelman, S., et al.: Review of particle physics. Phys. Lett. B **592**, 1 (2004)

Chapter 15
Neural Networks

15.1 Introduction

A neural network is a powerful tool for data classification: if we know a source
with input data, and also have knowledge of how the output data looks like, one can
obtain a relationship between inputs and outputs. This can be done totally numerically, without using any predefined function as in the case of curve fitting discussed
before. Once such a relationship is established, one can perform predictions based
on the input, assuming that the data set has the same characteristics as those used
for determination of the relationship during the so-called "training" step. Here we
will recommend several books [1, 2] with detailed information about the neural networks. Below we will touch only the basics of this subject.

But what exactly should be used to perform such classification of inputs? The entire concept of the neural-network approach tries to capture the essence of biological
neural systems. In particular, it tries to simulate the brain's way of processing information. A neural network consists of a number of interconnected "neurons" in oder
to simulate biological neural systems. We should remind that a human brain consists
of a very large number of interconnected neurons. Each neuron has an input and a
branching output structure. When a neuron is activated, it sends an electrochemical
signal via the synapses to other neurons which may, in turn, fire signals to inputs of
other neurons. The strength of the propagated signal depends on the efficiency of
the synapses.

Mathematically, a neural network consists of a set of interconnected units, called
nodes. Each node accepts a weighted set of inputs and responds with an output.
Each input has an associated weight representing the strength of that particular connection. A multilayer network with several layers of units is rather popular. For such
networks, the output from one layer serves as input to the next.

Below we will consider a *feedforward* neural network, in which the data from
input to output units is strictly feedforward. For this type of network, no feedback
connection exists. This is the most popular network in many practical applications.

The number of neurons in the input layer depends on the number of inputs, while
the number of neurons in the output layer depends on the number of desired outputs.

S.V. Chekanov, *Scientific Data Analysis using Jython Scripting and Java,*
Advanced Information and Knowledge Processing,
DOI 10.1007/978-1-84996-287-2_16, © Springer-Verlag London Limited 2010

The number of hidden layers and the number of neurons in each such layer must be defined empirically.

Each neuron performs accumulation of incoming pulses from its inputs using weights. All weights are sum up and the sum is passed into a non-liner activation function. We will use the most popular sigmoid activation function, $S(x) = 1/(1 + e^{-x})$. The output of such neutron is formed by applying the sigmoid function as: $S(d + \sum_{i=0}^{N} x_i w_i)$, where x_i are the input values from other neurons, w_i characterize the weights for input connections and d is a real number called the threshold. This number characterizes the neutron as whole.

The neural net should be trained to adjust w_i and d parameters. Once the input units have received the signal from outside and from output units with the resulting signal, one can adjust the weights in interconnected neurons such that one can perform predictions. During the learning process, the output vector of numbers is compared to the expected output. If the difference is not zero, the ("epoch") error on the prediction is calculated. The idea is to adjust the weights to minimize the difference between the neural-net output and the expected output. In this case, such networks can learn arbitrary associations between the input vector and the output.

15.1.1 Generating a Data Sample

Let us sketch a typical task which can be solved using the neural-net approach. Assume we have many events with the input vector of numbers. We also know some events in which the input vector produces a specific output. How to make predictions in case when only the input is available?

We will start with the preparation of a data sample, so later we will use it for our neural network studies. Let us assume that we have five variables generated in accordance with some distributions. We generate a new variable from the inputs using some functions. The vector with inputs and outputs will be stored persistently in a serialized file as shown below:

─────────── Creating a data set ───────────

```
from java.util import Random
from jhplot  import *
from jhplot.io import *

input=PND('Data')
out=PND('Output')
r= Random()
for i in range(1000):
    x1=10*r.nextDouble()
    x2=2**(10*r.nextDouble())
    x3=1/(r.nextDouble()+0.0001)
    x4=10*r.nextDouble()+0.0001
    input.add([x1,x2,x3,x4])
    out.add([x1/x4-x2+x3])
```

```
d={'input':input,'out':out}
name='data.ser'
Serialized.write(d,name)
print name+' is ready'
```

The code generates four variables and calculates the output y using the relationship $y = x_1/x_4 - x_2 + x_3$.

Of course, in reality, we do not know an analytic relationship between input and output. In fact, we are not interested in this form at all, since the whole idea of the neural network approach is to find a numerical relationship. In some cases, we even do not know the contribution of each input variables to the output, thus some contributions could be zero and the variable may have nothing to do with the actual output.

So, the task is seemingly clear. We have the input array and the output (which, in general, could be also an array). The question is how to establish the relationship between these variables pretending that, in reality, we know nothing about it.

15.1.2 Data Preparation

For neural network studies, it is advisable to standardize the input and to rescale the output to the range [0, 1] or [−1, 1], depending on the output activation function.

The standardization means that we rescale the data such that all inputs have the same weights. Without it, a variable with the largest scale may dominate, thus producing a bias towards the input of this variable. To circumvent this problem, each column should be standardized, i.e. transformed to:

$$S_i = (X_i - \bar{X})/\sigma$$

where X_i is the original value, $\bar{X}$ is the mean and σ is the standard deviation of the input data.

In order to standardize a PND, apply the following command:

```
>>>   pnd.standardize()
```

where 'pnd' is an input PND data container. The example below shows how to use this method:

```
>>> from jhplot import PND
>>> pnd=PND('pnd')
>>> pnd.add([0.02,10,3])
>>> pnd.add([0.01,6,1])
>>> pnd.add([0.03,12,5])
>>> print pnd.standardize()
```

```
  0.0    0.21    0.0
 -1.0   -1.09   -1.0
  0.99   0.87    1.0
```

In contrast, a normalization or rescaling means that each column is transformed such that all column values are either between [0, 1] or [−1, 1], depending on the range of the activation function. To normalize a PND, use the rescale() method. One should pass "0" as an argument in case if the rescaling should be done for the range [0, 1]:

$$S_i = (X_i - X_{min})/(X_{max} - X_{min})$$

In case of the range [−1, 1], one should apply this transformation:

$$S_i = 2 * (X_i - \text{midrange})/(X_{max} - X_{min}), \quad \text{midrange} = 0.5(X_{max} + X_{min})$$

which is implemented for the PND class using the method rescale(1). This method returns an array which then is used to transform the PND back to the original form. In this case, the output of the rescale method should be passed as an input to the rescale(obj) method. Consider the example:

```
>>> from jhplot import PND
>>> pnd=PND('pnd')
>>> pnd.add([0.02,10,3])
>>> pnd.add([0.01,6,1])
>>> pnd.add([0.03,12,5])
>>> a=pnd.rescale(0)
>>> print pnd
 0.5   0.66   0.5
 0.0    0.0   0.0
 1.0    1.0   1.0
>>> pnd.rescale(a)
>>> print pnd
 0.02   10.0   3.0
 0.01    6.0   1.0
 0.03   12.0   5.0
```

One can see that the rescale method has restored the original array, once you have passed the output from the first rescale(0) call which converts each column to the range [0, 1]. Analogously, if the rescale(i) method is called with the argument "1", all values will be moved to the range [−1, 1], and the conversion back will be done in exactly the same way as before.

In the next example we read the saved containers, standardize the input and normalize the output. Then we save the resulting container into a dictionary. We also save the array keeping the scaling factors, which can be used to restore the output later.

```
───────────── Standardizing data ─────────────
from jhplot.io import *

d=Serialized.read('data.ser')
input=d['input']
out=d["out"]

input.standardize()
scale=out.rescale(0)

d={'input':input,'out':out,'scale':scale}
name='data_scaled.ser'
Serialized.write(d,name)
print name+' is ready'
```

It should be noted that the standardization should be approached with caution because it discards certain information. Generally, there are no 'rules of thumb' that apply to all possible cases.

Now we are ready to use a fraction of the rescaled data for training of our neural network. But before, let us discuss the machinery behind the construction of a neural net.

15.1.3 Building a Neural Net

First, let us discuss how to build and visualize a neural network. In case of jHep-Work, a neural net can be constructed using the class HNeuralNet which is based on the Encog project [3]. In order to illustrate its usage, we will build a feedforward neural network with an input layer, two hidden layers and one output layer. The input layer will contain four neurons, two hidden layers will have five neurons each, and the output layer will have a single neuron:

```
>>> from jhplot import *
>>> net = HNeuralNet()
>>> net.addFeedForwardLayer(4)
>>> net.addFeedForwardLayer(5)
>>> net.addFeedForwardLayer(5)
>>> net.addFeedForwardLayer(1)
```

The code above simply adds layers, staring from input to the output, passing integer arguments and specifying the number of interconnected neutrons in each layer.

The next step is to reset all neuron thresholds to some (random) values. This should be done with the method reset().

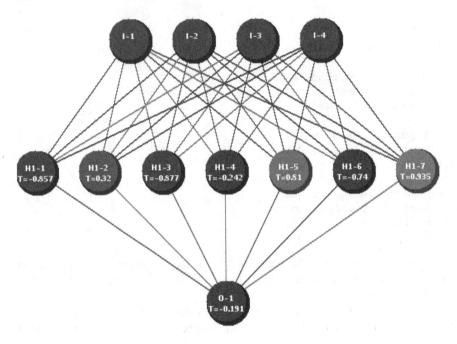

Fig. 15.1 A representation of a neural network with two hidden layers. The threshold values for the activation matrix is shown for each neutron. In this example, they are random numbers generated with the reset() method

The constructed network can be visualized with the method showNetwork(). Below we show how to initialize the network and display it on the screen:

```
>>> net.reset()
>>> net.showNetwork()
```

The result of the last line is a pop-up window which displays the structure of the network. This window is shown in Fig. 15.1. Each layer has the following notations: "I" is the input layer, "H" is the hidden layer and "O" is the output layer.

Once the network is created, we should set the input and the output data for training. This can be done with the method setData(input,output), where input and output are data holders of the type PND. One can look at the data by calling the editData() method which opens an editor with the neural-net input and output.

Next, the created network should learn how to adjust the weights and the neuron thresholds. This is done by applying the method:

```
>>> net.trainBackpopogation(b,max,learnRate,mom,err)
```

where:

b	is 1 (Java true) if a dialog window is required to show the learning rate and its errors. The smaller error on Y, the higher chances that the learning is successful. Make sure that the error value does not change with iteration (or epoch) number. If b is Java false ("0" in Jython), one cannot monitor the learning rate and error.
max	is the maximal number of epochs for learning. The learning continues until max is reached, but only if the epoch error is larger than errEpoch. If we reach a maximum count of iterations or epochs, this will mean that the training was not successful.
learnRate	is the rate at which the weight matrix will be adjusted based on learning (the so-called learning rate); This number is usually between 0.1 and 0.2.
mom	is the influence that previous iteration's training deltas will have on the current iteration. Usually it is set to 0.1–0.4.
err	is the "epoch" error for training at which the learning should be stopped. If the specified error is not reached during the learning, the program stops the learning after reaching "max" number of events.

It should be noted that the epoch error can also be obtained with the method net.getEpochError() without opening the control window.

It is always a good idea to call the showNetwork() method to look at the new threshold values. Once the network was trained, one can save the network in a file, so one can restore it later. This can be done as:

```
>>> net.save('test.eg','trainedNN','My trained NN')
```

The first string is the output file name ('test.eg' in this example), the second is the name of the network ('rainedNN') and the third argument is used for the network description.

One can restore the saved network at any time by calling the method:

```
>>> net.read('test.eg','trainedNN');
```

where 'test.eg' is an input file with the neural network.

15.1.4 Training and Verifying

The data after the standardization of the inputs and rescaling the outputs are ready for the next step—training. We will create a neural net with four neutrons in the input layer. The output layer will contain one neuron. Our hidden layer will contain seven neurons. After creating the net, we will reset all inputs and load the first half of our data (500 rows). The training will be done using 2000 epoches. During the training, we will open a frame displaying the value of the epoch error. Once the training is finished, we will save the results into the file `'test.eg'`, attributing `trainedNN` string to the name of our neural net. We also display the neural net after the training in order to verify that all thresholds indeed have been changed after the training. The code below shows the entire training process:

```
——————————————— Training a neural net ———————————
from jhplot import *
from jhplot.io import *

d=Serialized.read('data_scaled.ser')
input=d['input'].getRows('input',0,500)
out=d['out'].getRows('result',0,500)

net = HNeuralNet()
net.addFeedForwardLayer(4)
net.addFeedForwardLayer(7)
net.addFeedForwardLayer(1)
net.reset()

net.setData(input, out)
print net.trainBackpopogation(1,2000,0.01,0.02,0.005)

print "Epoch error=",net.getEpochError()
net.save("test.eg","trainedNN","My trained NN")
net.showNetwork()
```

Next, we will verify the performance of the network training using the second half of our generated sample. We will read the data and the neural net saved in the previous example, and generate predictions from the input. Then we will restore the input (remember, previously we had to scale the input data to the range [0, 1]) using the "scale" arrays, and apply the same recovery procedure to the predicted array. Then we will save restored output and neural net predictions in a separate file for the next analysis step. The code is given below:

```
——————————————— Neural net verification ———————————
from jhplot import *
from jhplot.io import *

d=Serialized.read('data_scaled.ser')

min,max =500,1000
input=d['input'].getRows('input',min,max)
out=d["out"].getRows('result',min,max)
scale=d["scale"]

net=HNeuralNet()
net.read('test.eg','trainedNN');
```

```
pred=net.predict(input)
pred.rescale(scale)
out.rescale(scale)

d={'predicted':pred,'expected':out}
Serialized.write(d,'data_verify.ser')
```

Finally, we can look at the outputs and the predicted values. We will print the values and also calculate the ratio of the predicted outputs to the original outputs. We will create a histogram with this ratio as shown below:

```
—————————— Neural network predictions ——————————
from jhplot import *
from jhplot.io import *

d=Serialized.read('data_verify.ser')
predicted=d['predicted']
expected=d['expected']
ratio=predicted.copy('ratio')
ratio.oper(expected,'/')

c1=HPlot()
c1.visible()
c1.setAutoRange()

h=H1D('ratio',50,-3,3.)
h.fill(ratio)
c1.draw(h)
c1.drawStatBox(h)

for i in range(100): # check first 100
    p=predicted.getRow(i)
    x=expected.getRow(i)
    d1=p.get(0)
    d2=x.get(0)
    print 'predicted=',d1,' expected=',d2
```

The script generates a canvas with the filled histogram. One can see that the histogram has a peak at unity as expected. The distribution has long tails indicating that, for some rear events, the predictions could still be far away from the original values.

15.2 Bayesian Networks

A Bayesian network (or "belief" network) is a graphical model for manipulating probabilistic relationships among variables of interest, and building decision scenarios requiring reasoning under uncertainty. Such networks are widely used in managing uncertainty in science, engineering, business and medicine [4, 5].

The Bayesian network is included into the jHepWork libraries using the JavaBayes package [6]. It calculates marginal probabilities and expectations, produces explanations and performs robustness analysis. In addition, it allows a user to import, create, modify and export networks.

The network editor can be called using the HBayes class:

```
>>> from jhplot import *
>>> HBayes()
```

This brings up the Bayesian-network editor and a console window. A user can follow the step-by-step instruction given in the console. The JavaBayes network is rather well documented and will not be discussed here further.

15.3 Self-organizing Map

In addition to the neural network algorithms which require a "learning" procedure, there is a class of unsupervised learning algorithms which attempt to find the most appropriate topological description of input data. Below we will consider a Bayesian self-organizing map (BSOM), which represents a method for estimating a probability distribution from input data on the basis of a Bayesian stochastic model. For this model, the data recognition algorithm tries to find relationships in high-dimensional data and converts this knowledge into a simple relationship using low-dimensions.

We will consider an example which can help to make the BSOM explanation more illustrative. For this, we will use the class HBsom based on the BSOM program [7].

The code below fills a histogram, converts it to a P1D container which is then used for the algorithm input. We use 30 interconnected units to analyze the topological shape in 2D, see the method setNPoints. The last line of the code brings up a window with visualized data points and interconnected units. Also, we save the generated input data (see the method toFile()), so we can use the same data for the example to be shown in the next subsection.

```
───────────── Interactive self-organizing map ─────────────
from java.util import Random
from jhplot  import *

h1 = H1D('Data',20,-100.0, 300.0)
r = Random()
for i in range(2000):
    h1.fill(100+r.nextGaussian()*100)

p1d=P1D(h1,0,0)
```

```
p1d.toFile('data.txt')

bs=HBsom()
bs.setNPoints(30)
bs.setData(p1d)
bs.visible()
```

To start the algorithm, one should set up the values α and β. Then press the button "learn". Initially the BSOM units are positioned randomly, then the algorithm calculates the most optimal positions for interconnected units. The resulting plot is shown in Fig. 15.2.

The BSOM model parameters, α and β, represent the strength of topological constraint and the noise level expected in data, respectively. α and $1/\beta$ correspond to the temperature of a physical system. For $\alpha \simeq 0$, topological constraint on the units is not imposed and BSOM can be viewed as a data clustering based on a spherical

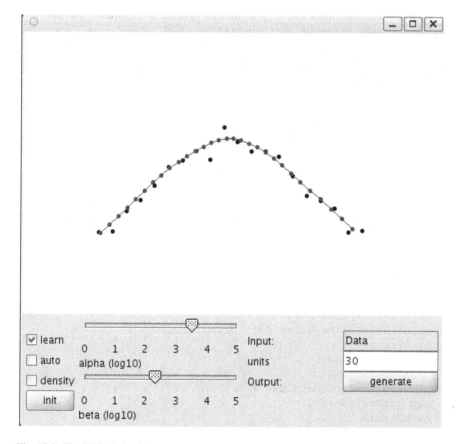

Fig. 15.2 The BSOM algorithm in action using an interactive mode

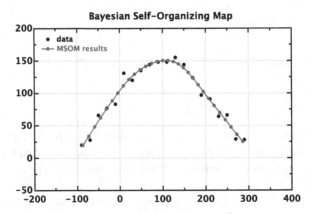

Fig. 15.3 Using the BSOM algorithm in a non-interactive mode

Gaussian mixture model. When β is infinitely large, BSOM is very similar to the K-means algorithm.

Running this algorithm will help to find an optimal configuration for data description. Usually, one should start learning from a high value of α and a low value of β. One should note that BSOM can automatically search for the optimal values of the parameters by pressing the 'auto' button. The 'density' button shows the estimated data density.

15.3.1 Non-interactive BSOM

It is more practical to run the BSOM program in a non-interactive mode. This can be done by removing the line with the `visible()` method and inserting the line `run()`. The results of the algorithm can be retrieved with the method `getResult()` with the output in the form of `P1D` array. Then we plot the results as in the code below:

```
————————————————— Non-interactive BSOM ————————

from jhplot  import *
from java.awt import Color

c1 = HPlot("Canvas")
c1.setGTitle("Bayesian Self-Organizing Map")
c1.visible()
c1.setAutoRange()

p1d=P1D("data","data.txt")
p1d.setErrToZero(1)
bs=HBsom()
bs.setNPoints(30)
bs.setData(p1d)
```

```
bs.run()
result=bs.getResult()
result.setStyle("pl")
result.setColor(Color.blue)
c1.draw(p1d)
c1.draw(result)
```

One should note that we set all statistical errors of the input P1D to zero. The resulting P1D is shown with symbols connected with the lines. The result of the above script is shown in Fig. 15.3.

Let us give more details about the use of the HBsom in the non-interactive mode. The method run() executes the learning algorithm. The learning stops if topological changes between each step are smaller than some parameter, which can be set via the method setDelta(). One can get the total number of iterations using the method getNiterations().

For more complicated topologies of data in 2D, it is important to adjust the initial α and β values using the method setAlphaBeta(a,b).

15.4 Neural Network Using Python Libraries

jHepWork packaged several libraries for neural-network studies implemented in Python. In particular, one can use the PyANN program [8] which was initially written for CPython. As you already know, numerical programs implemented in Python are slow. Therefore, we will not gain anything in terms of execution performance while running this package using Jython, compare to native Java or C++ libraries. But there are several other advantages in using the Jython approach: (1) The code is fully "transportable" and can be run using CPython; (2) There is a direct access to the original code if one needs to understand it and make necessary modifications; (3) Finally, you will get all benefits of the Python programming language, including its interactivity for easy debugging.

In this section, we will illustrate how to use the PyANN library using jHepWork. The corresponding library is located inside the directory pyann in python/packages. If you are using the jHepWork IDE, this directory is imported automatically. Below we will learn how to run a multilayer algorithm, which implements a feedforward, supervised algorithm for data classification.

As in the previous section, we will divide our work in several steps: (1) we will prepare a common module, to make our codding compact; (2) we will create files with input data for training and verifications. We will use the shelve module discussed in Sect. 2.16.6; (3) then we will build a neural net and train it. Finally, we will verify the performance of this algorithm.

Let us first make a common module which imports the necessary libraries from the PyANN package. Here we will also define several global parameters, as well functions for input and output. The example code is given below:

```
———————————— Common module "NNpython.py" ————————————
import sys,math,random,shelve

import pyann.mlp
from pyann.mlp.layer import *
from pyann.mlp.monitoring import *
from pyann.mlp.training import *
from pyann.mlp.training.backprop import *

WINDOW, MIN_VALUE, MAX_VALUE = 10,-1,1

def writeData(data,valid,pattern):
    sh=shelve.open('data.shelf')
    sh['data']=data
    sh['validate']=valid
    sh['pattern']=pattern
    sh.close()

def readData():
    return shelve.open('data.shelf')
```

Next, using the module above, we will create a file with the input data. We will use
sin(i) as an engine to generate our data. The data will have 73 rows, 63 will be
used for training and 12 for validation. Each row of data will be in a form of tuple
with 10 inputs, while we will have only one output (also in the form of tuple but
with one value).

```
———————————— Creating a data set ————————————
from NNpython  import *

data = [] # Build data set
for i in range(-360, 361, 10):
    data.append(math.sin(i))
print len(data)

start = len(data)-WINDOW-1
pattern=[] #  Build pattern
while len( data[start : start+WINDOW] ) == WINDOW:
   input = tuple(data[start : start+WINDOW])
   output = (data[start+WINDOW],)
   pattern.append(pyann.mlp.Pattern(input, output))
   start -= 1

valid = [] # validation pattern.  20% of data
for i in range(int(len(pattern)*0.2)):
  valid.append(pattern.pop(random.randrange(len(pattern))))
```

```
writeData(data,valid,pattern)
```

In the above example, use the method `type()` for debugging which helps to understand what the above code is doing. All data will be saved into a file.

Now we come to the training. First, we build a neural net with 10 inputs, 5 nodes for the hidden layer and one output node. Then we read the data and train the net using rather self-explanatory methods. The script prints the training ("epoch") errors. Finally, we will verify the net using our verification sample.

```
──────── Neural network training and predictions ────────
from NNpython  import *

layers = (SigmoidInputLayer(WINDOW),\
  SigmoidLayer(5),\
  SigmoidOutputLayer(1,minValue=MIN_VALUE,
  maxValue=MAX_VALUE))
net = pyann.mlp.Network( layers )

sh=readData()
data=sh['data']
validation=sh['validate']
patterns=sh['pattern']

monitor = VerboseMonitor()
stopOnMaxIter =MaxIterationsStopCondition(500)
stopOnMinError =MinErrorStopCondition(0.01)
stopOnVMinError =MinValidationErrorStopCondition(0.005)

trainer =BackpropagationTrainer(net, monitor)
trainer.setLearningRate(0.7)
trainer.setMomentum(0.5)
trainer.setRandomize(True)
trainer.setStopConditions((stopOnMaxIter,\
        stopOnMinError,stopOnVMinError),\
        joiner = 'or')

trainInfo = trainer.train(patterns, validation)
print """
Epochs: %(epoch)s
Final error: %(error)s
Validation set error: %(validationError)s
""" % { 'epoch': trainInfo.getIterationNumber(),
   'error': trainInfo.getError(),
   'validationError': trainInfo.getValidationSetError()}

for i in range(370, 725, 10):
    start = len(data) - WINDOW
```

```
predicted = net.classify( tuple(data[start:]) )[0]
data.append(predicted)
print 'sin(%s) = %s' % (i, predicted)
```

Run this code to see the result of the training. The code snippet prints out the predicted values, which are rather close to the true values of the sin(i) function.

References

1. Beale, R., Jackson, T.: Neural Computing: An Introduction. Institute of Physics Publishing, Bristol (1990)
2. Bharath, R., Drosen, J.: Neural Network Computing. McGraw-Hill, New York (1994)
3. Heaton, J.: Contributions: The ENCOG project. URL http://code.google.com/p/encog-java/
4. Neapolitan, R.E.: Learning Bayesian Networks. Prentice Hall, New York (2003)
5. Jensen, F.V., Nielsen, T.D.: Bayesian Networks and Decision Graphs, 2nd edn. Information Science and Statistics. Springer, Berlin (2007)
6. Cozman, F.G.: Javabayes: Bayesian networks in Java. URL http://www.cs.cmu.edu/~javabayes/
7. Utsugi, A.: Neural Comput. **9**, 623 (1997)
8. PyANN, a Python framework to build artificial neural networks. URL http://sourceforge.net/projects/pyann

Chapter 16
Steps in Data Analysis

In this chapter, we will consider rather common analysis techniques used to process and analyze data before final visualization. We will gain insight into how to transform data, reject unnecessary records or add additional information, remove duplicate records, sort data and many other common operations. We will also learn how to process data in parallel using multiple processors.

From the programming point of view, our discussion is heavily based on the knowledge collected in the previous chapters.

16.1 Major Analysis Steps

A first step towards understanding of a general analysis strategy is to abstract from a particular type of data and focus on generic steps for data processing, which can consist of

Measurement	A determination of numerical values of some observables and collection of experimental results into data files. A group of measured values forms an "event record", or simply, "events". Each event can contain a set of records, each consisting of a set of objects (strings, values, multi-dimensional arrays, histograms etc.). The event record may represent various characteristics of a single observation.
Transformation	Transformation of the event records into the most appropriate format for further manipulation. This can be done at any analysis step, depending on a concrete situation. If event records are too big for disk storage, they can be converted into a more compact format. But keep in mind that we may pay for this later with a lower processing speed during data analysis. If data requires a lot of processing and the file storage is not an issue, one can keep data in a less compact format.

S.V. Chekanov, *Scientific Data Analysis using Jython Scripting and Java,*
Advanced Information and Knowledge Processing,
DOI 10.1007/978-1-84996-287-2_17, © Springer-Verlag London Limited 2010

Checking integrity Furthermore, if data files have to be moved between different file storages using network, one should check data integrity. The main question is whether the data records have been corrupted during the process of copying. Data files can also be altered at various stages of data manipulation (intentionally or unintentionally), thus one should always verify if they are valid and contain the original information.

Data skimming Reduction of data volume by rejecting uninteresting events. First of all, one should reject events which do not pass selection requirements or do not convey useful information, events with duplicate records and so on. The reason for this is simple: run time for most data analysis algorithms is roughly proportional to the data volume. In addition, one can save disk space when saving the data persistently. The choice of skimming criteria is often a balance between many factors.

Data slimming Removing uninteresting information inside event records. Instead of dealing with complete event records, one may keep only the most interesting objects to be used for the final analysis. The reason is pretty much the same as for the skimming step.

Data thinning Reduction of information characterizing separate objects inside remaining event records after the skimming. Again, the reason for this step is the same as before: it guaranties that one deals with a meaningful part of data records in future.

Data inclusion Adding necessary information to the event records. This information can be derived from the remaining information stored in the event records (or from the rejected information before the previous "slimming" or "thinning" steps). For example, one can add new collections which represent statistical summaries of the event records by imposing some selection criteria.

Building metadata This is a process of constructing short event records that characterize either the entire event or objects inside event records. The metadata are usually necessary in order to quickly find necessary events without reading the entire events, in case if data should be processed more than once. The metadata information can be inserted either inside the event records (using the data inclusion process discussed above) or can be included into a separate file for easy manipulation.

Final preparation This is a step which is necessary for final data mining, statistical summaries and the final analysis. Typically, this stage includes a data projection onto lower dimensions for data visualization. It can also include data sorting, removal of duplicate entries or transformation to the most useful format for analysis.

The prepared event records can be used for meaningful predictions on the basis of discovered pattern. For instance, data can be visualized and used as inputs for the supervised or unsupervised learning algorithms discussed in the previous chapters. Finally, the selected data should be compared with a certain theory.

The data skimming, slimming and thinning are not new concepts in data analysis. However, in recent years, there has been growing interest in such analysis techniques during preparation to data taking at the Large Hadron Collider to be constructed by the international community in Geneva (Switzerland), where such concepts have been reshaped and refined [1].

In this book, our approach will remain to be the same: Instead of being abstract, we will illustrate the above analysis techniques using numerical examples.

16.2   Real Life Example. Analyzing a Gene Catalog

In this section will analyze a published gene catalog of human chromosome 11 and illustrate most of the analysis steps discussed above. Other aspects of data analysis will be discussed in the following sections.

Of course, we cannot discuss the actual measurement of the gene catalog; instead, we will copy the available data from a public domain and then perform necessary operations using Jython scripting. The analyzed data file can be copied from a web page given in [2]. For easy manipulation, we have transformed the original file into a CSV file (one can do this with any spreadsheet program).

Let us download a CSV file with the gene catalog and display it in the CSV browser.

```
_____ Getting the data _____
from jhplot.io.csv import *
from jhplot import *

http='http://projects.hepforge.org/jhepwork/'
file='nature04632-s16-2.csv'
wget(http+'examples/data/'+file)
r=CSVReader(file,',')
SPsheet(r)
```

As for the previous chapters, if the above web page location fails, use a mirror site by replacing the http address with the string:

```
http='http://jhepwork.sourceforge.net/'
or
http='http://jwork.org/jhepwork/'
```

The above code brings up a spreadsheet filled with the data from the gene catalog. A visual study of this file reveals that each gene symbol is characterized by several records. In fact, each row representing a gene can be considered an "event record" using the terminology adopted above. Each gene has its symbol, name, category, location, length and so on. We will use this file as a starting point for our examples.

16.2.1 Data Transformation

Now we will show how to transform the gene event records into a machine-readable form. For many practical applications, this is necessary in order to save disk space while storing the data, or for best performance while reading data.

In the example below we illustrate how to read the CVS file (see Sect. 11.7.1) and convert the event records, line by line, into a compressed serialized file (see Sect. 11.3):

```
───────────────── Transforming data records ─────────────────
from jhplot.io.csv import *
from jhplot.io import *

r=CSVReader('nature04632-s16-2.csv',',')
f=HFile('nature.ser','w')

i=0
while 1:
  line= r.next()
  if line == None: break
  if (i%100 == 0): print i
  f.write(line)
  i +=1
f.close()
r.close()
```

The code seemed very simple: We loop over all rows in the CSV file and serialize them into the file 'nature.ser' which keeps data in a compressed form. The size of the output file is typically three times smaller than the original file size.

16.2.2 Data Skimming

Let us turn to the skimming stage during which the data volume is reduced after rejecting uninteresting event records. For our example, we assume that some rows convey no useful information for further analysis. In particular, we would like to skip all genes with undefined gene names ("symbol"). By examining the CSV file using a spreadsheet, one can find that a gene symbol entry is not defined if the first column has the string 'undef'. So, let us rewrite the file by rejecting such entries:

```
───────────────── Skimming data records ─────────────────
from jhplot.io import *

f1=HFile('nature.ser','r')
f2=HFile('nature_skim.ser','w')
```

```
while 1:
  row=f1.read()
  if row == None:   break
  if row[0] == 'undef' or len(row)<18: continue
  f2.write(row)
f1.close()
f2.close()
```

The output will be saved into the file 'nature_skim.ser'. In the above code, we included a simple check which verifies the correct length (18) for the processed rows.

16.2.3 Data Slimming

This time we are not interested in storing all information withing each event record. For example, we would like to remove two last columns for each row, which do not convey any useful information for us, but keeping the same number of rows as before. It takes only a few lines of the code to perform this manipulation:

```
—————————————— Slimming data records ——————————————
from jhplot.io import *

f1=HFile('nature_skim.ser','r')
f2=HFile('nature_slim.ser','w')

while 1:
    row=f1.read()
    if row == None:   break
    s=len(row)
    del row[s-1]
    del row[s-2]
    f2.write(row)
f1.close()
f2.close()
```

16.2.4 Data Sorting

Next, we would like to sort the event records using some arbitrary criteria. Our event records are not very simple because they are constructed from several objects. Therefore, the standard use of the sort() method for Jython lists will not work: In case if each row contains several records, the sorting of rows should be done using

values stored in a certain column, thus we should use the standard Jython method
sort() in a somewhat different way.

Sorting of data will be performed in a computer memory to make the entire process fast. For this example, we will sort the records based on the first column. We
will read data into a list and use the sort() method together with the cmp() function which provides a sorting algorithm. After the sorting, we will write the event
records into a new file:

```
_____ Sorting records _____

from jhplot.io import *

f1=HFile('nature_slim.ser','r')
f2=HFile('nature_sorted.ser','w')

data=[]
while 1:
    row=f1.read()
    if row == None: break
    data.append(row)
f1.close()

def cmp1(a,b): return cmp(a[0],b[0])

data.sort(cmp1)
for k in data:
    f2.write(k)
f2.close()
```

In our example, the sort() method uses a custom cmt1() function, which
calls the standard Python cmt() function. However, one can build totally custom
cmp1(). For example, one can construct the cmp1() function used in the above
example as:

```
_____ A custom sorting function _____

def cmp1(a,b):
    if a[0]<b[0]:   return -1
    if a[0]==b[0]:  return  0
    if a[0]>b[0]:   return  1
```

One can define the function cmt1() before the sort() command, and include
any logic for sorting.

We do not need always to use the Python-type objects. One can also use pure-
Java collections to store and sort elements. In Sect. 2.7 we have shown how to use
the List class to store data and how to sort records using the Collection class.

16.2.5 Removing Duplicate Records

Removal of duplicate event records requires a small preparation step. First, let us write a simple function which accepts a list containing other lists and sorts the list in accordance with an input index. Then, we scan from the end of the list, deleting duplicate entries based on the input index as we go:

```
———————————— Module 'unique.py' ————————————
def unique(s,inx):
  def cmp1(a,b): return cmp(a[inx],b[inx])
  s.sort(cmp1)
  last = s[-1]
  for i in range(len(s)-2, -1, -1):
    if last[inx] == s[i][inx]:
        del s[i]
    else:
        last = s[i]
  return s
```

The function returns a list of elements without duplicate entries. We pass a variable inx which defines the column number used for the duplicate removal.

Now we should make a test of this function. Assume we have a list in which each element is another list. We will remove duplicate entries based on the first index (0), and then remove duplicates based on the second index:

```
———————————— Testing module ————————————
from unique import *

L=[[1,2],['a',10],['a',100],['b',1],['c',2]]
print unique(L,0)
print unique(L,1)
```

The output of this code is given below:

```
[[1, 2], ['a', 100], ['b', 1], ['c', 2]]
[['b', 1], ['c', 2], ['a', 100]]
```

As you can see, after the first call, we have removed ['a',10] (based on the first index), while the element [1,2] was rejected after the second call based on the second index.

Now we are ready to remove duplicate records in our example of the gene catalog. We will remove duplicate rows based on the second column (index = 1):

```
———————————— Removing duplicate entries ————————————
from jhplot.io import *
from unique import *
```

```
f1=HFile('nature_sorted.ser','r')
f2=HFile('nature_unique.ser','w')
data=[]
while 1:
    row=f1.read()
    if row == None: break
    data.append(row)
f1.close()

unique(data,1)
for k in data:
     f2.write(k)
f2.close()
```

16.2.6 Sorting and Removing Duplicate Records Using Java

We should also remind that there is no need for the Python-type approach for re-moving duplicate objects while working with Jython. On can always convert data into the Java collection Set discussed in Sect. 2.7. We have shown that one can use the class HashSet implementation of the interface Set which, by construc-tion, cannot contain duplicates. Moreover, one can use the class SortedSet that always maintains its elements in ascending order, so there is no need to worry about sorting and removing duplicates.

So, plan the analysis beforehand: if you think that the final data output should not contain the same elements in the output container, just find appropriate container to keep data records.

In the following subsections we will show two examples of how to work with event records using pure Java collections.

16.2.6.1 Processing Big Event Sample

You may find yourself in a typical situation: One needs to read a huge data sample and remove duplicate records based on some value. One cannot store all records in the computer memory, but one can still store values used for duplicate removal. In this case, there is no need for loading a full event into the computer memory: just read the file and write the event back as you go. Use the class HashSet to skip event records with exactly the same description (in our case, this is the first element). When you add a description into the HashSet object, it returns '1' (success) if no duplicate is found and '0' if a duplicate is found. Using this approach one should be able to process big data files without too much load on the computer memory.

16.2.6.2 Sorting and Removing Duplicate Records

In the next example we show how to sort and remove duplicate records using Java classes. In particular, the class `TreeMap` will be used for sorting and removing duplicate objects at the same time. This example is similar to that shown in Sect. 2.7.6. The only difference is that now we read and fill a file with the data using the `List` class to decouple the first record with the description from the rest of the event record. We use the description of genes as the keys for the `TreeMap` object which removes duplicates and sorts the keys as we fill it in. Finally, we loop over all sorted keys and restore the event records by combining keys with the key value.

```
————————————— Sorting and removing duplicates —————————————
from jhplot.io import *
from java.util import *

f1=HFile('nature_slim.ser','r')
map=TreeMap()
while 1:
    row=f1.read()
    if row == None: break
    ln=row.size()
    r=row.subList(1,ln)
    map.put(row[0], row.subList(1,ln))
f1.close()

f2=HFile('sorted_unique.ser','w')
for i in map:
    row=map[i]
    row.set(0,i)
    print row
    f2.write(row)
f2.close()
```

At this stage, we cannot use the gene catalog for our further examples. Our remaining gene records are not sophisticated enough to perform thinning or building a metadata file. The latter techniques will be illustrated using an appropriate data sample.

16.3 Using Metadata for Data Mining

We will continue with our discussion of data analysis techniques by diving into the notion of metadata. For this section, we cannot use the previous data sample as it lacks the necessary volume and complexity with which we usually have to deal with when the metadata records become necessary. The metadata is especially useful for large, data-intensive tasks in which one observation (or event) contains many data records.

As mentioned before, the metadata is a short record which captures the basic characteristics of the entire event or objects inside each event record. These characteristics appear to be useful when one needs to find a necessary record as fast as possible, without reading the entire data records.

The metadata concept is essential when one needs to find a few rear events, so there is no too much sense in reading the data again and again if one single processing can build a metadata file which can be used for searching. Especially, one can gain a lot in terms of a program performance when a multiple data processing is required. For example, one can benefit from the use of the metadata if data records have to be analyzed by many users.

The reason why the metadata can be useful for multiple processing is simple: in this approach, a program reads only a small ("metadata") record. Usually, this is neither IO nor CPU consuming. Of course, the metadata records should be first constructed using a priory knowledge about which data characteristics are important for future analysis. We should also note that the process of construction of the metadata files can be rather CPU consuming.

In this chapter we will illustrate the metadata concept using a few short code snippets and Jython scripting.

16.3.1 Analyzing Data Using Built-in Metadata File

First, let us create event records for our examples. Each event will consist of a string, representing the record number, and two arrays of the type POD. The size of both arrays is not fixed: it is determined by a Poisson distribution with the mean 500. One array is filled with a uniform random numbers, while the second array is filled with random numbers in the range between 0 and 1. We write 10,000 events into a serialized compressed file 'data.ser'

```
──────────── Building a data file ────────────
from jhplot.io import *
from jhplot import *

ps=math.Poisson(500)
def makeEvent(entry):
  p1,p2=POD('a'),POD('b')
  p1.randomUniform(ps.next(),0,1)
  p2.randomNormal(ps.next(),0,1)
  return [str(entry),p1,p2]

f=HFile('data.ser','w')
for i in range(10000):
  ev=makeEvent(i)
  if (i%100 == 0):
    print 'pocessed=',ev[0]
```

```
    f.write(ev)
f.close()
```

When the output file is ready, one question we may to ask themselves is this: how can we find events in which the sum of all elements inside both arrays is above some value cut? Of course, this can be any other data mining task. We just have picked up this one as it is simple to implement and, at the same time, it well captures the idea of using the metadata.

The code which reads the data file and counts all data records with the sum of all elements above cut=320 inside both arrays can look like this:

```
———————————————— Analyzing data ————————
from jhplot  import *
from jhplot.io import *
import time

f=HFile('data.ser')
start = time.clock()
i=0; cut=320
while 1:
    event=f.read()
    if event == None:
        print "End of events"
        break
    if int(event[0])%1000 == 0:
        print 'processed=',event[0]
    p1,p2=event[1],event[2]
    v=p1.getSum()+p2.getSum()
    if v>cut:
            i+=1
t = time.clock()-start
print 'Nr above '+str(cut)+' =',i, ' after time (s)=',t
f.close()
```

The above code also performs a very basic benchmarking. The output is listed below:

```
End of events
Nr above 320 = 30   after time (s)= 10
```

Of course, the number of selected events and the execution time can be different for your test.

Now, let us assume that we need to process the generated data file many times using different values of the variable cut for the event selection. This means that we should build a metadata by constructing an additional event record with the sum of all elements inside the arrays.

We are ready to design such code using the class HDataBase discussed in Sect. 11.3.6. As the key value to be associated with each event record, we will use a string which consists of two parts: the event number and a value representing the sum of all elements inside both arrays. The event number and the sum are separated by the underscore character.

```
_____ A database approach to store data and metadata _____
from jhplot  import *
from jhplot.io import *
import time

f=HFile('data.ser')
start = time.clock()

db=HDataBase('data.db','w')
while 1:
    event=f.read()
    if event == None:
        print 'End of events'
        break
    p1=event[1]
    p2=event[2]
    v=p1.getSum()+p2.getSum()
    db.insert(event[0]+"_"+str(v),event)
t = time.clock()-start
print 'Entries=',f.getEntries(), ' time (s)=',t
f.close()
db.close()
```

After execution of this script, our data will be converted into a small database 'data.db'. The time necessary to build such database is a factor two larger than that used in the previous example.

Now we can read this database by iterating over all keys. We extract the event number and the sum value from the string. Then we access only events which have the sum of all elements above the cut value cut.

```
_____ Data analysis using metadata _____
from jhplot  import *
from jhplot.io import *
import time

start = time.clock()
db=HDataBase('data.db')
keys=db.getKeys()
i=0; cut=320;
while keys.hasMoreElements():
    next = keys.nextElement();
    words = next.split('_')
```

```
            if int(words[0])%1000 == 0:
                print 'processed=',words[0]
            if float(words[1])>cut:
                i +=1
                event=db.get(next)
t = time.clock()-start
print 'Nr above '+str(cut)+' =',i, ' after time (s)=',t
db.close()
```

The above code prints:

```
End of events
Nr above 320 = 30  after time (s)= 1.5
```

Thus, the access time is significantly reduced compare to 10 seconds from the previous example when no metadata records were used.

16.3.2 Using an External Metadata File

The use of the metadata as a part of database keys is not the only possible approach. One can write metadata into an external file, thus completely decoupling the metadata from the actual data volume.

The example below shows how to write such external metadata file:

```
─────────── Rebuilding external metadata ───────────
from jhplot import *
from jhplot.io import *
import time

f=HFile('data.ser')
start = time.clock()

meta=open('data.meta','w')
db=HDataBase('data1.db','w')
while 1:
    event=f.read()
    if event == None:
        print "End of events"
        break
    p1,p2=event[1],event[2]
    v=p1.getSum()+p2.getSum()
    db.insert(event[0],event)
    meta.write(event[0]+' '+str(v)+'\n')
t = time.clock()-start
print 'Entries=',f.getEntries(), ' time (s)=',t
```

```
meta.close()
f.close()
db.close()
```

The file 'data.meta' contains rows with the event numbers and the sum of all elements inside the arrays in each event record. The database file has the keys based on the event number only.

During data processing, first load the metadata file and identify interesting events. Then read only selected event records using the keys:

```
                    Using external metadata records

from jhplot  import *
from jhplot.io import *
import time

start = time.clock()
db=HDataBase('data1.db')
i=0; cut=320
for line in open ('data.meta', 'rt'):
  key, sum = [x for x in line.split()]
  if int(key)%1000 == 0:
          print 'processed=',key
  if float(sum)>cut:
        event=db.get(key)
        i=i+1
t = time.clock()-start
print 'Nr above '+str(cut)+' =',i, ' after time (s)=',t
db.close()
```

The approach in which we decouple the metadata file from the actual data file is rather convenient as it does not require rebuilding the database with new metadata entries every time when we need to make modifications in our metadata definitions.

16.4 Multiprocessor Programming

Multiprocessor programming significantly increases program performance during data analysis. Essentially, this is just breaking up a single programming task into pieces, which can be executed in parallel. Then, the outputs will be combined upon completion of all pieces. The code segments should preferably be independent of each other. This allows to achieve the highest possible performance by taking advantage of modern multi-core systems and enables us to write very efficient programs that make maximum use of the CPU.

We remind that the Java language and, hence, Jython, have inherent multi-threading support which significantly simplifies parallel programming. In this section, we will illustrate how to write analysis programs that contain two or more parts that can run concurrently.

16.4.1 Reading Data in Parallel

One can achieve a better performance of a program using multiple threads when I/O bandwidth is not a bottleneck. This, obviously, depends on hardware and the actual calculations. Typically, numerical calculations can benefit from the use of parallel cores when the time necessary for processing events is larger than the time needed to access events from a file storage.

Let us illustrate the idea behind the parallel processing of data by looking at some concrete examples. As usual, first we will prepare two data samples to work with using a single thread. Both files contain a sequence of event records. Each event is represented by a POD array with 20 random numbers:

```
───────────────────── Generating data files ─────────────────────
from jhplot.io import *
from jhplot import *

f1=HFile('random1.ser','w')
f2=HFile('random2.ser','w')
for k in range(100000):
  p=POD()
  p.randomUniform(20,0,1.)
  f1.write(p)
  p.randomUniform(20,0,1.)
  f2.write(p)
  if k%1000==0: print 'done=', k
f1.close()
f2.close()
```

Next, we will create a small class which reads the files 'random1.ser' and 'random2.ser'. This class should accept a string with the file name, open this file and read the file record-by-record. We refill the records, constructed from single POD arrays, with 200 random numbers distributed in accordance with the normal distribution. We will use the Jython threads discussed in Sect. 2.13 to perform this task:

```
───────────────────── Module 'readthread.py' ─────────────────────
from jhplot.io import *
from jhplot import *
from threading import Thread

class testit(Thread):
   def __init__ (self,fin):
```

```
        Thread.__init__(self)
        self.fin = fin
        self.k =0
    def run(self):
        f1=HFile(self.fin,'r')
        self.k=0
        while 1:
            row=f1.read()
            if row == None:  break
            if self.k%10000==0:
                print self.fin+" done=",self.k
            row.sort()
            row.getStat()
            self.k=self.k+1
        f1.close()
```

Our new class inherits the Thread class, i.e. we sub-classing the Thread interface and overwrite its initialization and the method run(). In the method run(), we open the file to read its records. Then we sort the 1D array and evaluate its statistical characteristics.

In the example below we will read two data files using two independent threads. Then we perform a simple benchmarking by printing the total number of processed events and the timing.

```
—————————— Reading data in two threads ——————————
from readthread import *
import time

print 'Start at',time.ctime()
list=['random1.ser','random2.ser']
# list=['random1.ser']
t1 = time.clock()
tlist = []
for f in list:
    cu=testit(f)
    tlist.append(cu)
    cu.start()

for t in tlist:
    t.join()
    print 'From ',t.fin,' processed: ',t.k
t2=time.clock()-t1
print 'calculation takes',int(t2),' sec'
```

The code reads two files at the same time. But you can always read only one file by removing one file name from the list (we have commented out this case). If your computer has several computational cores, then what likely you will see is that the total processing time for two files will be less than the processing time for a single file times factor two. For the author's computer with two cores, 13 seconds were spent to read two files in parallel and 10 seconds to read one file.

The advantages of this approach are obvious. If data are distributed in several files, one can read these files in parallel. We were able to parallelize our program without any effort, since Jython and Java take care of how to perform such parallelization.

We should note that there is no too much point in using more threads than the number of available cores: if the number of cores is less than the number of threads, then you will end up in an illusion that you are still reading data in parallel; this will not be quite true, since threads will share the same computational cores and you will not gain a higher performance in comparison with the case when the number of threads is equal to or less than the number of available processing cores.

16.4.2 Processing a Single Input File in Parallel

In the above example we have considered the case when parallel processing of data is achieved by dividing data into several files. In principle, it is not too difficult to parallelize a program designed to read a single input file. As before, we should note that this approach makes sense for CPU intensive calculations which require input data.

In Chap. 11 we have considered the `PFile` and `EFile` classes which can store data in sequential oder using the Protocol Buffers format. One important feature of these classes is that one can access a particular event record using the method `read(inx)`, where `inx` specifies a position of an event record inside the file. This is important feature for parallel calculations, since one can jump to a necessary record within a file without uncompressing the entire file.

Let us use the class `PFile` for our next example. We will create a file with 1D arrays using the same approach as before:

```
———————— Data writer. Module 'pwrite.py' ————————

from jhplot.io  import *
from jhplot  import *
import time

f=PFile('test.pbu','w')
start = time.clock()
for i in range(90000):
   if (i%1000 == 0):
      print 'pocessed=',i
   p0= P0D('event'+str(i))
   p0.randomNormal(1000,0.0,1.0)
   f.write(p0)
print 'PFile  time (s)=',time.clock()-start
f.close()
```

Execution of this script creates an input file 'test.pbu' (be careful: this file is rather large, typically it has the size of 650 MB).

Next, we will prepare a module which takes a file name, initial and final position of records within the input file and a name of the thread which calls this class:

```
——————————— Module 'readthread2.py' ———————————
from jhplot.io import *
from jhplot import *
from threading import Thread

class testit(Thread):
    def __init__ (self,fin,process,i1,i2):
        Thread.__init__(self)
        self.fin = fin
        self.process=process
        self.k =0
        self.i1,self.i2 = i1,i2
    def run(self):
        f1=PFile(self.fin,'r')
        self.k=0
        for j in range(self.i1,self.i2):
            row=f1.read(j)
            row.sort()
            row.getStat()
            if row == None:   break
            if self.k%1000==0:
                print self.process+' done=',self.k
            self.k=self.k+1
        f1.close()
```

As before, the program reads a 1D array and performs sorting and calculation of all major statistical characteristics.

Finally, let us check our program. We will read all event records in one single thread as shown below:

```
——————————— Data reader using 1 thread ———————————
from readthread2 import *
import time

print 'Start at',time.ctime()
t1 = time.clock()
tlist = []

cu=testit('test.pbu','cu',1,90000)
tlist.append(cu); cu.start()
cu.start()

for t in tlist:
```

```
   t.join()
   print 'From ',t.fin,' processed: ',t.k
t2=time.clock()-t1
print 'calculation takes',int(t2),' sec'
```

For calculations in parallel on multiple cores, we will create three threads. Each thread will read a fraction of data, such the total number of processed events will be exactly as in the single-thread example:

─────────────── Data reader using 3 threads ───────────────
```
from readthread2 import *
import time

print 'Start at',time.ctime()
t1 = time.clock()
tlist = []

cu = testit('test.pbu','cu1',1,30000)
tlist.append(cu); cu.start()

cu = testit('test.pbu','cu2',30001,60000)
tlist.append(cu); cu.start()

cu = testit('test.pbu','cu3',60001,90000)
tlist.append(cu); cu.start()

for t in tlist:
    t.join()
    print 'From ',t.fin,' processed: ',t.k
t2=time.clock()-t1
print 'calculation takes',int(t2),' sec'
```

Run this code and compare the answers. For a computer with four cores, the second program with three threads is typically 40% faster. This is not due to faster disk I/O (which is still the same!), but due to the fact that the program can perform calculations in parallel. One can easily check this by removing our 1D array calculations from the file 'readthread2.py'.

16.4.3 Numerical Computations Using Multiple Cores

Now let us move on and consider calculation-intensive programs which do not require access to data. Thus, instead of reading and processing objects using extensive I/O, we will perform numerical calculations without any input data.

Again, we will prepare a class which inherits the properties of the Thread class. This time, we pass a two-dimensional function F2D discussed in Sect. 3.4. We cal-

culate the sum of all values in a double loop from 0 to 2000 inside the method
run():

```
————————————— Module 'functhread.py' —————————————

from jhplot.io import *
from jhplot import *
from threading import Thread

class functhread(Thread):
    def __init__ (self,func):
        Thread.__init__(self)
        self.func = func
        self.sum=0
    def run(self):
        for i in range(2000):
            for j in range(2000):
                self.sum += self.func.eval(i,j)
```

Next we will perform numerical evaluations of both input functions in parallel:

```
——————————— Parallel numerical computations ———————————

from functhread import *
import time

print "Start at",time.ctime()

f1=F2D('cos(x*y)*x+y')
f2=F2D('cos(x*y)*y-x')
list=[f1,f2]
# list=[f1]
t1 = time.clock()
tlist = []
for f in list:
    current = functhread(f)
    tlist.append(current)
    current.start()

for t in tlist:
    t.join()
    print 'From ',t.func.getTitle(),' processed: ',t.sum

t2=time.clock()-t1
print 'calculation takes=',int(t2),' sec'
```

Run this script using two functions specified in the Jython list as given in this example. Then, make a new test: Uncomment the line with the list containing only one
function. You will see that, for two functions, the calculation time is rather close to
the time necessary to evaluate a single function. Again, this can be seen if one uses
a computer with multiple cores.

The conclusion of this section is simple: take advantage of multiple cores that make maximum use of your computer. First, you should think about how to split up your program into independent segments to increase efficiency of your calculations. Then, write a custom class based on the `Thread` class and use the `run()` method to include the necessary calculations.

16.5 Data Consistency and Security. MD5 Class

While your data analysis code grows, one may start to worry about consistency and integrity of data objects which are either stored in the computer memory (non-persistent data) or saved in files (stored persistently). Surely, any data container can easily be modified inside an analysis code. In order to prevent deliberate (or accidental) modifications of data, one should find fingerprints of the objects that keep data, and compare them with those generated during object initializations. This should be done at runtime, i.e. during the code execution.

More importantly, it is often necessary to check integrity of data files, to make sure that files have not been modified by someone after posting them on the Web. Again, what you will need is a fingerprint or some sort of unique signatures for such files.

16.5.1 MD5 Fingerprint at Runtime

Maintaining data consistency is not too difficult: what you really need to do is to use the well known MD5 algorithm. The usual procedure is following: a publisher should provide a MD5 signature of an object or a file. A user confirms it by calculating the signature again and comparing it to the one provided by the publisher. If the signatures do not match, then this will indicate that the object or the file was altered since it was created.

Below we will generate a MD5 hash of a `P1D` object at runtime. This is simply a string with a hexadecimal number which reflects the counts of each byte of your object. Any recreation or modification of this object at runtime leads to a modification of this string.

We use the class `MD5` which accepts any Java object as an argument. After generating a MD5 string, we regenerate it again after modifying the `P1D` container by adding a row of data:

```
>>> from jhplot.security import *
>>> from jhplot import *

>>> p1=P1D('test')
>>> p1.add(10,20)
>>> md5=MD5(p1)
```

```
>>> print md5.get()
da14f51d302dd6756457f16f989b8eb9

>>> # modify a object and check MD5
>>> p1.add(20,30)
>>> md5=MD5(p1)
>>> print md5.get()
c5671fff0834bcc09489798ec1f3f752
```

The method get() returns a hex-string representing the array of bytes of the input object. It should be noted that the printed strings in your case could be rather different from those shown here, since every new creation of objects leads to a different fingerprint. The important thing to remember is that they do not change at runtime.

One can perform a similar fingerprint using the Jython module md5. This approach works in case of jHepWork objects which are convertible into strings (usually, this means that the method toString() is implemented for such objects). The syntax is somewhat different since now we pass strings to the md5 instance:

```
>>> from jhplot import *
>>> import md5
>>>
>>> p1=P1D('test')
>>> p1.add(10,20)
>>> # check MD5
>>> m=md5.new()
>>> m.update(p1.toString())
>>> print m.hexdigest()
c5671fff0834bcc09489798ec1f3f752
```

As before, the string returned by the method hexdigest() can be rather different in your case.

16.5.2 Fingerprinting Files

Analogously, one can calculate a MD5 checksum for any file using the class MD5. In case of files, a publisher should provide a MD5 signature and a user must confirm it by calculating the signature again and comparing it to the one provided by the publisher. If the signatures do not match then this will indicate that the file was altered after it was published.

This time the fingerprint string is exactly the same as long as a file is not modified on the file system. Below we calculate a MD5 checksum for the file 'file.txt' and print it.

```
>>> from jhplot.security import *
>>> from java.io import *
>>>
>>> md5=MD5( File("file.txt") )
>>> print md5.get()
```

The string returned by the method `get()` is unique and can be used for detecting file corruptions or deliberate modifications, since the chances of accidentally having two files with identical MD5 checksums are very small.

References

1. Cranshaw, J., et al.: A data skimming service for locally resident analysis data. J. Phys. (Conf. Ser.) **119**, 072011 (2008)
2. Taylor, T., et al.: Human chromosome 11 DNA sequence and analysis including novel gene identification. Nature **440**, 495 (2006)

Chapter 17
Real-life Examples

Now we have all the machinery that is necessary to perform a realistic data analysis. Unlike the previous chapters with the programming recipes, this chapter features extensive real-research examples based on Java scripting.

Here we will learn how to simulate data and how to use data from the real world. We will show how to perform a full-scale data analysis to make a first step in extraction of knowledge about the underlying nature of data. The self-contained examples given in the following sections can provide the basis to conduct your own research.

17.1 Measuring Single-particle Densities

In this chapter, we will analyze particle distributions, starting from a single-particle distribution and then moving towards studies of multi-particle densities, correlations and fluctuations.

As usual, before doing actual analysis, first we should prepare an event sample to be used for our research. We will create it in almost exactly the same way as scientists usually do when they need to model real-world data: we will *simulate* a data sample using some mathematical principles. The simulated samples usually help us to understand and explain the underlying dynamics in the observed data, especially if there are missing data or data distorted by experimental apparatus. In this book, the simulated data will serve for an illustration of programming aspects of data analysis.

We will consider the following experiment: Particles created or observed in an experiment are counted in certain phase-space intervals. "Particles" could be real elementary particles produced by an accelerator machine or by cosmic rays, photons produced by a light source and so on. We do not need to go into such details, as well as details of the underlaying particle production mechanism: We will use this notion in a very generic way. For example, one can think about particles as people coming to a store during a certain time interval, cars entering a highway, telephone calls per time unit and so on.

S.V. Chekanov, *Scientific Data Analysis using Jython Scripting and Java,*   407
Advanced Information and Knowledge Processing,
DOI 10.1007/978-1-84996-287-2_18, © Springer-Verlag London Limited 2010

Each particle can be characterized by same continues characteristics. The measured number (N) of particles will depend on the time interval or/and the size of some spatial region where particles are detected.

17.1.1 Preparing a Data Sample

To be more specific, let us consider a Rutherford-like experiment, when a beam of particles from a radioactive source (or an accelerator) strikes a target (say, a thin foil). A special detector counts all particles after interactions with the target and measures the scattering angle θ. The distribution for $\cos(\theta)$ is expected to be proportional to

$$\frac{1}{(1 - \cos(\theta))^2} \tag{17.1}$$

This is the famous Rutherford formula for scattering of particles from a given nucleus. This form of the scattering angle is a signature for scattering off a point-like object without structure (all such point-like objects are inside the thin foil). A deviation of the measurement from this function is an indicator of structure inside the target.

The code below was constructed from the code snippets which have already been discussed in the previous chapters. We create a Poisson distribution with the mean 10 (can be any number!), and distribute particles in each event using our expected functional form in (17.1). In addition, we will need to store extra information about the incoming particle intensity, which was set to 1000 (using some arbitrary units). To do this, we create a dictionary at the very beginning of the file record.

```
———— Creating a data set using the Rutherford formula ————
from jhplot  import *
from jhplot.io import *
from jhplot.math.StatisticSample  import *

Xmin,Xmax=0.1, 1
f=F1D('0.01/(1-cos(x))^2',Xmin,Xmax)
p=f.getParse()
max=f.eval(Xmin)
f=HFile('events.ser','w');
pos=math.Poisson(10)
f.write( {'Intensity':1000} )
for i in range(10000):
    if (i%1000 == 0):  print 'Event=',i
    a=randomRejection(pos.next(),p,max,Xmin,Xmax)
    f.write(a)
f.close()
```

We write events in a loop. Each event consists of a random number of particles with the scattering angles distributed as described above. Then we create 10,000 events in a loop, writing each event into a file. In this form, our program is well scalable since we do not leave any objects in the computer memory, unlike the case when first we create a list of events and then write such list into a file.

17.1.2 Analyzing Data

Now we come to an analysis of the event sample generated in the previous section. We remind that the file 'events.ser' contains everything we need: the intensity of the incoming beam of particles and the event records for multiple observations. Each event contains a certain number of particles characterized by one-dimensional quantity (in our example, this is the cosine of the scattering angle).

Our task is to read all information from the data file and plot what is called a differential cross section calculated as:

$$\frac{d\sigma}{d\theta} = \frac{N}{\Delta \cdot L}$$

where L is the intensity of the incoming particles and Δ is the width of the histogram bin where particles are counted. As you can see, the distribution above is nothing but a probability density discussed in Sect. 8.1.1 scaled by the intensity value.

The distribution above conveys information about: (1) the shape of the particle distribution. Note that the distribution itself does not depend on the chosen bin size since we divide the counted number of particles in each bin by the bin size; (2) the intensity with which events are happening. This is due to the scaling of the density distribution by the intensity number, L.

Integration of the distribution must give a total cross section or a total rate of interactions, σ, which is simply $\sigma = N_{tot}/L$, where N_{tot} is the total number of counted particles after the interaction.

The code below shows how to extract the differential distribution and then visualize it:

```
───────── Analyzing a particle distribution ─────────
from jhplot import *
from jhplot.io import *

f=HFile('events.ser')
a=f.read()
Intensity=a['Intensity']
print 'Intensity=',Intensity
h=H1D("Rutherford Scattering",40,0.1,1)
while(1):
    p=f.read()
    if p == None:
            print 'End of events'
            break
    n=f.getEntries();
```

```
        ntot=len(p)
        if (n%1000 == 0):
            print 'Event=',n, ' particles=',ntot
        h.fill(p)
f.close()

c1 = HPlot('Scattering')
c1.visible()
c1.setAutoRangeAll()
c1.setLogScale(1,1)
c1.setNameX('cos &theta; [rad]')
c1.setNameY('d&sigma; / d cos &theta;')
h.scale(1/(h.getBinSize()*Intensity))
h.setFill(1)
c1.draw(h)
```

The code reads the data file and then extracts the intensity of the beam L using the known key. Then we read the event records one by one, using the loop over all entries until the end of the file is reached. For each event, we extract the array of numbers representing the particle angles and then use this array to fill a histogram. Note that the method `fill(obj)` accepts arrays (not single values!). Then we scale the histogram using the bin width and the intensity value. The resulting plot is shown in Fig. 17.1.

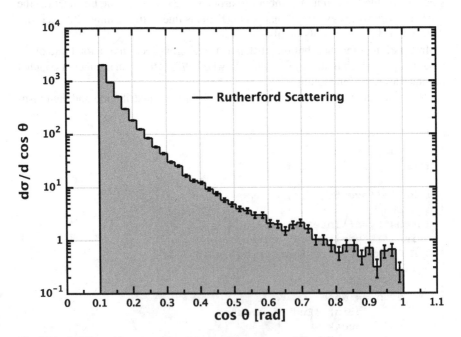

Fig. 17.1 A differential cross section extracted from a Java serialized file

One can use the methods of the H1D class (see Sect. 8.1) to access a detailed information about this distribution or even perform a fit as discussed in Sect. 14.2 to make sure that we indeed observe the famous scattering-angle cross section given in (17.1).

17.2 Many-particle Densities, Fluctuations and Correlations

Now we will discuss correlations between particles. The notion of correlations is directly related to the question of fluctuations inside small intervals [1], and both phenomena ultimately imply the existence of forces between particles.

In the previous example we distributed particles independent of each other, ignoring the fact that this may not be true in real situations. This time, our plan is to model a data sample by introducing interdependence between particles. Then, using this simulation, we will try to measure the strength of correlations/fluctuataions using several statistical tools included in jHepWork.

17.2.1 Building a Data Sample for Analysis

We will assume that each particle can be characterized by a continues value y. For example, this can be a particle velocity, or a momentum, or the time at which a particle is detected. First, we will consider a sample without any correlations in the variable y, and then we will introduce correlations between particles.

17.2.1.1 Independently Produced Events

In the simplest case, we assume that the total number of particles falling into the measured interval $\Delta y = y_{max} - y_{min}$ is distributed in accordance with a Poisson distribution:

$$P_n = \frac{\lambda^n \exp(-\lambda)}{n!}$$

where λ defines the average number of particles in the counting interval, Δy. This distribution naturally arises when the number of occurrences of rare events is measured in a long series of trials. Inside the interval Δy, particles are distributed independent of each other.

Let us prepare such data sample and write it to an output file:

```
──────────────── Creating a data set I ────────────────
from jhplot  import *
from jhplot.io import *
from cern.jet.random.engine import *
from cern.jet.random import *
```

```
Ymin,Ymax=0,1
file=HFile('sample1.ser','w')
ps=Poisson(50,MersenneTwister() )
for i in range(10000):
    if (i%1000 == 0): print 'Event=',i
    p=POD('Event='+str(i))
    ntot=ps.nextInt()
    p.randomUniform(ntot,Ymin,Ymax)
    p.sort()
    file.write(p)
print 'Entries written =', file.getEntries()
file.close()
```

This script assumes 10000 generated events. Each event record contains a certain number of particles arriving inside the interval [0, 1]. The global distribution is given by a Poisson with the average 50. We distribute particles uniformly inside the interval [0, 1] using the method randomUniform(). Each event is put into a POD array which is then written into a serialized file.

To understand what is written, it is instructive to write the data into a XML file using the HFileXML() class, instead of HFile(). One can view such XML file using any favorite editor, but to store such file is not recommended since its size is a factor eight larger than in the case of the compressed serialized file.

17.2.1.2 Including Interactions Between Particles

So far we have considered an idealized situation when particles are produced totally independent of each other. For each event, particles produces at certain region do not know anything about the presence of other particles.

Actually, this almost never happens in reality: there are always interactions between particles which can be detected by looking at particle correlations at any given event. For example, for particle or cosmic-ray physics, each particles can interact with other particles in accordance with some underlying dynamics.

Let us consider another example from the real life. Assume that our particles are shop customers counted in certain time intervals. Usually, people come to stores in groups of relatives and friends, thus customers are never independent of each other. Of course, such bunching of people in certain time intervals is due to a social mechanism which can be studied using various statistical tools.

There are two types of interactions: one type is attraction which can be characterized by a positive correlation. In fact, this is the most frequent situation in our example with the number of visitors per time interval. The second type of interaction is a repulsion, which features anti-correlations. Probably, you have already guessed what does it mean: this corresponds to a hypothetical town where everyone hates each other and people avoid meeting with other people in public places.

Now let us introduce interactions between particles. Positive correlations will be modeled by shifting particles closer to each other, while in case of anti-correlation, we will increase the distance between particles. Before doing this, let us determine the distribution of distances between particles. This can be done with this modified code:

```
——————————— Distances between particles ———————————
from jhplot  import *
from cern.jet.random.engine import *
from cern.jet.random import *

Ymin,Ymax = 0,1
h=H1D('delta',100,0,0.2)
ps=Poisson(50,MersenneTwister() )
for i in range(10000):
    if (i%1000 == 0):  print 'Event=',i
    p=P0D('Event='+str(i))
    ntot=ps.nextInt()
    p.randomUniform(ntot,Ymin,Ymax)
    p.sort()
    for j in range(p.size()-1):
        delta=p.get(j+1)-p.get(j)
        h.fill(delta)
c1 = HPlot('Delta')
c1.visible()
c1.setAutoRange()
c1.draw(h)
c1.drawStatBox(h)
```

The code above calculates the distances between particles in an event and plots the corresponding histogram, see Fig. 17.2. The distribution has exponentially falling shape with the mean (and RMS) being around 0.02.

Try to fit this histogram using the method discussed in Sect. 14.5. This can be done in the above code snippet by appending the lines:

```
a=HFit(c1,'c1',h,'h')
a.addFunc('User1', 'Tooltip', 'b* exp(-a*x[0])','a,b')
```

and rerun the script. You will see a dialog window for interactive curve fitting. Select with the mouse the "User1" function (left list), and click on the "Add" button. Then click on the "Fit" button. You will see that the histogram with the distances is well described by the function:

$$1.8 \cdot 10^4 \cdot x^{31}$$

Here we did not discover anything new: the exponential distributions are known to describe distances between events with the uniform distribution in time or space.

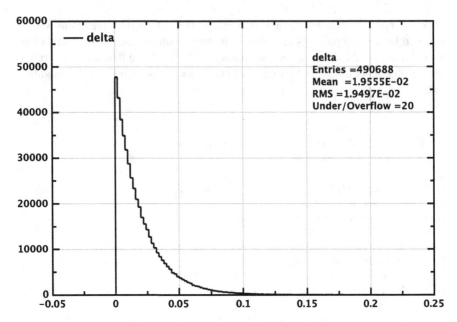

Fig. 17.2 The distribution of distances between any two particles

The distribution above is proportional to $\lambda \exp(-n\lambda)$, with the mean $1/\lambda$ and variance $1/\lambda^2$. It is obvious that λ depends on the full phase-space and on the average number of particles per event.

Our task is not to go into the details of this distribution, but try to introduce interactions between particles by either shifting them closer to each other (correlations), or moving them away from each other (anti-correlations).

Let us write a small function for this operation. We assume that the input for such function is a POD object which keeps the information about the original particle locations in an event. The output is a new POD with a modified distance between particles. We define R as a radius of interactions and will modify the distance between particles only if the size between any two particles is smaller than R. In this case, we will set a new inter-particle distance as:

```
———————————— Creating a data set II ————————————
from jhplot  import *
from jhplot.io  import *
from cern.jet.random.engine import *
from cern.jet.random import *

Ymin,Ymax=0,1
file=HFile('sample2.ser','w')
ps=Poisson(50,MersenneTwister() )
R=0.01
for i in range(10000):
    if (i%1000 == 0):  print 'Event=',i
```

```
    p=POD('Event='+str(i))
    ntot=ps.nextInt()
    p.randomUniform(ntot,Ymin,Ymax)
    p.sort()
    for j in range(p.size()-1):
        delta=p.get(j+1)-p.get(j)
        if delta<R:
            newP=p.get(j)+R/4
            p.set(j+1,newP)
    file.write(p)
print 'Entries written =', file.getEntries()
file.close()
```

The code above is sufficiently simple: The value newP is used to substitute the original particle position if the distance delta is smaller than the interaction radius.

Let us prepare two samples with distinct interactions: in one case, we will use $R = 0.01$ and change the particle distances using the algorithm newP=p.get(j)+ R/4, as given in the example above. This case corresponds to two-particle correlation, since particles will arrive in pairs if the additional term $R/4$ is sufficiently small.

The second sample is prepared using the algorithm newP=p.get(j)+R. This sample contains anti-correlations, since normally particles are separated by a smaller distance than R and now we require that they will always be separated by R. We will write our modified event records into the file 'sample3.ser'.

Let us emphasize again that the examples above are just toy models. We made simple interactions between two particles, ignoring the fact that interactions can also happen between three or more particles. In the above case, we have introduced only two-particle correlations. Also, we did not change the functional shape of the distribution used to parametrize distances between objects.

17.2.2 Analyzing the Data

Now we are equipped with three data samples in the files 'sample1.ser', 'sample2.ser', 'sample3.ser'. The first file, 'sample1.ser', contains fully uncorrelated events and two other files contain data sets with toy-like interactions characterized by the radii $R = 0.01$ and $R = 0.1$. How can we say what is inside such files, pretending that we do not know anything about their content?

In general terms, the main question is following: assume we have a data sample which represents occurrence of some events at specific interval (time or some phase space), what tools should be used to understand interdependence of events? What is the strength of such interactions? Is it attraction ("correlations") or repulsion (anti-correlations)? What is the mechanism behind such interaction?

The last question is difficult to answer, and, by no means, we will try to offer a solution for such question. Our task is rather modest: we will try to understand

whether events are independent or not. If we will find that events are heavily inter-
dependent, then we will try to determine the type of such dependence (attraction or
repulsion). Furthermore, we will try to estimate the strength of interdependence.

We should also add that we are not going to offer a universal and comprehensive
approach to deal with such kind of problems. Our task is to illustrate what can be
done using Java scripting and the tools included in the jHepWork package.

17.2.3 Reading the Data and Plotting Multiplicities

First, let us read the prepared data files and plot a few basic distributions. We have
all necessary equipment for doing this. The code below shows how to read the data
files and plot multiplicity distributions inside a small region [0–0.5]. We remind that
our original points are distributed inside the phase space window [0–1]:

——————————— Multiplicity distributions ———————

```
from jhplot  import *
from jhplot.io import *
from java.awt import Color
import os

Ymax=0.5
h=[]
files=['sample1.ser','sample2.ser','sample3.ser']

for f in range(len(files)):
    name=files[f]
    if os.path.exists(name):
        file=HFile(name)
        print 'Open: '+name
        h.append(H1D(name,25,0,50))
        while(1):
            p=file.read()
            if p == None:
                print 'End of events'; break
            i=file.getEntries();
            if (i%1000 == 0):  print 'Event=',i
            n=0
            for j in range(p.size()):
                if p.get(j)<Ymax: n=n+1
            h[f].fill(n)
        file.close()
c1 = HPlot('Multiplicity')
c1.visible()
c1.setAutoRange()

for i in range(len(h)):
```

```
    h[i].setFill(1)
    color=Color.getHSBColor(0.3*i, 1.0, 1.0);
    h[i].setColor(color)
  h[1].setPenDash(2)
  h[2].setPenDash(6)
  c1.draw(h)
```

Running this script will bring a canvas with three distributions, each of them has a Poisson-like shape. They are all very similar, so there is no way to say about the difference we have introduced for two-particle distances. So, we should find a better solution to tackle the problem of correlations. One solution would be to study fluctuations inside small phase space windows.

Since we have introduced two-particle correlations, we should expect that the multiplicity distributions inside small regions should deviate from a Poisson distribution: in case of interactions, it should be broader than the standard Poisson (large fluctuations). In case of repulsion between particles, the distribution should be narrower.

When fluctuations of separate particles are measured, it is convenient to transform a multiplicity distribution $P_n(\delta)$ inside a region δ to the following observables:

$$\text{BP:} \quad \eta_q(\delta) = \frac{q}{q-1} \frac{P_q(\delta) P_{q-2}(\delta)}{P_{q-1}^2(\delta)} \tag{17.2}$$

$$\text{NFM:} \quad F_q(\delta) = [n(\delta)]^{-q} \sum_{n=q}^{\infty} \frac{n!}{(n-q)!} P_n(\delta) \tag{17.3}$$

where the abbreviations denote the bunching parameters (BP) or the normalized factorial moments (NFM). The parameter q defines the order of the BP or NFM. Note that the experimental definitions of the BPs [2, 3] and the NFM [4, 5] are different from those given above since, in reality, one should scan all phase space regions to increase statistics when looking into ever smaller regions.

These two quantities measure deviations of a multiplicity distribution $P_n(\delta)$ from a Poisson distribution for which $F_q(\delta) = \eta_q(\delta) = 1$. Note that such deviations are measured differently by these three methods (for a review see [1]).

Uncorrelated particle production inside δ leads to the Poisson statistics, thus deviations of the NFM and BP from unity indicate correlations (interactions) between particles leading to dynamical fluctuations (i.e. non-Poisson type of statistics). If $\eta_q(\delta)$ or $F_q(\delta)$ are larger than one, one can say about positive correlations. This case is expected for the first sample generated without including the toy interactions. In contrast, if both quantities are smaller than unity for ever smaller δ regions, one can say about anti-correlations (this, naturally, should be the case for our second sample with defined repulsion).

jHepWork contains a Java library for calculations of both $\eta_q(\delta)$ and $F_q(\delta)$. We will consider the class called `BunchingParameters`, and also a similar class `FactorialMoments` from the package `jhplot.stat`.

We can initialize our calculations as:

```
>>> bp=BunchingParameters(MaxOrder,Bins,Step,Min,Max)
```

where `MaxOrder` is the maximum order of BP, `Bins` is the number of bins used
for the calculations, `Step` is any integer number to increase the binning step. The
bins are defined as $1 + i * \text{Step}$, where i runs from 2 to `Bins`. This number of bins
is used to divide the total phase space defined by `Min` and `Max`.

Once the object `'bp'` is created, we can use the method `run(array[])` inside
the event loop, passing the array with all particles in the event. After the end of the
event loop, we should call the method `eval()` to evaluate the BPs values. To access
the values for BPs as a function of the number of phase-space divisions, one should
call the method `getBP(order)`, which returns a P1D array with the BP values
and their statistical errors for a given `order` of BP.

Now we are ready. Let us read all three samples prepared in the previous section
and calculate the BP up to the third order. The code is shown below:

```
────────── Studies of fluctuations ──────────
from jhplot import *
from jhplot.stat import *
from jhplot.io import *
Ymax=0.5
bp2,bp3=[],[]
files=['sample1.ser','sample2.ser','sample3.ser']

for i in range(len(files)):
    name=files[i]
    if os.path.exists(name):
        file=HFile(name)
        print "Open: "+name
        bp=BunchingParameters(3,20,5,0,Ymax)
        while(1):
            p=file.read()
            if p == None:
                    print 'End of events'; break
            n=file.getEntries();
            if (n%1000 == 0):  print 'Event=',n
            bp.run(p.getArray())
        bp.eval()
        result=bp.getBP(2)
        result.setTitle('BP_{2} for '+name)
        result.setSymbol(3+i)
        bp2.append(result)
        result=bp.getBP(3)
        result.setTitle('BP_{3} for '+name)
        result.setSymbol(3+i)
        bp3.append(result)
        file.close()
```

```
c1 = HPlot('BP',800,400,2,1)
c1.visible()
c1.setAutoRangeAll()
c1.cd(1,1)
c1.draw(bp2)
c1.cd(2,1)
c1.draw(bp3)
```

As you can see, we do this rather complicated calculations using a few lines of the code. We build two lists which keep the results for η_2 and η_3 for all three processed samples and then run over each of this sample. The result of this code is shown in Fig. 17.3. As expected, the BPs for the sample without interactions are indeed independent of $\delta = (Ymin - Ymax)/Bins$ and they are all equal to unity ("Poisson statistics"). The sample with the interactions, when particles arrive in bunches, has a sharp increase for the calculated $\eta_2(\delta)$ with decreasing δ. The $\eta_3(\delta)$ does not show too much variations with the number of bins, indicating that we are dealing with two-particle correlations. For the third sample, $\eta_2(\delta)$ decreases with δ (anticorrelations) and stays constant when δ reaches some δ_{min}, which roughly corresponds to anti-correlation radius in our code (the parameter "R").

Now, let us calculate the NFMs. This can be done by replacing the class name `BunchingParameters` with the name `FactorialMoments`. To retrieve the results, we should call `getNFM()` instead of `getBP()`. The resulting plot is shown in Fig. 17.4. One can see a similar result as that shown on Fig. 17.3. The only notable difference is that F_3 for the sample with interactions still shows some increase with the bin size. This is not totally surprising since F_3 reflects not only three-particle

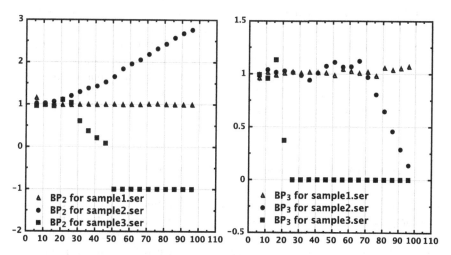

Fig. 17.3 Bunching parameters η_2 (*left*) and η_3 (*right*) as functions of the number of bins used to divide the total phase space [0–0.5]. The calculations were done for three data samples: with no correlations ('sample1.ser') with correlations ('sample2.ser') and with anti-correlations ('sample3.ser')

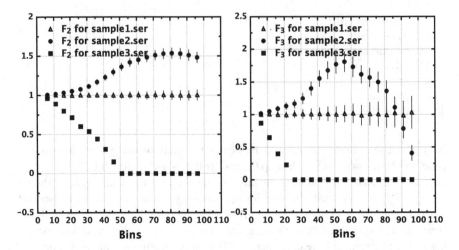

Fig. 17.4 The normalized factorial moments F_2 (*left*) and F_3 (*right*) as functions of the number of bins used to divide the total phase space. The calculations were done for the three data samples: with no correlations (`'sample1.ser'`), with correlations (`'sample2.ser'`) and with anti-correlations (`'sample3.ser'`)

correlations but also contains a contribution from two-particle correlations which (we know!) have been included when the sample was generated. One can find more details about the use and properties of BP or NFM in the review [1].

At the very end, you may still wonder why not just to look at the distribution of distances between two particles if we want to know how particles are distributed with respect to each other. We know that inter-particle distances should be distributed in accordance with the exponential distribution in case of the Poisson statistics. In principle, this can be done as well, but then we should find a method to compare such distributions with a reference distribution which is known to have exactly the same properties (multiplicities, shapes of single-particle distributions) but without any correlations between particles. This is not too easy and is outside the scope of our example: the use of the BPs, NFMs or other tools is very handy since they do not require construction of any reference sample [1].

17.3 Analyzing Macro Data: Nearby Galaxies

In this section, we give another real-life example: this time we plan to analyze data on nearby galaxies. Our intention is not to be too scientific, nor to discuss in detail topics which require special knowledge. What we want to do is to illustrate the power of scientific scripting: in a few lines of the code we will perform almost complete data analysis. It will consist of reading a data file, fetching necessary data records and plotting one type of values against another. The readers who are interested in science will immediately find areas where the discussed code snippets can be reused for their own research or even to make a detailed study of the data file discussed in this section.

What we are going to do now is to analyze the catalog of Southern spirals galaxies (Mathewson+ 1996) publicly accessible from the Astronomical Data server [6]. To simplify our task, we have prepared a file with the catalog to be used for this tutorial.

Let us first copy the catalog with the galaxy's data and open it in the jHepWork editor in order to understand its format:

```
────────────── Getting data from the web ──────────────
http='http://projects.hepforge.org/jhepwork/'
file='Mathewson1996.tsv'
wget(http+'examples/data/'+file)
view.open(file, 0  )
```

The script above fetches the file 'Mathewson1996.tsv' and opens it in a new editor tab. If you are using this script outside jHepWork, do not forget to import webustils package from the directory macros/system.

If the script above fails, try several other mirrors by replacing the http value with the string:

```
http='http://jhepwork.sourceforge.net/'
or
http='http://jwork.org/jhepwork/'
```

By looking at the file structure, one can see that each entry in the row is separated by a space. Thus, if we want to read this file line by line, one should find a way to split each line of this file. Also, one can see that comment lines always start with the symbol "#". Such lines have to be ignored. Finally, we split the strings into pieces and convert such sub-strings into float values.

We are dealing with a typical CSV file discussed in Sect. 2.16.4, thus it can be opened using the class CSVReader. This class may not be flexible ehough, but the good thing is that to write a small Jython code which parses a CSV file is very easy, perhaps as easy as when using a ready-to-go third-party library!

So, let us write a small Jython code which reads a CSV file and performs the necessary conversions. Our module will be designed for the following task: It splits a string into n-size pieces using an arbitrary string as the delimiter and ignores the lines which start from the symbols "#" or "*" (this will be necessary for our next example). Finally, we will need to prepare a function which converts a string into a number. Normally, this can be done with the float() method (in case if we expect a real number). But we should take care of the situations when we need to convert the strings such as "1/2", "2/3" etc. into float values.

Let us create a Jython module file called 'reader.py' with the following code:

```
——————————— 'reader.py' module ———————————
def get(line, delim=None):
  s=line.strip()
  if s.startswith('*') or s.startswith('#'): return None
  if delim == None:
     return s.split()
  else:
     return s.split(delim)

def conv(s):
  try:
    return float(s)
  except ValueError:
    try:
      num, denom = s.split('/')
      return float(num) / float(denom)
    except ValueError:
      return None
```

The function called 'get()' takes any string and splits it into peaces using the delim as a delimiter string. If the delim is not used, we assume that the string should be split using a white space. In addition, we included the function 'convert()' which accepts any string and converts it into a float value. For example, if we pass "1/2" to this function, it will return the float 0.5. If no proper conversion is possible, the function will raise an exception (see Sect. 2.15) and return None.

Now let us read this file and build a profile histogram 8.5. We would like to plot the average value of rotation velocities versus the face-on diameter of the spiral galaxies. By studying the file, one can easily see that we need to plot values in the column 18 against values in the column 14. So, let us write the following simple code:

```
——————————— Analyzing galaxies ———————————
from jhplot  import *
from reader import *

c1 = HPlot('Canvas')
c1.setGTitle('Southern Spiral Galaxies')
c1.visible()
c1.setRange(-1,4,100,200)

h=HProf1D('2447 galaxies (Mathewson+ 1996)',20,0.05,3.0)
for line in open('Mathewson1996.tsv'):
  tab = get(line)
  if tab == None or len(tab)<18: continue
  x=conv(tab[13])
  y=conv(tab[17])
```

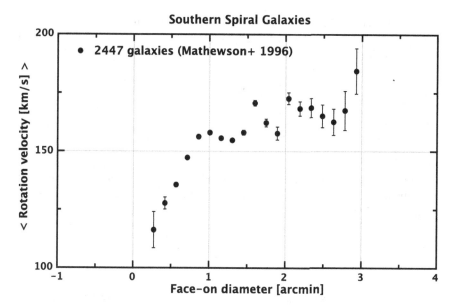

Fig. 17.5 A dependence of the rotational velocities on the galaxy size

```
    if x != None and y!=None: h.fill(x,y)

c1.setNameX('Face-on diameter [arcmin]')
c1.setNameY('< Rotation velocity [km/s] >')
h1=h.getH1D()
h1.setStyle('p')
c1.draw(h1)
```

The example shows how to open a CVS file and read it line by line. We split each line and make sure that we have the correct number of pieces (18). Then we fill the profile histogram.

The resulting image is shown in Fig. 17.5. As we can see, there is rather obvious dependence of the average velocity on the galaxy radius. I am sure some readers will find this plot familiar and will connect it to some physics phenomena. If the reader is not one of them, then this would be your entry to the world of science.

17.4 Analyzing Micro Data: Elementary Particles

In this section we plan to study elementary particles from the Particle Data Book [7]. A typical size of the galaxies analyzed in the previous example is 10^{20} meters, while the proton size is 10^{-15} meters, so the examples in this book span more than 35 orders of magnitude in distances! Sure, the universe we leave in is a big place, and we have to find a way to study it.

Let us first copy a file with particle characteristics from the Particle Data Group web page [7]. To simplify our task, we have prepared an easy-to-copy file. We will open this file in the jHepWork editor for a visual examination:

```
————————————————— Reading a data file ———————
http='http://projects.hepforge.org/jhepwork/'
file='mass_width_2008.csv'
wget(http+'examples/data/'+file)
view.open(file, 0  )
```

As before, the script above fetches the data and opens the file in a new editor tab. If the script above fails, try another mirror sites by replacing the value http with the string:

```
http='http://jhepwork.sourceforge.net/'
http='http://jwork.org/jhepwork/'
```

Now let us show how to read this file and how to understand correlations between different values in this file. Let us plot the squared mass of the hadron on the J value, which is the total angular momentum. We will use the same module 'reader.py' developed in the previous section. But, for this example, we will do something new: We will attempt to read not only masses, but also errors on the masses and put all of this into a P1D container (see Sect. 5.1) for plotting. We will use the delimiter ";" for separation of values.

```
—————————— Analyzing elementary particles ————————
from jhplot  import *
from reader import *

p1=P1D('hadrons')
file=open('mass_width_2008.csv')
for line in file:
   tab = get(line,',')
   if tab == None : continue
   id=conv(tab[12])
   J=conv(tab[8])
   if id>100 and id<100000 and J != None:
     mass=conv(tab[0])
     er1=conv(tab[1])
     er2=conv(tab[2])
     p1.add(J,mass,er1,er2)

p2=p1.copy()
p1.oper(p2,'*')
c1 = HPlot('Canvas')
c1.setGTitle('Hadrons')
c1.setNameY('M^{2} [MeV]')
```

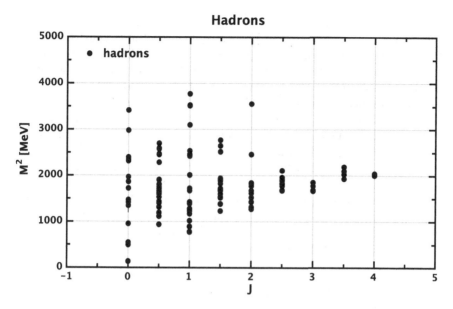

Fig. 17.6   A dependence of the hadron masses on the angular momenta

```
c1.setNameX('J')
c1.visible()
c1.setRange(-1,5,0,5000)
c1.draw(p2)
```

Figure 17.6 shows the result.

As you can see, there is a correlation between M^2 and J. In fact, what we have discovered here using a few lines of the code is a reflection of the well-known observation that hadrons from the same family lie on special trajectories (the so-called Chew-Frautschi conjecture [8–10]). Such a relationship between J and M^2, also known as the principle of exchange degeneracy, is usually interpreted as a manifestation of the linear potential of the strong forces between constituent quarks inside the proton. One can see clear regularities using a rigorous linear regression fit and the most recent data in this articles [11, 12].

Here we will stop discussing any further scientific issues: the example above is just an invitation to science via the offered Java scripting and the jHepWork program.

17.5 A Monte Carlo Simulation of Particle Decays

In this section we will build a simple Monte Carlo simulation of particle decays based on the class `HepLorentzVector` discussed in Sect. 7.6. We will consider two-body decays of a particle with a mass M which can be written as

$M \rightarrow M_1 + M_2$, where M_1 and M_2 are masses of daughter particles. When a particle decays, one should keep in mind the usual four-momentum conservation. The calculation itself can be organized in two steps: (1) a calculation of decays in center-of-mass frame of daughter particles and; (2) transformation of four-momenta of daughter particles to the laboratory frame using the Lorentz transformation.

In our numerical calculation we will be interested in the distribution of the maximum angle between daughter particles with respect to the initial particle. We will look at this on a statistical bases by introducing some randomness of decay angles in the center-of-mass frame of the original particle.

First, we define an initial particle to be a "top" quark, while daughter particles will be called W and b. It should be said that this corresponds to quite realistic situation in particle physics when a top quark decays into a W boson and b quarks. At this stage, however, we will be more interested in the program implementation of our task, rather than in physics interpretation of our results.

Let us take a look at the code below:

```
─────────────── Particle decays ───────────────
from jhplot  import *
from java.awt import Color
from java.util import Random
from hephysics.particle import *
from math import *


P=100
Top=LParticle('Top quark',170)
Top.setV3( 0,0,P )
W,b =LParticle('W',80),LParticle('b',5)

r = Random()
h1 = H1D("&Theta;_{max} angle",50, 0,3)
for i in range(10000):
    if (i%1000==0): print 'event=',i
    Top.twoBodyDecay(W, b,1);
    theta =acos(2.0*r.nextDouble()-1.0 )
    phi   =6.28*r.nextDouble()
    Top.setThetaPhiP(theta,phi,P)
    W.boost( Top ); b.boost( Top )
    p=P0D('max angle')
    p.add(Top.angle(W))
    p.add(Top.angle(b))
    pmax=p.getMax()
    h1.fill(pmax)

c1 = HPlot('Canvas')
c1.setGTitle('Max angle t &rarr; W b')
c1.visible()
c1.setAutoRange()
c1.setNameX('&Theta;_{max} [rad]')
```

```
c1.setNameY('Events')
h1.setFill(1)
h1.setFillColor(Color.yellow)
c1.draw(h1)
```

Let us make some comments. First, we have defined three particles using the LParticle class discussed in Sect. 7.6. The particles have the masses 170, 80, 5, respectively (using some arbitrary units). We have assumed that the total momentum (P) of the original particle is 100 (again, in arbitrary units). Then we have generated 10,000 events, introducing a randomness in the azimuthal (ϕ) and polar (θ) angle, since angles defined in the center-of-mass frame can have any values covering the full solid angle. The line 6.28*r.nextDouble() simply randomizes the azimuthal angle in the range [0, 2π]. We redefine four-momenta of the original particle using the input ϕ and θ. Then we have performed the Lorentz boost and calculated angles between the original particle and the daughter particles. Finally, we have determined the maximum angle for each decay and filled a histogram.

Run this code and look at its output shown in Fig. 17.7. We can do the following simple experiment with this code: change the initial momentum from 100 to a higher value. We will see that the shape of this distribution will change depending on this value.

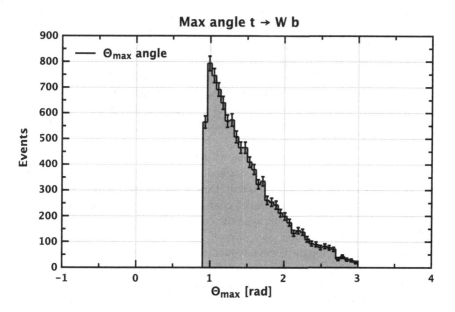

Fig. 17.7 A distribution of angles between daughter particles and the original parent particle

17.6 Measuring the Speed of Light Using the Internet

Yes, this is correct. What we going to do now is to measure the speed of light. It is not going to be the most precise measurement, probably it will be the most imprecise among other imprecise measurements, but it still be a measurement which can give you ideas about this fundamental physical constant, and the way how it can be extracted. At least, we guarantee that the value of the speed obtained in this chapter will be of the same order of magnitude as the established value, 299,792 km/c. The good thing about this measurement is that we do not need to have any apparatus: a home computer with a reasonable Internet bandwidth should be sufficient for our purpose.

We will only warn you before going into details of this experiment: for our measurement, Linux or Mac computers are best suited as they usually have the "ping" command. The program to be discussed below was tested using Ubuntu Linux, a community developed Linux-based operating system. But one can use any operating system with the "ping" program, as long as it is installed. We will use jHepWork in a combination with the ping program.

Now let us come to the main idea. The "ping" command is usually used for testing connectivity and response time over networks. For example, when one types `ping -c1 www.google.com`, it will test the response of the Google server. The execution of this program prints the time necessary for a package to reach a server and return to your PC.

We know that the signal in a vacuum travels with the speed of light, which appears to be very important physical constant with a value of 299,792 kilometers per second. It travels a little bit slower in fiber optics, but we will ignore this for the moment since we are not too interested in such precision. The distance, between two sufficiently separated points, say between Chicago and Berlin, is around 7,000 km. Thus, if one pings a server in Berlin from Chicago, we should get the response at least in $2 * 23$ milliseconds, where the factor two is introduced to take into account the round signal trip. In the real word, the response time will be longer than 46 ms, since the Internet signal does not travel along a straight line, and various switches and routers slow down the transfer speed. If the signal is transported by a fiber, it bounces around along the path. Therefore, it is a reasonable assumption that the signal needs 40–50% more time to travel.

Now we can try to get a rough estimate of the speed of light on a statistical basis. We will ping few servers in Europe, USA and Asia. Then we will make histograms with the response time for each continent. Depending on the location, one can estimate an approximate distance between continents by analysing the histograms. If you know what is the distance between the continents, one can measure the speed of light with a certain precision.

Below we will develop the necessary modules to perform our measurement. The example we are going to show is rather informative: we will illustrate how to use Jython lists, how to call an external program, how to process data and use a random number generator. Finally, the example will help you to develop a genuine feeling for the quantities and concepts involved.

17.6.1 Getting Host Names in Each Continent

First, we should obtain the lists of hosts in each continent. For this, we will use again the Web site of Ubuntu with a list of mirror servers located in different parts of the world. We will use the same method as in the previous examples: to make things easier, we will download three prepared files containing the computer host names in a HTML form:

```
──────────────── Getting data from the web ────────────────
http='http://projects.hepforge.org/jhepwork/'
list=['ubuntu_usa.html',\
      'ubuntu_europe.html',\
      'ubuntu_asia.html']
for file in list:
    wget(http+'examples/data/'+file)
```

We remind that the script above fetches files from the Web. If you are using this script outside jHepWork, do not forget to import webustils package from the directory macros/system. If the script above fails, try another mirror by replacing the http value with the strings:

```
http='http://jhepwork.sourceforge.net/'
or
http='http://jwork.org/jhepwork/'
```

Upon download completion, open one of such files in the jHepWork editor. You will see that this is a HTML file with the server names. Our task is to make a small Jython module which reads the names of the servers from such files. The names usually start with the string http://, and end after a quotation mark. We want to make a list of servers from each file, and pick up only several random servers from each list.

```
─────────── Getting host names. The module 'get_host.py' ───────────
from random import *

def get_host(file,number):
    r=Random()
    f = open(file)
    list=[]
    for line in f.readlines():
        inx1=line.find('//')
        if inx1>0:
            ss=line[inx1+2:len(line)]
            ss=ss[0:ss.find("/")]
            list.append(ss)
```

```
      list=r.sample(list,3)
      return list
```

Let us test this module, assuming that it was saved in a file "get_host.py". We will print a list with four hosts in USA. The host names will be drawn at random using the random module discussed in Sect. 2.8.

```
─────────────── Testing the module 'get_host.py' ───────────────

from get_host import *

print get_host("ubuntu_usa.html",4)
```

This script prints the list with the server names. If everything looks correct, we can proceed to the next step.

17.6.2 Checking Response from Servers

Next we want to test the ping response for a certain server. We will write a small module for this which will be put into the file 'get_ping.py'. The input to this module is a list with the server names. For each remote server, the program calls the external command "ping". Each server will be 'pinged' exactly three times. (see the option -c3). The module returns a list with the response from each "ping" call.

```
─────────── Ping hosts. The module 'get_ping.py' ───────────

import os

def get_ping(list):
    answer=[]
    for i in list:
        pingaling = os.popen('ping -c3 '+i,'r')
        while 1:
            line = pingaling.readline()
            if not line: break
            line=line.strip()
            if not line: continue
            answer.append(line)
    return answer
```

17.6.3 Creating Histograms with the Response Time

The module above returns the output from the ping command. Each output has a string in the form time=Number ms, where Number is the response time in

milliseconds for each server. We should identify this number using the standard
Python string method (like the method find() and plot each number on a graph.

Our next module will be called "get_h1d.py". It accepts the list with the server
names from the previous example, and identifies the response time. Then, a histogram with the response time will be filled for further investigation. We will also
pass the color for this histogram and the title for better visualization. Once the histogram is filled, the module will return the filled histogram.

```
───────────── Getting H1D using 'get_h1d.py' ─────────────
from jhplot import H1D

def get_h1d(answer,title,color):
  h1=H1D(title,50,0,500)
  h1.setFill(1)
  h1.setFillColor(color)
  h1.setColor(color)
  for m in answer:
    inx1=m.find('time=')
    if inx1>0:
        ss=m[inx1+5:len(m)]
        ss=ss[0:ss.find('ms')]
        d=float(ss)
        h1.fill(d)
  return h1
```

17.6.4 Final Measurements

Now we are ready to collect all pieces together. First, we will import all functions
from the files we created so far. Then, we will prepare a list with the input files from
which the server names will be taken. Then, for each file, we will build a list with the
servers, ping at random several servers from each continent and fill the histograms
with response time. In addition, we will print the mean values of the server response
for each continent. This can be done using the method mean(). Here is our code
example:

```
───────────── Response-time measurement ─────────────
from get_host import *
from get_ping import *
from get_h1d import *
from jhplot import *
from java.awt import Color


sites=['europe','usa','asia']
colors=[Color.red,Color.blue,Color.green]
```

```
all=[]
for i in range(len(sites)):
    list=get_host('ubuntu_'+sites[i]+'.html',4)
    answer=get_ping(list)
    h1d=get_h1d(answer,sites[i],colors[i])
    all.append(h1d)
    print sites[i],' mean=',h1d.mean()

c1 = HPlot('Canvas')
c1.setGTitle('Response time (ms)')
c1.visible(1)
c1.setRange(0.0,300,0,10)
c1.draw(all)
```

The execution of this script prints the mean values and brings up a canvas with three filled histograms. Each histogram is shown with different color and has appropriate annotations.

For the author's computer, the execution of the above script prints:

```
europe  mean= 142.0
usa   mean   = 54.1
asia  mean   = 252.8
```

Let us try to make sense from these numbers: The author of this book is in USA, and the average response time to a random server inside USA is 54 ms. The average travel time to European servers is 142 ms. The signal needs to travel in one direction roughly $(142 - 54)/2 = 44$ ms. This is a upper limit, since it does not take into account how much time is lost in European servers, assuming that 54 ms is the time which was lost in USA servers.

So, a lower limit on the speed of light is 159,090 km/s. It is twice lower than expected, since we did not take into account many factors discussed above (server response time, travel of light inside fibers rather than in a vacuum, large geographical spread of servers). But we kept our promise: we are close to the correct answer, at least the order of magnitude for the extracted speed is correct.

Let us give a final comment: we would not recommend to increase the number of servers and/or the number of pings for each server. This will not help you to increase the precision of the measurement. If you really want to do this measurement with a high precision, one should pick up a certain server and assess all the effects discussed above. To do this is not easy. But what we can say is this: all such effects are minimal if the ping test is done over very long distances. So, the advice is to come back to this approach when you will be a scientist or seriously interested in this type of measurements. Or, when you will be able to ping Internet servers on the Moon or Mars from your home (in not too distant future!).

References

1. De Wolf, E.A., Dremin, I.M., Kittel, W.: Scaling laws for density correlations and fluctuations in multiparticle dynamics. Phys. Rep. **270**, 1–141 (1996)
2. Chekanov, S.V., Kuvshinov, V.I.: Bunching parameter and intermittency in high-energy collisions. Acta Phys. Pol. B **25**, 1189 (1994)
3. Chekanov, S.V., Kittel, W., Kuvshinov, V.I.: Generalized bunching parameters and multiplicity fluctuations in restricted phase-space bins. Z. Phys. C **74**, 517 (1997)
4. Bialas, A., Peschanski, R.B.: Moments of rapidity distributions as a measure of short range fluctuations in high-energy collisions. Nucl. Phys. B **273**, 703 (1986)
5. Bialas, A., Peschanski, R.B.: Intermittency in multiparticle production at high-energy. Nucl. Phys. B **308**, 857 (1988)
6. ADC, Astronomical data center, access to astronomy data and catalogs. URL http://adc.astro. umd.edu/
7. Eidelman, S., et al.: Review of particle physics. Phys. Lett. B **592**, 1 (2004)
8. Chew, G.F., Frautschi, S.C.: Principle of equivalence for all strongly interacting particles within the s matrix framework. Phys. Rev. Lett. **7**, 394–397 (1961)
9. Chew, G.F., Frautschi, S.C.: Regge trajectories and the principle of maximum strength for strong interactions. Phys. Rev. Lett. **8**, 41–44 (1962)
10. Frautschi, S.: Regge Poles and S-Matrix Theory. Benjamin, New York (1968)
11. Anisovich, V., et al.: Quark Models and High Energy Collisions. World Scientific, Singapore (2004)
12. Chekanov, S.V., Levchenko, B.B.: Regularities in hadron systematics, Regge trajectories and a string quark model. Phys. Rev. D **75**, 014007 (2007)

Index of Examples

S.V. Chekanov, *Scientific Data Analysis using Jython Scripting and Java,*
Advanced Information and Knowledge Processing,
DOI 10.1007/978-1-84996-287-2, © Springer-Verlag London Limited 2010
435

Index